Surveys of Modern Mathematics
Volume 14

Introduction to
Moduli Spaces of Riemann Surfaces and Tropical Curves

Lizhen Ji
Eduard Looijenga

International Press
www.intlpress.com

HIGHER EDUCATION PRESS

Surveys of Modern Mathematics, Volume 14

Introduction to Moduli Spaces of Riemann Surfaces and Tropical Curves

Lizhen Ji (University of Michigan)
Eduard J. N. Looijenga (Tsinghua University and Universiteit Utrecht)

ISBN: 978-1-57146-353-1

Printed in the United States of America.

Surveys of Modern Mathematics

www.intlpress.com/smm

Preface

Riemann introduced in 1851 in his thesis *Grundlagen für eine allgemeine Theorie der Functionen einer veränderlichen complexen Grösse* the surfaces we have now named after him. In this thesis he set up a new foundation for complex analysis of one variable and thus completely changed the prevailing perspective on that field. For example, he pointed out that the natural domain of a holomorphic function need not be contained in the complex plane $\mathbb{C}$, but lies in general on a Riemann surface. This groundbreaking work was followed a few years later (1857) by another very influential paper of his, *Theorie der Abel'schen Functionen*, in which he developed systematically the topology of Riemann surfaces, discussed integrals of holomorphic and meromorphic 1-forms and proved the crucial Riemann inequality in the Riemann-Roch theorem, thereby establishing the equivalence between compact Riemann surfaces and plane algebraic curves over the complex number field. Moreover, in order to classify compact Riemann surfaces or rather function fields of algebraic curves, he introduced the notion of "moduli" for such surfaces, and for the purpose of solving the Jacobi inversion problem for a general compact Riemann surface, he developed the theory of (what we now call) Riemann theta functions. Although analysis played a crucial role, this paper also initiated the algebraic geometry (in particular the birational geometry) of algebraic curves. So Riemann was fully aware of the rich world of connections between his surfaces and algebraic curves.

Much has happened since, and both Riemann surfaces and their moduli spaces have come to assume a central role in several areas in mathematics. The moduli spaces in question are considered to be among the most important algebraic varieties, but have also a prominent place in string theory. The subject is still very much vibrant as the focus of current research, with many of its central questions still being unanswered.

A milestone in the study of moduli spaces of Riemann surfaces was the introduction of Teichmüller spaces endowed with the action of a mapping class group, with the orbit space then yielding the moduli space. This view point provided an analogy with locally symmetric spaces, which by definition are obtained as orbit space of symmetric spaces endowed with an action of an arithmetic group. This analogy, though imperfect, has been very fruitful in the past decades, as it suggested both problems and solutions.

As we mentioned, compact Riemann surfaces can also be regarded as algebraic curves over the complex numbers. In the past few years, there has been an explosion of work on their tropical counterparts, i.e., algebraic curves over the tropical semifield. These have appeared in many unexpected topics and make currently a very active area of study. It turns out that the moduli spaces of tropical curves also fit into the above analogy, in the sense that these are the orbit spaces of tropical Teichmüller spaces by outer automorphism groups of free groups. The outer automorphism groups of free groups are among the most important groups in geometric group theory, and the tropical Teichmüller spaces were studied as spaces of marked metric graphs before the subject of tropical geometry (and its name) existed.

Given the rich history of Riemann surfaces and their moduli spaces, and their generalizations, applications and connections with other subjects, it seems helpful to provide an accessible and timely introduction to all the above topics, while emphasizing the underlying connections. The current book is an attempt towards this goal. It consists of two parts. The first part deals with mapping class groups of surfaces, Teichmüller spaces and their applications to moduli spaces of Riemann surfaces, whereas the second part deals with tropical analogues and some applications in geometric group theory. Though these parts were conceived independently, together they cover both basic and essential results in these subjects as well as some recent developments. Although there exist several works on some of the above topics, we hope and believe that the nature of the subject justifies our offering of our own perspective on this wonderful world.

Lizhen Ji

Eduard Looijenga

September 2016

Contents

Part I

Moduli Spaces of Riemann Surfaces

by Eduard Looijenga

This is the write up of a course I taught at Tsinghua University in the Fall of 2011, while being supported by the there-based Mathematics Sciences Center.

One can enter the subject via algebraic geometry, complex analysis, (combinatorial) topology and even homotopy theory. My aim was not to confine myself to just one of these approaches, but rather to give the students a sense how they not only supplement, but sometimes also reinforce each other. That sounds rather ambitious and so I hasten to add that the course's content was not only subject to time constraints, but also to my probable bias and the limitations of my knowledge. Having said that, I believe that the implementation of this philosophy not only exhibits the intertwined nature mentioned above, but also allows occasionally for shorter proofs, at least shorter than the ones I found in the literature.

In the version prepared during the course, references were absent. This has now been remedied, to the extent that I provided them where I believed the text demanded this. Of course, this does not do the literature justice and the reader is in this regard probably best served by the references list in the recent book by Arbarello-Cornalba-Griffiths [1]. This magnum opus of almost a thousand pages is also an excellent reference for much of the material that is discussed here and I highly recommend to use it on the side and for further study.

I knew from the outset that this 'multidisciplinary approach' would be rather demanding on the students. But my audience, which even counted a few undergraduates, proved to be very motivated and gave me plenty of feedback (which occasionally led to a correction or a more detailed discussion). The course was a joy to give.

In these notes a *surface* always means an oriented 2-manifold admitting a countable base. We denote the unit circle if $\mathbb{C}$ of complex numbers of norm 1 by $\mathbb{C}_1$.

1 Mapping class groups and Dehn twists

Mapping class groups

Let S be a surface. For any subset $P \subset S$. We denote by $\mathrm{Diff}(S,P)$ the group of diffeomorphisms of S onto S that are the identity on P (but usually write $\mathrm{Diff}(S)$ when $P = \emptyset$). We write $\mathrm{Diff}^\circ(S,P) \subset \mathrm{Diff}(S,P)$ for its identity component—this is a normal subgroup. Two elements $h, h' \in \mathrm{Diff}(S)$ are called *isotopic relative to* P if they lie in the same path component of $\mathrm{Diff}(S,P)$, in other words, if $h^{-1}h'$ in $\mathrm{Diff}^\circ(S,P)$.

The group of orientation preserving diffeomorphisms in $\mathrm{Diff}(S,P)$, denoted by $\mathrm{Diff}^+(S,P) \subset \mathrm{Diff}(S,P)$, contains $\mathrm{Diff}^\circ(S,P)$ as a normal subgroup. We call the quotient $\mathrm{Mod}(S,P) := \mathrm{Diff}^+(S,P)/\mathrm{Diff}^\circ(S,P)$ (which comes endowed with the discrete topology) the *mapping class group* of (S,P). As these notions only depend on the closure of P in S, we may as well assume that P is already closed in S.

Dehn twists

We now assume $P \subset S$ closed. The collection of embedded circles in $S \smallsetminus P$ is acted on by $\mathrm{Diff}(S,P)$. Two embedded circles α, α' are said to be isotopic relative to P if they have the same $\mathrm{Diff}^\circ(S,P)$-orbit. So $\mathrm{Mod}(S,P)$ acts on the set of isotopy classes of embedded circles in $S \smallsetminus P$.

An embedded circle $\alpha \subset S \smallsetminus P$ leads to a so-called Dehn twist $D_\alpha \in \mathrm{Mod}(S,P)$: let $\phi : (-1,1) \times \mathbb{C}_1 \to S \smallsetminus P$ be an open, *orientation preserving* embedding such that α is the image of ϕ_0 and let $f : (-1,1) \to [0,2\pi]$ be a smooth function which is constant 0 on $(-1,-\frac{1}{2})$ and constant 2π on $(\frac{1}{2},1)$. Let $h : S \to S$ be the identity outside the image of ϕ and be such that $h(\phi(t,u)) = \phi(t, ue^{-\sqrt{-1}f(t)})$. Then h is a diffeomorphism and one can check that its image in $\mathrm{Mod}(S,P)$ only depends on the isotopy class relative to P of α. In particular, it does not depend on an orientation of α. We call this element of $\mathrm{Mod}(S,P)$ the *Dehn twist* associated to α and denote it by D_α.

Exercise 1.1. Prove that D_α is trivial when α bounds a disk in S which meets P in at most one point.

We will be mostly interested in the case when S is a closed surface and $P \subset S$ is finite subset. We then call the pair (S,P) a *P-pointed surface*. We then often write S° for $S \smallsetminus P$. If S is connected and g is the genus of S, then we sometimes denote $\mathrm{Mod}(S,P)$ by $\mathrm{Mod}_{g,P}$, or if the elements of P have been effectively numbered: $P = \{p_1, \dots, p_n\}$, by $\mathrm{Mod}_{g,n}$.

It can be shown that $\mathrm{Mod}_{g,P}$ is finitely presented with Dehn twists as generators. We shall see that $\mathrm{Mod}_{0,3}$ is trivial and that $\mathrm{Mod}_{1,1} \cong \mathrm{SL}(2,\mathbb{Z})$.

Now assume that S is a closed connected surface of genus g and that P is finite. The $\mathrm{Diff}^+(S,P)$-orbits of embedded circles in S° can then be topologically distinguished: a *nonseparating* embedded circle $\alpha \subset S^\circ$ is by definition such that its complement $S \smallsetminus \alpha$ connected. That complement is then the interior of a compact connected surface of genus $g-1$ with two boundary components. Otherwise α is *separating* and splits S into two connected components S', S'', each of which has as its closure a compact subsurface with α as boundary. If the genera of these surfaces are denoted by g' and g'', then $g = g' + g''$. The members of P will divide themselves in $P' := P \cap S'$ and $P'' := P \cap S''$. The diffeomorphism type of a compact surface with boundary is completely given by its genus and its number of boundary components and this is still true in the relative situation where the surface has been equipped with a finite subset in its interior. From this we easily deduce:

Proposition 1.2. *The* non-separating *embedded circles in S° make up a single orbit under* $\mathrm{Diff}^+(S,P)$ *and for a* separating *embedded circle its* $\mathrm{Diff}^+(S,P)$*-orbit is characterized by the unordered pair of pairs $\{(g',P'),(g'',P'')\}$ defined above. This gives rise to a corresponding characterization of the* $\mathrm{Mod}(S,P)$*-conjugacy classes of Dehn twists.*

The notion of a mapping class group does not change if we pass to the topological setting: any closed topological surface admits a differentiable structure, and for (S,P) as above the inclusion $\mathrm{Diff}^+(S,P) \subset \mathrm{Homeo}^+(S,P)$ induces an isomorphism on their path component groups $\mathrm{Mod}(S,P) \xrightarrow{\cong} \mathrm{Homeo}^+(S,P)/\mathrm{Homeo}^\circ(S,P)$. This is even true if we descend to the homotopy category: the natural map from $\mathrm{Mod}(S,P)$ to the group $\mathrm{Htp}(S,P)$ of homotopy equivalences $(S,P) \to (S,P)$ relative to P is an isomorphism. That property amounts to a characterization of $\mathrm{Mod}(S,P)$ as a group of *outer automorphisms* of the fundamental group of S°, which we describe the next subsection. Before we do so, we mention that there is an intermediate structure which is quite useful in Teichmüller theory, namely that of a quasi-conformal structure. We will not give the definition, but just mention that the connected component group of the group of automorphisms of that structure which preserve orientation and fix P pointwise also maps isomorphically to $\mathrm{Mod}(S,P)$.

Fundamental groups and mapping class groups

We will here describe without proof a characterization of a mapping class group in terms of its (outer) action on the fundamental group. We begin with general discussion of outer actions of groups.

Recall that given a group π, the group $\mathrm{Aut}(\pi)$ of group automorphisms of π contains the group $\mathrm{Inn}(\pi)$ of its inner automorphisms as a normal subgroup. The quotient group $\mathrm{Aut}(\pi)/\mathrm{Inn}(\pi)$ is called the group of its *outer automorphisms* and is denoted $\mathrm{Out}(\pi)$. Since $\mathrm{Inn}(\pi)$ acts trivially on the (co)homology of π, the group $\mathrm{Aut}(\pi)$ acts on this (co)homology always via $\mathrm{Out}(\pi)$. If $B \subset \pi$ is a $\mathrm{Inn}(\pi)$-invariant subset, in other words, a union conjugacy classes of π, then we can form the subgroup $\mathrm{Aut}(\pi, B) \subset \mathrm{Aut}(\pi)$ of automorphisms which preserve each conjugacy class in B and as this this contains $\mathrm{Inn}(\pi)$, we also have defined the quotient group $\mathrm{Out}(\pi, B) := \mathrm{Aut}(\pi, B)/\mathrm{Inn}(\pi)$.

Let X be a path connected space. Its *fundamental groupoid* π_X is the category whose objects are the points of X and for which a morphism from $p \in X$ to $q \in X$ is a homotopy class of arcs from p to q. It is a groupoid since every morphism is an isomorphism. The fundamental group based at $p \in X$, $\pi(X, p)$, then appears as the group of π_X-endomorphisms of p (this presupposes that elements of $\pi(X, p)$ are read from right to left). Given $p, q \in X$, then by assumption there exists a π_X-morphism $\gamma : \pi(X, p) \cong \pi(X, q)$ and any two such differ by an element of $\pi(X, q)$. So the resulting isomorphism $\mathrm{Out}(\gamma) : \mathrm{Out}(\pi(X, p)) \cong \mathrm{Out}(\pi(X, q))$ is independent of the choice of γ. We denote by $\mathrm{Out}(\pi_X)$ this common group. Or if we insist on treating all the points of X on an equal footing: an element of $\mathrm{Out}(\pi_X)$ is the subgroup of $\prod_{p \in X} \mathrm{Out}(\pi(X, p))$ of elements whose components are related by the isomorphisms $\mathrm{Out}(\gamma)$. This renders evident the observation that a homotopy equivalence $h : X \to X$ induces an element of $\pi(h) \in \mathrm{Out}(\pi_X)$ and that thus is defined a group homomorphism from the group $\mathrm{Htp}(X)$ of homotopy classes of self homotopy equivalences $X \to X$ to $\mathrm{Out}(\pi_X)$, to which one often refers as an *outer action* (of $\mathrm{Htp}(X)$) on the fundamental group.

Let us return to S, a closed connected oriented surface. If $o \in S$ is a base point, then it is well-known that $\pi(S, o)$ has a (standard) presentation with generators $\alpha_{\pm 1}, \dots, \alpha_{\pm g}$, and relation $[\alpha_g, \alpha_{-g}] \cdots [\alpha_1, \alpha_{-1}] \equiv 1$.

The outer action of $\mathrm{Diff}^+(S)$ on $\pi_S = \pi(S, o)$ defines a group homomorphism

$$\mathrm{Mod}(S) \to \mathrm{Out}(\pi_S).$$

The theorem alluded to above asserts that for genus $g > 0$ this is an isomorphism onto a subgroup $\mathrm{Out}^+(\pi_S)$ of $\mathrm{Out}(\pi_S)$ of index 2. (The preservation of orientation can be expressed in terms of π_S, because $H^2(S)$ can be understood as group cohomology: $H^2(S) = H^2(\pi_S)$; we give another description below which avoids this language.) For $g = 0$, $\mathrm{Mod}(S)$ is trivial. The above result implies among other things that $\mathrm{Mod}(S)$ is finitely generated and even finitely presented. We will later give another proof of this fact (Corollary 7.3).

For a cofinite subgroup of π_S (i.e., a subgroup of finite index), the collection of its $\mathrm{Aut}(\pi_S)$-conjugates is finite in number (for π_S is finitely generated) so that the intersection of these is still cofinite in π_S. It is clear that for such an $\mathrm{Aut}(\pi_S)$-invariant cofinite subgroup π of π_S is normal in π_S and that we have an action of $\mathrm{Aut}(\pi_S)$ on the finite group π_S/π. The kernel of this action produces a cofinite subgroup of $\mathrm{Aut}(\pi_S)$ and its intersection with $\mathrm{Mod}(S)$ will be cofinite in the latter. A subgroup of $\mathrm{Mod}(S)$ which contains the kernel of such a homomorphism is called a *congruence subgroup* of $\mathrm{Mod}(S)$. Such a subgroup is clearly cofinite in $\mathrm{Mod}(S)$. It is still an open question whether the converse holds:

Question 1.3 (Grothendieck)**.** Is every cofinite subgroup of $\mathrm{Mod}(S)$ a congruence subgroup?

We assume $P \subset S$ finite nonempty and write $S^\circ := S \smallsetminus P$ as before. We also assume that when $g = 0$, $|P| \geq 2$ (we will see that $\mathrm{Mod}(S,P)$ is trivial otherwise). Choose $o \in S^\circ$. In order to give a presentation of $\pi_{S^\circ} = \pi(S^\circ, o)$, we first number the points of P so that $P := \{p_1, \dots, p_n\}$. Then we choose for $k = 1, \dots, n$, $\beta_k \in \pi(S^\circ, o)$ representing a simple positive loop in S° around p_k based at o such that $\beta_n \beta_{n-1} \cdots \beta_1$ is represented by a positive loop which encircles P and bounds a disk in S. A presentation π_{S° has now generators $\alpha_{\pm 1}, \dots, \alpha_{\pm g}, \beta_1, \dots, \beta_n$ that are subject to the relation

$$\beta_n \beta_{n-1} \cdots \beta_1 [\alpha_g, \alpha_{-g}] \cdots [\alpha_1, \alpha_{-1}] \equiv 1.$$

Since this relation allows us to eliminate β_n, this group is in fact freely generated by $\alpha_{\pm 1}, \dots, \alpha_{\pm g}, \beta_1, \dots, \beta_{n-1}$. However it has some additional structure: the conjugacy class $B_k \subset \pi_{S^\circ}$ of β_k is invariantly defined as it consists of *all* the simple positive loops in S° around p_k based at o. If we divide the group $\pi(S^\circ, o)$ out by the normal subgroup generated by B_k, then we get the fundamental group of $S^\circ \cup \{p_k\}$. We put $B := B_1 \cup \cdots \cup B_n$ and observe that we have an evident map

$$\mathrm{Mod}(S,P) \to \mathrm{Out}(\pi_{S^\circ}, B).$$

Another basic result asserts that this map is also an isomorphism. (There is no orientation issue, for an orientation reversing diffeomorphism maps B_1 to B_1^{-1}, which differs from B_1.) Now consider the exact sequence of groups

$$1 \to \mathrm{Inn}(\pi_{S^\circ}) \to \mathrm{Aut}(\pi_{S^\circ}, B) \to \mathrm{Out}(\pi_{S^\circ}, B) \to 1.$$

Since π_{S° has trivial center (it is a free group), we may identify it with its group of inner automorphisms. The middle term may be identified with $\mathrm{Mod}(S, \tilde{P})$, where $\tilde{P} := P \sqcup \{o\}$ and we thus find the *Birman exact sequence*

$$1 \to \pi_{S^\circ} \to \mathrm{Mod}(S, \tilde{P}) \to \mathrm{Mod}(S,P) \to 1.$$

This is in fact also valid when $P = \emptyset$: $\mathrm{Mod}(S)$ can be identified with group of outer automorphisms h of $\pi(S,o)$ that, in terms of the above presentation, can be lifted to an automorphism $\tilde{h}$ of the free group on the generators $\tilde{\alpha}_{\pm 1}, \dots, \tilde{\alpha}_{\pm g}$ which preserves the conjugacy class of $[\tilde{\alpha}_g, \tilde{\alpha}_{-g}] \cdots [\tilde{\alpha}_1, \tilde{\alpha}_{-1}]$.

2 Conformal structures and a rough classification

Conformal structures

Let T be a real vector space of dimension two. A *conformal structure* on T is an inner product on T given up to scalar multiplication. If T is endowed with an orientation, then a conformal structure yields a notion of angle (rotation over an angle $\theta \in \mathbb{R}/2\pi\mathbb{Z}$ is defined) and thus turns T into a complex vector space of complex dimension one: scalar multiplication by $\sqrt{-1}$ is rotation over $\pi/2$. In fact, giving a complex structure on T is equivalent to giving an orientation plus a conformal structure.

The space of positive definite quadratic forms on T makes up a convex cone in the space of quadratic forms on T (if T has coordinates (x,y), then a quadratic form $ax^2+2bxy+cy^2$ is positive definite if and only if $b^2 < ac$ and $a > 0$). So the set of conformal structures on T makes up a disk $\mathrm{Conf}(T)$ in the real projective plane associated to the space of quadratic forms on T. This disk has a convex structure, but has no natural center (it is rather like the upper half plane which we will later discuss in some detail).

Let S be a surface. The spaces $\mathrm{Conf}(T_pS)$, $p \in S$, make up a disk bundle $\mathrm{Conf}(S)$ over S. A *conformal structure on* S is a C^∞-section of this bundle. The space of such sections has also a convex structure: two conformal structures are naturally joined by an interval. In particular, the space $\mathrm{Conf}(S)$ of conformal structures on S is contractible.

Since S is oriented, a conformal structure on S defines a complex structure J_p on each T_pS. If $U \subset S$ is open, then we say that a C^∞-map $f: U \to \mathbb{C}$ is *holomorphic* if $D_pf : T_pU \to T_{f(p)}\mathbb{C} = \mathbb{C}$ is $\mathbb{C}$-linear for all $p \in U$. If $S = \mathbb{C}$ (so that U open in $\mathbb{C}$), then this means that f satisfies the Cauchy-Riemann equations and so is holomorphic in the usual sense. It is a classical but nontrivial fact that a Riemann metric on a surface is locally conformal to a Euclidean metric. This implies that S admits an atlas consisting of holomorphic charts. Since the coordinate changes of such an atlas are holomorphic, the conformal structure turns S into a Riemann surface. Conversely, a Riemann surface has an underlying orientation and conformal structure.

For a closed surface S and a finite subset $P \subset S$, the group $\mathrm{Diff}^+(S,P)$ acts on $\mathrm{Conf}(S)$. The *Teichmüller space* of (S,P) is the space of conformal structures on S given up to isotopy relative to P:

$$\mathscr{T}(S,P) = \mathrm{Diff}^\circ(S,P)\backslash\mathrm{Conf}(S).$$

If S is connected of genus g, we may write $\mathscr{T}_{g,P}$ for $\mathscr{T}(S,P)$, and if in addition the elements of P have been numbered: $P = \{p_1,\dots,p_n\}$, then we also write $\mathscr{T}_{g,n}$. We will see that $\mathscr{T}(S,P)$ has the structure of a finite dimensional complex manifold. Notice that $\mathrm{Mod}(S,P)$ acts on $\mathscr{T}(S,P)$. Assuming now that S is connected of genus g, then we usually suppress S in the notation of the orbit space:

$$\mathscr{M}_{g,P} := \mathrm{Diff}^+(S,P)\backslash\mathrm{Conf}(S),$$

as the ambiguity has then essentially been eliminated (we write $\mathscr{M}_{g,n}$ if P has been numbered): a point of $\mathscr{M}_{g,P}$ can be understood as an isomorphism class of pairs $(C, \iota : P \to C)$, with C a closed Riemann surface of genus g and ι an injection of the abstract finite set P in C. We will see that this has structure of a complex-algebraic variety.

The following theorem is central in the theory of Riemann surfaces. We will take it for granted.

Theorem 2.1 (Uniformization theorem). *Every simply connected Riemann surface is isomorphic to the Riemann sphere* $\mathbb{P}^1$*, the affine line (= complex plane)* $\mathbb{C}$ *or the upper half plane* $\mathbb{H}$.

The automorphism groups of these surfaces are as follows.

$\mathrm{Aut}(\mathbb{P}^1)$ is the *Möbius group* $\{z \mapsto \frac{az+b}{cz+d}\}$, where $\left(\begin{smallmatrix} a & b \\ c & d \end{smallmatrix}\right) \in \mathrm{SL}(2,\mathbb{C})$. This is $\mathrm{SL}(2,\mathbb{C})/\{\pm 1\} = \mathrm{PSL}(2,\mathbb{C})$. Notice that it is simply transitive on ordered triples of distinct points: if $z_0, z_1, z_\infty \in \mathbb{P}^1$ are distinct, then there is unique automorphism of $\mathbb{P}^1$ taking them to resp. $0, 1, \infty$.

$\mathrm{Aut}(\mathbb{C})$ is the $\mathrm{Aut}(\mathbb{P}^1)$-stabilizer of ∞, i.e., the affine group $\mathrm{Aff}(\mathbb{C}) := \{z \mapsto az+b\}$, where $a \in \mathbb{C}^\times$ and $b \in \mathbb{C}$.

$\mathrm{Aut}(\mathbb{H})$ is the $\mathrm{Aut}(\mathbb{P}^1)$-stabilizer of $\mathbb{H}$, i.e., the group of real Möbius transformations $\{z \mapsto \frac{az+b}{cz+d}\}$, where $\left(\begin{smallmatrix} a & b \\ c & d \end{smallmatrix}\right) \in \mathrm{SL}(2,\mathbb{R})$. This case is special because $\mathbb{H}$ comes with a metric, the *Poincaré metric* $y^{-2}(dx^2+dy^2)$, that is invariant under $\mathrm{Aut}(\mathbb{H})$. This metric is also complete and has constant curvature -1 (more about this later).

If C is a connected Riemann surface, then its universal cover $\tilde{C}$ is isomorphic to one of the three above. If Π denotes the covering group of the universal cover $\tilde{C} \to C$, then Π acts freely and properly discretely on $\tilde{C}$. So Π can be identified with a subgroup of $\mathrm{PSL}(2,\mathbb{C})$, $\mathrm{Aff}(\mathbb{C})$ or $\mathrm{PSL}(2,\mathbb{R})$ acting freely and properly discretely on $\mathbb{P}^1$, $\mathbb{C}$, $\mathbb{H}$ respectively, and C may be identified with its orbit space. We thus obtain the following rough classification into spherical, flat and hyperbolic cases.

Spherical cases

Every element of $\mathrm{SL}(2,\mathbb{C})$ has an eigen vector in $\mathbb{C}^2$ and so every element of $\mathrm{PSL}(2,\mathbb{C})$ has a fixed point in $\mathbb{P}^1$. Hence the only possible subgroup $\Pi \subset \mathrm{PSL}(2,\mathbb{C})$ that acts freely on $\mathbb{P}^1$ is the trivial group. It follows that we have an isomorphism $f : C \cong \mathbb{P}^1$. Notice that we can then pick 3 distinct points p_0, p_1, p_∞ in C and ask that f maps these to $0, 1, \infty$, respectively. This makes $f : C \to \mathbb{P}^1$ unique.

Corollary 2.2. *The Teichmüller space* $\mathscr{T}_{0,3}$ *is a singleton and the mapping class group* $\mathrm{Mod}_{0,3}$ *is trivial.*

Proof. Let $S = S_0$ and $P = \{p_0, p_1, p_\infty\}$. The assertion that $\mathscr{T}_{0,3}$ is a singleton means that any two conformal structures on S are isotopic relative to P. To see this, let J be a conformal structure on S. As we observed above, there exists a unique orientation preserving diffeomorphism $f : \mathbb{P}^1 \cong S$ that is conformal and is such that $f(k) = p_k$ for $k = 0, 1, \infty$. Let J' be another conformal structure on S. By the convexity property of conformal structures, we can connect J with J' by a path of complex structures $\{J_t\}_{0\le t\le 1}$ (with $J=J_0$ and $J'=J_1$). This gives a path of isomorphisms $f_t:\mathbb{P}^1\cong(S,J_t)$ with $f_t(k)=p_k$ ($k=0,1,\infty$). Since $h_t:=f_tf_0^{-1}$ takes J to J_t and preserves P, the family of diffeomorphisms $\{h_t\}_{0\le t\le 1}$ accomplishes an isotopy between J and J'.

This argument also shows that $\mathrm{Diff}^+(S,P)=\mathrm{Diff}^0(S,P)$. To see this, we may just as well assume that $S=\mathbb{P}^1$ and $p_k=k$ for $k=0,1,\infty$. Given $h\in\mathrm{Diff}^+(\mathbb{P}^1,P)$, and denoting by J the standard conformal structure J on $\mathbb{P}^1$, then arguing as above we get a path $\{h_t\}_{0\le t\le 1}$ in $\mathrm{Diff}^+(\mathbb{P}^1,P)$ with $h_0=1$ and $h^*J=h_1^*J$. Since $h_1^{-1}h$ respects J and fixes $0,1,\infty$, the uniqueness property implies that $h=h_1$. It follows that $h\in\mathrm{Diff}^0(\mathbb{P}^1,P)$. □

Remark 2.3. This implies that for $n=0,1,2$, $\mathscr{T}_{0,n}$ is also a singleton and $\mathrm{Mod}_{0,n}$ is trivial. For $n\ge 4$ this is no longer the case: $\mathscr{T}_{0,n}$ can then be identified with the configuration space of pairwise distinct $(n-3)$-tuples $(z_4,\dots,z_n)$ in $\mathbb{P}^1\smallsetminus\{0,1,\infty\}=\mathbb{C}\smallsetminus\{0,1\}$ (and the mapping class group $\mathrm{Mod}_{0,n}$ is the quotient of the colored braid group on $(n-1)$ strands modulo its center).

Flat cases

Every element $z\mapsto a+bz$ *in* $\mathrm{Aff}(\mathbb{C})$ with $a\ne 0$ has the fixed point $z=b/(1-a)$ unless $a=1$. In other words, it is without fixed point if and only if it is a translation $z\mapsto z+b$ ($b\ne 0$). So we may regard $\Pi\subset\mathrm{Aff}(\mathbb{C})$ as an additive subgroup of $\mathbb{C}$. This subgroup is discrete and hence has rank at most two and we have $C\cong\mathbb{C}/\Pi$. Notice that $\mathbb{C}/\Pi$ is not just a Riemann surface, but also an abelian group. Any automorphism σ of $\mathbb{C}/\Pi$ as a Riemann surface lifts to its universal cover $\mathbb{C}$. It must be then of the form $\tilde\sigma:z\mapsto az+b$ with $a\in\mathbb{C}^\times$. This transformation has to commute with the covering transformations, i.e., with the group Π and this just means that multiplication by a maps Π onto itself. The full translation group of $\mathbb{C}$ is normalized by $\tilde\sigma$ (conjugation with $\tilde\sigma$ acts as multiplication by a). To phrase this intrinsically: C is a natural manner a torsor of a connected abelian complex Lie group $\mathrm{Tr}(C)$ of dimension one (its *translation group*) that is isomorphic to the abelian group $\mathbb{C}/\Pi$ (the choice of any $o\in C$ gives C the structure of that group having o as its identity element) and $\mathrm{Aut}(C)$ is the semi-direct product of this translation group $\mathrm{Tr}(C)$ and the automorphism group of $\mathrm{Tr}(C)$.

If $\Pi=\{0\}$, then we get $\mathrm{Tr}(C)\cong\mathbb{C}$ and $\mathrm{Aut}(C)\cong\mathrm{Aff}(\mathbb{C})=\mathbb{C}^\times\ltimes\mathbb{C}$.

If $\Pi=\mathbb{Z}\omega_0$, with $\omega_0\ne 0$. Then $z\mapsto e^{2\pi iz/\omega_0}$ identifies $\mathrm{Tr}(C)$ with the multiplicative group $\mathbb{C}^\times$ of $\mathbb{C}$ and $\mathrm{Aut}(C)\cong\{\pm 1\}\ltimes\mathbb{C}^\times$.

If Π is of rank 2, then $\mathrm{Tr}(C)\cong\mathbb{C}/\Pi$ is a complex torus ($g=1$). In this case, it is convenient to fix an oriented basis (e_0,e_1) for $\pi_1(C,o)$ (any orientation preserving diffeomorphism $C\cong\mathbb{R}^2/\mathbb{Z}^2$ yields one). If $\varepsilon_i:[0,1]\to C$ is a loop based at o that represents e_i and $\tilde\varepsilon_i:[0,1]\to\tilde C$ the lift of ε_i which begins in $\tilde o$, then the end point $\tilde\varepsilon(1)\in\tilde C$ only depends on e_i and we denote that point by $e_i(1)$. Then under the isomorphism $\tilde C\to\mathbb{C}$, $\tilde p$ is mapped to 0 and $(e_0(1),e_1(1))$ is mapped to an oriented basis (ω_0, ω_1) of $\mathbb{C}$, when regarded as a real vector space. Then after multiplication by ω_0^{-1}, we get an isomorphism $\tilde{f} : \tilde{C} \cong \mathbb{C}$ with $\tilde{f}(\tilde{o}) = 0$ and $\tilde{f}(e_0(1)) = 1$. This is unique, for any automorphism of $\mathbb{C}$ which fixes both 0 and 1 must be the identity. Notice that $\tau := \tilde{f}(e_1(1))$ lies in $\mathbb{H}$ (for $(1, \tau)$ must be an oriented basis) and so we get a unique isomorphism $\mathrm{Tr}(C) \cong \mathbb{C}/(\mathbb{Z} + \mathbb{Z}\tau)$ which on fundamental groups sends $e_0(1)$ to the class represented by the loop defined by $[0,1] \subset \mathbb{C} \to \mathbb{C}/(\mathbb{Z} + \mathbb{Z}\tau)$. The group $\mathrm{Aut}(\mathrm{Tr}(C))$ contains $\{\pm 1\}$. It is in fact the group of $\lambda \in \mathbb{C}^\times$ which preserve the lattice $\mathbb{Z} + \mathbb{Z}\tau$; in general this is just $\{\pm 1\}$, but for special values of τ this is the group of sixth roots of unity or the group of fourth roots of unity (e.g., $\tau = \exp(2\pi\sqrt{-1}/3)$ or $\tau = \sqrt{-1}$).

Corollary 2.4. *The Teichmüller space $\mathscr{T}_{1,1}$ can be identified with $\mathbb{H}$ and the mapping class group $\mathrm{Mod}_{1,1}$ with $\mathrm{SL}(2,\mathbb{Z})$ (acting as a group of fractional linear transformations).*

Proof. If $S = S_1$ and $P = \{o\}$, then choose an oriented basis (e_0, e_1) for $\pi_1(S, o)$ as above. The preceding construction finds for every conformal structure J on S a unique $\tau \in \mathbb{H}$ and a *unique* isomorphism

$$f : (S, J) \cong \mathbb{C}/(\mathbb{Z} + \mathbb{Z}\tau)$$

with the property that it maps p to 0 and e_0 to the class of the loop $[0,1] \subset \mathbb{C} \to \mathbb{C}/(\mathbb{Z} + \mathbb{Z}\tau)$.

Suppose now J' is another conformal structure on S and assume that the above construction produces the same $\tau \in \mathbb{H}$. We show that J' is isotopic to J relative to o. Since the space of conformal structures on S is contractible, there exists a path $\{J_t\}_{0 \leq t \langle 1}$ of such structures with $J_0 = J$ and $J_1 = J'$. This yields a continuous family of isomorphisms $f_t : (S, J_t) \cong \mathbb{C}/(\mathbb{Z} + \mathbb{Z}\tau_t)$ as above with $\tau_0 = \tau_1 = \tau$. Let $\tilde{h}_t : \mathbb{C} \to \mathbb{C}$ be the orientation preserving $\mathbb{R}$-*linear* isomorphism defined by $\tilde{h}_t(1) = 1$ and $\tilde{h}_t(\tau_t) = \tau$. This induces an orientation preserving diffeomorphism $g_t : \mathbb{C}/(\mathbb{Z} + \mathbb{Z}\tau_t) \to \mathbb{C}/(\mathbb{Z} + \mathbb{Z}\tau)$ that sends 0 to 0. Notice that $g_0 = g_1 = 1$. Consider now the diffeomorphism $f_t' := g_t f_t : S \to \mathbb{C}/(\mathbb{Z} + \mathbb{Z}\tau)$. If we replace J_t by the image of the conformal structure J_t' on $\mathbb{C}/(\mathbb{Z} + \mathbb{Z}\tau)$ under $f_t'^{-1}$, then we get a new family of conformal structures connecting J with J' for which the corresponding curve τ_t' in $\mathbb{H}$ is constant equal to τ. This means that J_t' is the image of the given complex structure J under $h_t := f_t' f_0'^{-1}$. So $J' = J_1'$ is isotopic to $J = J_0'$ relative to o.

The last part of the argument also shows that τ only depends on the isotopy class of J. So we have a well-defined map $\mathscr{T}_{1,1} \to \mathbb{H}$ and this map is injective. It is also onto: we hit every $\tau \in \mathbb{H}$ by choosing an orientation preserving diffeomorphism $f : S \to \mathbb{C}/(\mathbb{Z} + \mathbb{Z}\tau)$ with $f(o) = 0$ and which induces on $\pi_1(S, o)$ the map which sends e_0 resp. e_1 to the loops defined by 1 resp. τ.

The isomorphism $\mathscr{T}_{1,1} \cong \mathbb{H}$ depends on the choice of oriented basis (e_0, e_1). Another choice yields in general another isomorphism. Since the oriented bases of $\pi_1(S, o) = \mathbb{Z}e_0 + \mathbb{Z}_1$ are transitively permuted by $\mathrm{SL}(2,\mathbb{Z})$, we get an identification of $\mathrm{Mod}_{1,1}$ with $\mathrm{SL}(2,\mathbb{Z})$. We show that this induces the familiar action of $\mathrm{SL}(2,\mathbb{Z})$

of $\mathbb{H}$. If (e'_0, e'_1) is another oriented basis of $\pi_1(S,o)$, then write (additive notation) $e'_1 = ae_1 + be_0$ and $e'_0 = ce_1 + de_0$ with $a,b,c,d, \in \mathbb{Z}$. We must have $\sigma := \left(\begin{smallmatrix} a & b \\ c & d \end{smallmatrix}\right) \in \mathrm{SL}(2,\mathbb{Z})$. This takes the pair $(\tau, 1)$ of generators of $\mathbb{Z}\tau + \mathbb{Z}$ to the pair $(a\tau + b, c\tau + d)$. If we compose with scalar multiplication by $(c\tau + d)^{-1}$ it acquires the desired form $(\sigma(\tau), 1)$. □

Since $\mathrm{SL}(2,\mathbb{Z})$ acts on $\mathbb{H}$ through the (faithful) action of $\mathrm{PSL}(2,\mathbb{Z})$, it follows that $\mathrm{Mod}_{1,1}$ acts on $\mathscr{T}_{1,1}$ with a kernel of order 2.

Hyperbolic cases

These are the remaining cases. It follows from the uniformization theorem that all such surfaces are naturally endowed with a complete metric of curvature -1 (also called a *hyperbolic structure*). If C is isomorphic to a connected closed Riemann surface of genus g minus an n-element subset, then $(g,n) = (0,0)$ is excluded for then $C \cong \mathbb{P}^1$ by Remark 2.3. Similarly are excluded $(g,n) = (0,1), (0,2), (1,0)$, for according to Corollary 2.4 we are then in the flat case (C is isomorphic to $\mathbb{C}$, $\mathbb{C}^\times$ and a complex torus respectively). Notice that the excluded cases are precisely the ones for which the Euler characteristic $\chi(C)$ of C, which is $2 - 2g - n$, is ≥ 0. We conclude that Riemann surfaces of the above type are hyperbolic if and only if $\chi(C) < 0$.

3 Geometry of the upper half plane

We begin with observing that certain subsets of $\mathbb{P}^1 = \mathbb{C} \cup \{\infty\}$ merit to be called a circle: this can either be an ordinary circle in $\mathbb{C} \subset \mathbb{P}^1$ having an equation $|z - a| = r$ with $r > 0$ or the union of a straight line in $\mathbb{C}$ and ∞. The justification is provided by

Exercise 3.1. Prove that $\mathrm{PSL}(2,\mathbb{C})$ transitively permutes the circles in $\mathbb{P}^1$. (A conceptual proof consists in showing that a circle can also be given as the zero set of a nondegenerate Hermitian form on $\mathbb{C}^2$ of indefinite signature $(1,1)$. Then deduce the transitivity property from the fact that such a Hermitian form can be brought into a standard diagonal form $|z_1|^2 - |z_0|^2$.)

We notice that the boundary $\partial\mathbb{H}$ of $\mathbb{H}$ in $\mathbb{P}^1$ is a circle, namely the real axis together with $\{\infty\}$, or more canonically, the real projective line $P^1(\mathbb{R})$.

Lemma 3.2. *Let $p_0, p_1, p_\infty \in \partial\mathbb{H}$ be distinct. Then there is a unique $g \in \mathrm{PSL}(2,\mathbb{R})$ with $g(p_0) = 0$, $g(p_1) = 1$ and $g(p_\infty) = \infty$. The group of $g \in \mathrm{PSL}(2,\mathbb{R})$ which fix both 0 and ∞ consists of the homotheties $z \mapsto az$ with $a > 0$. The group of $g \in \mathrm{PSL}(2,\mathbb{R})$ which fix ∞ only is the affine group $z \mapsto az + b$ with $a > 0$ and $b \in \mathbb{R}$.*

The proof is left as an exercise.

We fix some terminology regarding geodesics. Let γ be a map from an interval I to a Riemannian space M. We say that γ is a *shortest geodesic parameterized by arc length* if for every pair $t,t' \in I$, $d(\gamma(t),\gamma(t')) = |t-t'|$. We recall that given a $p \in M$ and a unit vector $v \in T_pM$ there is a shortest geodesic parameterized by arc length $\gamma : (-\varepsilon,\varepsilon) \to M$ with $\gamma(0) = p$ and $\dot\gamma(o) = v$. This is unique in the sense that any other such geodesic coincides with this one where both are defined. The map $\gamma : I \to M$ is called a *geodesic parametrized by arc length* if it is locally a shortest geodesic parametrized by arc length and the image of such a map is called a geodesic. We use the adjective *complete* if $I = \mathbb{R}$. A complete geodesic parametrized by arc length which is periodic (so that it is compact) is said to be *closed*.

Lemma 3.3. *Let $p,q \in \partial\mathbb{H}$ be distinct. Then there is a unique circle in $\mathbb{P}^1$ through p and q which meets $\partial\mathbb{H}$ in these points perpendicularly. The intersection of this circle with $\mathbb{H}$ is a complete geodesic which is minimal for each pair of points lying on it and all complete geodesics are thus obtained.*

Proof. The proof of the first assertion is left to you. To prove that the half circle defined by p and q is a geodesic, it is enough, in view of Lemma 3.2, to do the case $p = 0$ and $q = \infty$. We prove that $\gamma : \mathbb{R} \to \mathbb{H}$, $\gamma(t) := \sqrt{-1}e^t$, is a complete geodesic parametrized by arc length. We have $\dot\gamma(t) = \sqrt{-1}e^t = \gamma(t)$ and so this is a unit vector for the Poincaré metric $y^{-2}(dx^2+dy^2)$. If $\delta(t) = x(t)+\sqrt{-1}y(t) : [0,1] \to \mathbb{H}$ is a path from $\sqrt{-1}y_0$ to $\sqrt{-1}y_1$ with $y_0 < y_1$, then we have the following lower bound for its length:

$$L(\delta) = \int_0^1 \sqrt{\frac{\dot x(t)^2+\dot y(t)^2}{y(t)^2}}\,dt \ge \int_0^1 \frac{|\dot y(t)|}{y(t)}\,dt \ge \int_0^1 \frac{\dot y(t)}{y(t)}\,dt = L(\gamma|_{[\log y_0,\log y_1]}).$$

If we have equality, then we must have $x(t) \equiv 0$ and $\dot y(t) \ge 0$, which means that δ traverses in a monotonous manner a piece of γ. So γ is a complete geodesic.

The last assertion follows from the fact that every unit tangent vector of $\mathbb{H}$ is tangent to a circle that is perpendicular to $\partial\mathbb{H}$. □

Exercise 3.4. Prove that the group $\mathrm{PSL}(2,\mathbb{R})$ acts transitively on the set tangent vectors of $\mathbb{H}$ of unit length.

Proposition 3.5. *Any $1 \ne g \in \mathrm{Aut}(\mathbb{H}) = \mathrm{PSL}(2,\mathbb{R})$ has either a fixed point in $\mathbb{H}$ (and is then called* elliptic*) or has two distinct fixed points in $\partial\mathbb{H}$ (and is then called* hyperbolic*) or has a single fixed point in $\partial\mathbb{H}$ with multiplicity 2 (and is then called* horocyclic or parabolic*) and is then conjugate to one of the following transformations*

elliptic: *g has in $\mathbb{H}$ the fixed point $\sqrt{-1}$ and is a hyperbolic rotation around it over an angle θ (defined by the image of $\begin{pmatrix} \cos(\theta/2) & -\sin(\theta/2) \\ \sin(\theta/2) & \cos(\theta/2) \end{pmatrix}$ in $\mathrm{PSL}(2,\mathbb{R})$),*

hyperbolic: *g has on* $\partial\mathbb{H}$ *the fixed points* 0 *and* ∞ *and is given by the homothety* $z \mapsto az$ *with* $1 \neq a > 0$,

parabolic: *g has in* $\partial\mathbb{H}$ *the unique fixed point* ∞ *and is given by the translation* $z \mapsto z+1$.

The proof is left as an exercise.

Let M be a Riemannian manifold, $N \subset M$ a submanifold and $p \in M - N$. Then it is well-known (and not hard to see) that any geodesic $\gamma : [0,a] \to M$ which realizes the distance between p and N (so $\gamma(0) = p$, $\gamma(a) \in N$ and $a = d(p,M)$) is perpendicular to N ($\dot\gamma(a) \perp T_{\gamma(a)}N$).

Lemma 3.6. *Let* $\ell \subset \mathbb{H}$ *be a complete geodesic which joins* $p,q \in \partial\mathbb{H}$. *Then each orbit of the group of homotheties in* $\mathrm{PSL}(2,\mathbb{R})$ *which fix both* p *and* q *is the intersection of a circle passing through* p *and* q *with* $\mathbb{H}$. *Such an orbit consists of points having a fixed distance,* $d \geq 0$ *say, to* ℓ *and any geodesic which meets* ℓ *perpendicularly also meets each of these orbits perpendicularly.*

Proof. It is clear that if $z \in \mathbb{H} \smallsetminus \ell$, and $g \in \mathrm{PSL}(2,\mathbb{R})$ fixes both p and q, then $d(gz,\ell) = d(z,\ell)$. So the locus in question is invariant under the group of homotheties. For the rest of the assertion we may assume that $p = 0$ and $q = \infty$. The orbit of $\tau \in \mathbb{H}$ is then a ray $\{r\tau\}_{r>0}$ and the distance from τ to ℓ is realized by the circle arc from τ to ℓ that has its center in 0. The assertion easily follows from this. □

For what follows we need the following

Proposition 3.7. *Let* $D \subset \mathbb{H}$ *be a right-angled hexagon and enumerate its vertices in their natural counterclockwise cyclic order:* $\{p_i\}_{i\in\mathbb{Z}/6}$ *(this means that* D *is a compact domain in* $\mathbb{H}$ *bounded by the 6 geodesic arcs* $\{p_ip_{i+1}\}_{i\in\mathbb{Z}/6}$ *with* $p_{i-1}p_i$ *meeting* p_ip_{i+1} *in* p_i *at a right angle). Then* D *is determined up to isometry by the lengths of the three disjoint geodesic arcs* p_0p_1, p_2p_3 *and* p_4p_5.

Proof. Let us write $d_{i,i+1}$ for the length of p_ip_{i+1}. We show how D can be reconstructed up to isometry from d_{01}, d_{23}, d_{45}. We begin with choosing $p_2, p_3 \in \mathbb{H}$ such that $d(p_2,p_3) = d_{23}$. If p_2', p_3' is another such pair, then there is a unique $\sigma \in \mathrm{PSL}(2,\mathbb{R})$ with $\sigma(p_2) = p_2'$ and $\sigma(p_3) = p_3'$. So it is enough to show that from the pair (p_2,p_3) and the positive numbers d_{01} and d_{45} we can uniquely reconstruct the other p_i's.

To see this, we first draw the complete geodesic ℓ_{34} through p_3 which makes a right angle with p_2p_3 and denote by ℓ_{34}^+ the part of ℓ_{34} that begins in p_3 and 'goes to the left'. Denote by ℓ_5' the circular arc of points that have distance d_{45} to ℓ_{34}, lie in the same component of $\mathbb{H} \smallsetminus \ell_{34}$ as p_2p_3 and lie in the same connected component of $\mathbb{H} \smallsetminus \ell_{23}$ as ℓ_{34}^+. Do similarly at the other side and find ℓ_{12}, ℓ_{12}^+ and ℓ_0'. It now suffices to show that there is a unique complete geodesic ℓ_{50} which is tangent to both ℓ_5' and ℓ_0' and is such that both ℓ_5' and ℓ_0' lie in the same connected component of $\mathbb{H} \smallsetminus \ell_{50}$ at p_2p_3. For if ℓ_{50} meets ℓ'_5 in p_5 and ℓ'_0 in p_0, then we take for p_4 the point of ℓ^+_{34} that is closest to p_5, and for p_1 the point on ℓ^+_{21} that is closest to p_0. This then reproduces D. (See Fig. 1).

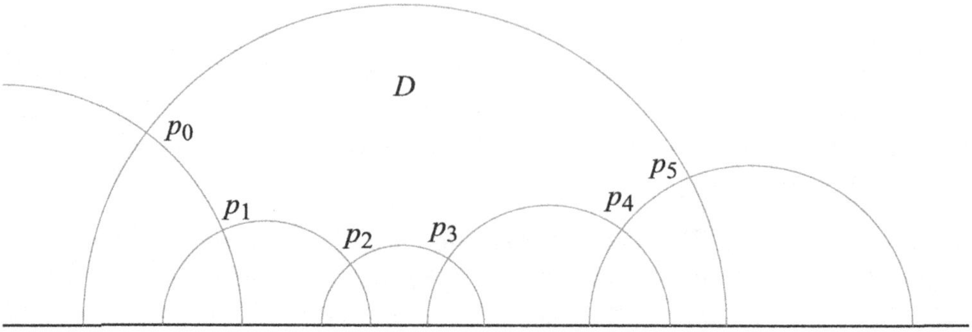

Fig. 1 A typical right-angled hyperbolic hexagon

The existence of the asserted ℓ_{50} is based on elementary planar geometry: consider the locus of points that are the center of a circle which is tangent to ℓ'_5 and ℓ'_0 and contains both these arcs in its interior. This is part of hyperbola which meets $\partial\mathbb{H}$ in a unique point. So there is a circle centered on $\partial\mathbb{H}$ that is tangent to both ℓ'_5 and ℓ'_0. Its intersection with $\mathbb{H}$ is the desired geodesic. □

Remark 3.8. The values of d_{01}, d_{23}, d_{45} can be arbitrary positive numbers. But this proposition also holds (with essentially the same proof) if some of them are zero, although its statement then needs to be properly interpreted. For instance $d_{01} = 0$ must be understood as: $p_0 = p_1$ and this vertex is *improper*, meaning that it lies on $\partial\mathbb{H}$. The extremal case we get when $d_{01} = d_{23} = d_{45} = 0$ is what is called an *ideal triangle*.

Remark 3.9. A well-known formula states that the area of a hyperbolic geodesic triangle (measured with the hyperbolic measure $y^{-2}dxdy$), is given by its angle excess: π minus the sum of its interior angles, regardless whether vertices are proper or not. This implies that the area of any right-angled hexagon is π (which is also the value that we get for an ideal triangle).

4 Hyperbolic surfaces

Closed geodesics on a hyperbolic surface

Let C be a connected Riemann surface of genus g and of hyperbolic type. We choose a universal cover $\tilde{C} \to C$ and denote by Π its covering group. An isomorphism $\tilde{C} \cong \mathbb{H}$ identifies Π with a discrete subgroup of $\mathrm{PSL}(2,\mathbb{C})$ that acts properly and freely on $\mathbb{H}$ so that C is identified with $\Pi\backslash\mathbb{H}$.

Let $\gamma \subset C$ be a complete geodesic. This lifts to an oriented geodesic $\tilde{\gamma}$ in $\mathbb{H}$, joining $p \in \partial\mathbb{H}$ with $q \in \partial\mathbb{H}$. If γ is closed, i.e., is an embedded circle, then there must exist an element of $g \in \Pi$ which preserves $\tilde{\gamma}$ and which is such that γ is obtained as the quotient of $\tilde{\gamma}$ by the action of g on it. In particular, g is a hyperbolic transformation and hence determines $\tilde{\gamma}$. If $\tilde{p} \in \tilde{\gamma}$, then a parametrization of the arc in $\tilde{\gamma}$ from $\tilde{p}$ to $g\tilde{p}$) projects to a parametrization of the closed geodesic γ. So g is then the covering transformation associated to this loop. Notice that after composing the chosen isomorphism $\tilde{C} \cong \mathbb{H}$ with an element of $\mathrm{PSL}(2,\mathbb{R})$ we can arrange that $\tilde{\gamma}$ joins 0 with ∞ (this makes the isomorphism unique up to multiplication by a positive scalar). Then $\tilde{\gamma}(t) = \sqrt{-1}e^t$ is a parametrization by arc length and g is given by multiplication by a scalar $\lambda > 1$. This scalar has the property that the length ℓ of γ must be equal to $\int_1^\lambda y^{-1}dy$, and so $\lambda = e^\ell$. For later purposes it is useful to have a concrete parametrization of a neighborhood of γ.

Lemma 4.1. *There is a holomorphic isometry of a neighborhood of the closed geodesic γ onto an annulus $\exp(-\delta/r) < |w| < \exp(\delta/r)$ with γ mapping to the circle $|w| = 1$, where $r = \ell_\gamma/2\pi$ and the annulus is endowed with the metric $\frac{r^2|dw|^2}{|w|^2\cos^2(r\log|w|)}$. This isometry sends every geodesic perpendicular to γ to an interval on a ray emanating from* 0.

Proof. For this purpose we parametrize $\mathbb{H}$ by means of polar coordinates: $z = \exp(\rho + \sqrt{-1}(\theta + \pi/2))$ with $\rho \in \mathbb{R}$ and $|\theta| < \pi/2$. Then the hyperbolic metric takes the form $(d\rho^2 + d\theta^2)/\cos^2(\theta)$. This shows in particular that an ε-neighborhood of the positive imaginary axis is a sector of the form $|\theta| < \delta$ (this also follows from the fact that this axis is an orbit under the multiplicative subgroup of isometries consisting of positive real scalars).

Let $U_\varepsilon(\gamma)$ denote the set of points in C of having distance $< \varepsilon$ to γ. For sufficiently small $\varepsilon > 0$, the preimage $\tilde{U}_\varepsilon(\gamma)$ of $U_\varepsilon(\gamma)$ in $\mathbb{H}$ is a sector as above, i.e., of the form $|\arg(z) - \pi/2| < \delta$ for some $\delta \in (0, \pi/2)$. We then recover $U_\varepsilon(\gamma)$ by taking the g-orbit space of this sector. We put $r := \ell/2\pi$ and then identify this orbit space with the annulus of the lemma using the holomorphic map

$$f : z \in \tilde{U}_\varepsilon(\gamma) \mapsto \exp\Big(\frac{\pi/2 + \sqrt{-1}\log z}{r}\Big) \in \mathbb{C}.$$

Here we take the logarithm of z with argument in $(0, \pi)$. In terms of our polar coordinates, $f(z) = \exp((-\theta + \sqrt{-1}\rho)/r)$. We see from this expression that f induces an isomorphism of $U_\varepsilon(\gamma)$ onto the annulus $\exp(-\delta/r) < |w| < \exp(\delta/r)$ with γ mapping onto the circle $|w| = 1$. It sends geodesic arcs perpendicular to γ to rays. It is straightforward to check that this becomes an isometry if we endow this annulus with the given metric. □

Geodesic shear

In the situation of this lemma we have the notion of *shear* along the closed geodesic γ over an angle $u = e^{\sqrt{-1}\theta} \in \mathbb{C}_1$, with $\theta \in \mathbb{R}$. This amounts to cutting C open along γ (in such a manner that we get a surface with two boundary components, each of which lies over γ) and then glueing back the two boundary components after we have displaced the opposite component over a distance $\theta\ell_\gamma$. Let us make this more precise with the help of the model furnished by Lemma 4.1. Denote by U_γ^+ resp. U_γ^- the closed subset of U_γ that maps to the part of the annulus defined by $|w| \leq \exp(-\pi/2r)$ resp. $|w| \geq \exp(-\pi/2r)$. Then we rotate U_γ^+ over an angle u to produce a new Riemann surface $C_\gamma(u)$ with a *new* underlying topological space: an open subset O of $C_\gamma(u)$ is given as a pair open subsets $O^\pm \subset U_\gamma^\pm$ such that $O^+ \cap \gamma$ is obtained by rotating $O^- \cap \gamma$ over v (= translating along γ over rv). Similarly, a complex valued function on this open set will be holomorphic for $C_\gamma(u)$ if it is given as a pair of holomorphic functions $\kappa_\pm : O^\pm \to \mathbb{C}$ so that $\kappa^+|O^+ \cap \gamma$ precomposed with rotation over v yields $\kappa^-|O^- \cap \gamma$. Notice that from whatever side we approach γ on C, if we want to cross γ, but want to stay in $C_\gamma(u)$, then the continuation on the other side must take place over a rotation v further down on γ 'on the left'. This formulation only uses the orientation of C; not a particular one of γ. So $C_\gamma(u)$ only depends on γ as a closed (unoriented) geodesic.

We may express this as follows: the family $\{C_\gamma(u)\}_{u\in\mathbb{C}_1}$ makes up a 3-manifold fibered over the circle: $\mathscr{C}_\gamma \to \mathbb{C}_1$ such that (i) the fibers are hyperbolic (Riemann) surfaces, (ii) each fiber $C_\gamma(u)$ comes with a closed geodesic $\gamma(u)$ of length $\ell(\gamma)$ and (iii) the complements $\{C_\gamma(u) = \gamma(u)\}_{u\in\mathbb{C}_1}$ form in a canonical way a trivial family. If we pull back this family along universal cover of the base $\theta \in \mathbb{R} \mapsto \exp(\sqrt{-1}\theta)$, then the resulting family $\mathscr{C}_\gamma \to \mathbb{R}$ is C^∞-trivial and the trivialization can be chosen to be the identity outside an ε-neighborhood of γ. Such a trivialization yields a diffeomorphism $h_v : C \to C_\gamma(\exp(\sqrt{-1}v))$ that depends continuously on v. For $v = 2\pi$ it gives a Dehn twist in C along γ. In coordinates: if $\phi : [\pi/2-\delta, \pi/2] \to [0, 2\pi]$ is constant 0 near $\pi/2-\delta$ and constant 2π near $\pi/2$, then define $h_v : C \to C_\gamma(u)$ by letting it be the identity outside U_γ^- and on U_γ^- be given by

$$h_v f(e^{\rho+\sqrt{-1}\theta}) = f(e^{\rho+\sqrt{-1}(\theta - v\phi(\rho))}), \quad |w| \leq e^{-\pi/2)/r}.$$

This is indeed a diffeomorphism and is even complex-analytic outside U_γ^-. Other choices of parametrization may lead to another diffeomorphism, but that diffeomorphism will always lie in the same connected component of $\mathrm{Diff}(C, C_\gamma(u))$, in other words, will differ from h_θ by an isotopy. Since the isotopy class of each h_v is independent of choices, assigning to v the conformal structure of $C_\gamma(\exp(\sqrt{-1}v))$ pulled-back with h_v gives a well-defined map $\mathbb{R} \to \mathscr{T}(S)$ that sends 0 to the isotopy class defined by C. The following proposition essentially sums up the preceding discussion, but states the conclusion somewhat differently:

Proposition-definition 4.2 (Shearing action). *Let S be a connected surface of (finite) negative Euler characteristic and $\alpha \subset S$ an embedded circle given up to iso-*

topy that is not homotopically trivial. If J is a conformal structure on S (so that we have defined a hyperbolic structure on S, a geodesic representative γ of α and an element $[J] \in \mathscr{T}(S)$), then the pull-back of the conformal structure of $C_\gamma(u)$ along h_θ yields a new element of $\mathscr{T}(S)$ that we will denote by $\theta_\alpha[J]$. This defines an action of the additive group $(\mathbb{R},+)$ on $\mathscr{T}(S)$ with the property that $2\pi \in \mathbb{R}$ induces the action of the Dehn twist $D_\alpha \in$ Mod(S).

Geodesic representation

Let now, conversely, be given an embedded circle $\alpha \subset C$ and suppose α is not homotopically trivial in C. Choose $p \in \alpha$ and $\tilde{p} \in \tilde{C} \cong \mathbb{H}$ over p. Then α is covered by a copy $\tilde{\alpha}$ of $\mathbb{R}$ which passes through $\tilde{p}$. With these choices the loop α defines a covering transformation $g \in \Pi$ which generates the group of covering transformations of the universal covering $\tilde{\alpha} \to \alpha$. If we let α vary in its isotopy class, say by a continuous one parameter family $\{\alpha_t\}_{0 \le t \le 1}$, then we can let p and $\tilde{p}$ follow that family and as the corresponding family g_t can only move in the discrete set Π, it will be constant. The isotopy class will contain an embedded circle γ of minimal length. This will be a geodesic which admits a lift $\tilde{\gamma}$ invariant under g so that we are in the situation considered above. In particular, $g \in \mathrm{PSL}(2,\mathbb{R})$ is a hyperbolic transformation and hence determines $\tilde{\gamma}$. It follows that γ is the unique geodesic representative of α.

Disjunction of geodesics

Let $\alpha' \subset C$ be an embedded circle that is not isotopic to α. By the above discussion its associated geodesic γ' is then distinct from γ. We claim that if α' is disjoint with α, then γ' is disjoint with γ. For if that were not the case, you can argue (using the isotopies that makes them disjoint) that there must then exist geodesic intervals γ_1 on γ and γ_1' on γ' with identical end points that are path homotopic. But then these geodesic intervals can be lifted to $\mathbb{H}$ such that they have the same end points. We then end up with two distinct geodesic intervals in $\mathbb{H}$ with the same end points and this is impossible.

Pair of pants

A *pair of pants* is a compact surface with boundary of genus zero whose boundary has three connected components. A *hyperbolic pair of pants* is a pair of pants endowed with a hyperbolic structure for which the three boundary components are geodesics.

Notice that a pair of pants has the homotopy type of a one point union of two circles (a figure eight) and so has Euler characteristic -1.

Lemma 4.3. *A hyperbolic pair of pants M is determined up to isometry by the circumference of its three boundary components: if we are given another hyperbolic* *pairs of pants M' and a bijection $\pi_0(\partial M) \cong \pi_0(\partial M')$ which preserves circumference, then there is a unique orientation preserving isometry $M \cong M'$ inducing this bijection. In particular (take $M' = -M$), M has a unique involution which reverses orientation; this involution has on each boundary component exactly two fixed points and its orbit space is a right-angled hexagon.*

Proof. Let M be hyperbolic pair of pants and let $\{\partial_i M\}_{i\in\mathbb{Z}/3}$ be a cyclic numbering of the boundary components. The shortest path from $\partial_i M$ to $\partial_{i+1}M$ will be realized by a geodesic arc. It goes from $p_i \in \partial_i M$ say, to $q_{i+1} \in \partial_{i+1}M$ say, and meets these boundary components there perpendicularly. So if we cut M open along these three arcs we get two right-angled hexagons. These two right-angled hexagons have three of their alternating sides in common and so by Proposition 3.7 they must be congruent, or rather (if we take their orientation into account), must be mirror images of each other. This implies that the two arcs on $\partial_i M$ joining its two points p_i and q_i must have the same length. In particular, this length is half the circumference of $\partial_i M$. But then another application of Proposition 3.7 shows that these circumferences completely determine the two hexagons up to congruence and hence determine M.

The last assertion follows essentially from the above construction. □

Pants decompositions

Let (for now) S be a closed connected surface of genus $g \geq 2$. A *pants decomposition* of S is the isotopy class of a closed one dimensional submanifold $A \subset S$ (so every connected component of A is then an embedded circle) with the property that each connected component of $S \smallsetminus A$ is diffeomorphic to a sphere with three punctures.

Notice that the mapping class group $\mathrm{Mod}(S)$ acts on the collection of pants decompositions of S. There are in general several $\mathrm{Mod}(S)$-orbits of pants decompositions. This is best understood in terms of a graph associated to the pair (S,A) that is defined as follows.

The Euler characteristic of S is $2-2g$, that of a pair of pants is -1 and that of a circle is 0, and so the additive properties of the Euler characteristic imply that if $A \subset S$ defines a pants decomposition, then $S \smallsetminus A$ must have $2g-2$ connected components. If α is a connected component of A, then for any orientation of α, there is a unique connected component of $S \smallsetminus A$ such that α with this orientation appears as the oriented boundary of that component (with its orientation inherited from S). In this way we hit every connected component of $S \smallsetminus A$ three times and so the number of connected components of A must be $\frac{3}{2}(2g-2) = 3g-3$. The topological type of the pair (S,A) is conveniently encoded by a connected graph $G(S,A)$ with vertex set $X_0(S,A) := \pi_0(S \smallsetminus A)$ and edge set $X_1(S,A) := \pi_0(A)$, with the obvious incidence relation.

Exercise 4.4. Let $X(S,A)$ be the set of boundary components of the connected components of $S \smallsetminus A$ (so we have $6g-6$ of these) and let $X(S,A) \to X_1(S,A)$ and $X(S,A) \to X_0(S,A)$ be the evident forgetful maps whose fibers have cardinality 2 resp. 3. Reconstruct $G(S,A)$ as a quotient of the product $[-1,1] \times X(S,A)$ (where $X(S,A)$ has the discrete topology) solely in terms of these maps.

So we have $2g-2$ vertices, all of degree 3 (one then says that the graph is *trivalent*) and has $3g-3$ edges (in particular, its Euler characteristic of the graph is half that of S, namely $1-g$). Notice that if a connected component of A happens to have the same component of $S \smallsetminus A$ at both sides, then it defines a loop.

Clearly, two pants decompositions that lie in the same $\mathrm{Mod}(S)$-orbit define homeomorphic graphs. In fact, it is not difficult to see that the graph is a complete invariant of the $\mathrm{Mod}(S)$-orbit and that every connected trivalent graph of Euler characteristic $1-g$ thus occurs.

Exercise 4.5. Draw the possible graphs for genus 2. Then try genus 3.

Exercise 4.6. Let G be the graph attached to a pants decomposition (S,A). By altering just one member of A, we may get a different graph G' whose vertex set may be identified with that of G. Describe the possible graphs G' in terms of G.

Fenchel-Nielsen coordinates

Let us fix a pants decomposition of S, defined by $A \subset S$, say. A conformal structure J on S is subordinate to a natural hyperbolic structure on S. With respect to that structure every connected component α of A is uniquely represented by a closed geodesic. According to the discussion at the beginning of this section these geodesics are disjoint. Denote by $\ell_\alpha(J)$ the length of the geodesic representing α. Since the connected components of A label the vertices of $G(S,A)$, we may view this also as defining a metric on $G(S,A)$: the edge of $G(S,A)$ labeled by α is then given the same length as its associated geodesic. Since this only depends on the isotopy class of the conformal structure we have thus defined a function

$$\ell_\alpha : \mathscr{T}(S) \to \mathbb{R}_{>0}.$$

Notice that we have attached to J a decomposition into hyperbolic pants and according to Proposition 3.7 the metric $\ell(J) : \alpha \in \pi_0(A) \mapsto \ell_\alpha(J) \in \mathbb{R}_{>0}$ on $G(S,A)$ completely determines each of these hyperbolic pairs of pants up to isometry. However, this does not enable us yet to reconstruct (S,J), for this gives us only the building blocks (namely, the hyperbolic pairs of pants) and although the graph tells us which pairs of boundary components should be welded onto each other, it does not determine (S,J) yet, for any such welding requires an angular parameter. To be precise, two conformal structures J,J' on S for which $\ell(J) = \ell(J')$ will differ by geodesic shears along the geodesic representatives of the connected components of A. This shows that $\mathscr{T}(S)$ is as a topological space locally like $\mathbb{R}^{6g-6}$.

Theorem 4.7 (Fenchel-Nielsen parametrization). *The map ℓ is a principal $\mathbb{R}^{\pi_0(A)}$-fibration: the action of the vector group $\mathbb{R}^{\pi_0(A)}$ on $\mathscr{T}(S)$ by geodesic shears is free, each orbit is a fiber of $\ell : \mathscr{T}(S) \to \mathbb{R}_{>0}^{\pi_0(A)}$ and ℓ admits a section so that we have a homeomorphism of $(\mathbb{R}_{>0} \times \mathbb{R})^{\pi_0(A)}$ onto $\mathscr{T}(S)$. In particular, $\mathscr{T}(S)$ is contractible and the Dehn twists $(D_\alpha)_{\alpha \in \pi_0(A)}$ commute with each other and are independent so that they generate in* $\mathrm{Mod}(S)$ *a free abelian subgroup D_A of rank $|\pi_0(A)|$.*

Proof. Let us write X_1 for $X_1(S,A) = \pi_0(A)$. We begin with constructing a section $j : \mathbb{R}_{>0}^{X_1} \to \mathscr{T}(S)$ of ℓ. Let $l \in \mathbb{R}_{>0}^{X_1}$. This gives for us for every connected component P of $S \smallsetminus A$ a hyperbolic pair of pants $(P, J_{\bar{P},l})$. Such a hyperbolic pair of pants comes with a natural involution $\iota_{\bar{P},l}$. Choose on every boundary component a fixed point of this involution (that is for every boundary component a choice of one out of two) and then assemble these hyperbolic pair of pants into a hyperbolic surface C_l as prescribed by (S,A), by welding boundary component on boundary component in such a manner that the marked fixed points get identified as well. The resulting hyperbolic surface C_l comes with an evident decomposition Γ_l into hyperbolic pair of pants. It has the the additional remarkable property that the involutions in the parts come from a global involution (it is as if C_l were defined over $\mathbb{R}$), but this interesting feature is not so relevant here: we just wanted to give a construction that is sufficiently rigid so that we can say that we have produced a family of hyperbolic surfaces with hyperbolic pants decompositions that depend continuously on l. There is no obvious diffeomorphism between two incongruent right angled hexagons, so neither between the various C_l's, but our construction is such that they appear as the fibers of a locally trivial fibration $\mathscr{C} \to \mathbb{R}_{>0}^{X_1}$ of manifolds (with local trivializations also respecting the pants decompositions). This local triviality implies that any two fibers can be identified by means of a diffeomorphism which preserves both orientation and decomposition. That diffeomorphism is not unique, but its isotopy class is. So once we have chosen for some $l_o \in \mathbb{R}_{>0}^{X_1}$ an orientation preserving diffeomorphism $S \to C_{l_o}$ such that the pull-back of Γ_{l_o} is in A, then we have a well-defined isotopy class of such diffeomorphisms $S \to C_l$ for every $l \in \mathbb{R}_{>0}^{X_1}$. If we denote the pull-back of the conformal structure on C_l to S along this isotopy class by $j(l) \in \mathscr{T}(S)$, then we have obtained a continuous section $j : \mathbb{R}_{>0}^{X_1} \to \mathscr{T}(S)$ of ℓ.

The shearing actions defined by the given closed geodesics now produce an $\mathbb{R}^{X_1}$-equivariant map

$$J : \mathbb{R}^{X_1} \times \mathbb{R}_{>0}^{X_1} \to \mathscr{T}(S), \quad (t,l) \mapsto t(j(l)),$$

where on the right, $t = (t_\alpha)_\alpha$ acts by shearing the geodesic associated to α over t_α. It is clear that its composite with $\ell : \mathscr{T}(S) \to \mathbb{R}_{>0}^{X_1}$ is the obvious projection $(t,l) \in \mathbb{R}^{X_1} \times \mathbb{R}_{>0}^{X_1} \mapsto l \in \mathbb{R}_{>0}^{X_1}$. Notice that if $J(t,l) = J(t',l')$, then $l = l'$ (clear), but we also have $t_\alpha \cong t'_\alpha \pmod{\pi\mathbb{Z}}$ for every α, because the two fixed point pairs of the involution on the associated boundary components must be in the same relative position. It follows from the preceding that J is a homeomorphism onto its image, when restricted to $I \times \mathbb{R}^{X_1}_{>0}$, where $I \subset \mathbb{R}^{X_1}$ is a product of open intervals of length $\leq \pi$. In fact, J is a covering map, for if we let $I \subset \mathbb{R}^{X_1}$ be a product of open intervals of length $\leq \pi/2$, then it follows that $J^{-1}J(I \times \mathbb{R}^{X_1}_{>0})$ is a collection of copies of $I \times \mathbb{R}^{X_1}_{>0}$, each of which is obtained by a translating $I \times \mathbb{R}^{X_1}_{>0}$ over a vector, which if not lying in $(2\pi\mathbb{Z})^{X_1}$, then certainly belongs to $(\pi\mathbb{Z})^{X_1}$, and hence are pairwise disjoint.

A covering map going from a connected space to a simply connected space is always a homeomorphism. So the first assertion of the theorem follows if we show that $\mathscr{T}(S)$ is simply connected, i.e., that loop $t \in [0,1] \to u_t \in \mathscr{T}(S)$ is null-homotopic. To see this, lift u to a path $t \in [0,1] \to \tilde{u}_t \in \mathrm{Conf}(S)$. We can arrange $\tilde{u}$ to be a loop: as $\tilde{u}_0$ and $\tilde{u}_1$ are isotopic, there exists a path $t \in [0,1] \mapsto h_t \in \mathrm{Diff}(S)$ with $h_0 = 1$ and such that $h_{1*}\tilde{u}_1 = \tilde{u}_0$, and then $t \mapsto u'_t := h_{t*}\tilde{u}_t$ is a loop that lifts u. Assume now that $\tilde{u}$ is a loop. Since $\mathrm{Conf}(S)$ is contractible, $\tilde{u}$ will be then be null-homotopic and a contraction of $\tilde{u}$ gives one for u.

The last clause of the theorem is now seen as follows: it is clear that under the shearing representation, the subgroup $(2\pi\mathbb{Z})^{X_1} \subset \mathbb{R}^{X_1}$ maps onto D_A. Since the former acts faithfully on $\mathbb{R}^{X_1} \times \mathbb{R}^{X_1}_{>0}$ by translations, the map $(2\pi\mathbb{Z})^{X_1} \to D_A$ is an isomorphism. □

Hyperbolic surfaces with cusps

We here no longer assume that $g \geq 2$, but we suppose given a subset $P \subset S$ of n elements, such that the Euler characteristic of $S^\circ = S \smallsetminus P$ is negative, in other words, that $2g - 2 + n > 0$. In order to deal in a similar fashion with the Teichmüller space $\mathscr{T}(S,P)$ we make some observations about quotients of $\mathbb{H}$. We know that an element $g \in \mathrm{PSL}(2,\mathbb{R})$ which has no fixed points in $\mathbb{H}$ is either hyperbolic or parabolic. We show that this dichotomy is still manifest if we pass to the orbit space $\mathbb{H}_g$ of $\mathbb{H}$ by the subgroup of $\mathrm{PSL}(2,\mathbb{R})$ generated by g. First observe that the isomorphism type of $\mathbb{H}_g$ only depends on the conjugacy class of g in $\mathrm{PSL}(2,\mathbb{R})$: if $g' = hgh^{-1}$, then $h : \mathbb{H} \to \mathbb{H}$ induces an isomorphism $\mathbb{H}_g \cong \mathbb{H}_{g'}$. So if g is hyperbolic, then we may assume it is scalar multiplication by a positive real number $\neq 1$, and as we have seen in Lemma 4.1, $\mathbb{H}_g$ is then biholomorphic to an annulus $r_1 < |w| < r_2$ with $0 < r_1 < r_2$. When g is parabolic, then we may assume it is given by $z \to z+1$ and $w := e^{2\pi\sqrt{-1}z}$ then identifies $\mathbb{H}_g$ with the punctured unit disk $0 < |w| < 1$.

An annulus is not holomorphically equivalent to a punctured disk: a bounded holomorphic function on a punctured disk extends, by the Riemann extension theorem, holomorphically to the unit disk and so has a well-defined limit in the puncture. But for an annulus as above, the coordinate w is bounded and holomorphic and has the property that if we approach one of the boundary circles, it does *not* go to a constant value. We also notice a metric difference: for g hyperbolic, the evident isotopy class of embedded circles on $\mathbb{H}_g$ has a unique geodesic representative which minimizes length. But if $g(z) = z+1$, then $t \in [0,1] \mapsto t + \sqrt{-1}y$ defines a loop on $\mathbb{H}_g$ around the missing point of length $1/y$, and so this length can be made arbitrarily small in its isotopy class.

Suppose we are given a conformal structure J on S. Since $2g-2+n>0$, its restriction to S° underlies a complete hyperbolic metric, so that we can identify S° with $\Pi\backslash\mathbb{H}$ for a subgroup $\Pi\subset \mathrm{PSL}(2,\mathbb{R})$. Let $p\in P$. Choose a closed disk $D\subset S$ having p in its interior such that D meets no other point of P. So $D\smallsetminus\{p\}$ is closed in S°.

Let $q\in\partial D$ and let α the loop based at q which traverses ∂D. We identify Π with $\pi_1(S^\circ,q)$, so that α determines an element $g\in\Pi$. We begin with observing that α cannot be homotopically trivial. For then the function $w:=\frac{z-\sqrt{-1}}{z+\sqrt{-1}}$ (which maps on $\mathbb{H}$ isomorphically onto the unit disk $\mathbb{D}$) can be regarded as a holomorphic embedding of $D\smallsetminus\{p\}$ in $\mathbb{D}$. It extends holomorphically to $w:D\to\mathbb{C}$ (by the Riemann extension theorem) and the maximum principle forces $w(p)\in\mathbb{D}$. But since $D\smallsetminus\{p\}$ is closed in S°, its image in $\mathbb{D}$ must be closed as well and we get a contradiction.

In a similar fashion we can exclude that g is hyperbolic. For then $D\smallsetminus\{p\}$ would embed in an annulus $r_1<|w|<r_2$. By the Riemann extension theorem such an embedding extends holomorphically to D and the maximum principle then forces p taking a value in the interior of the annulus. But then we contradict the fact the image of $D\smallsetminus\{p\}$ in the annulus must be closed.

We conclude that g is parabolic so that $\mathbb{H}_g$ is isomorphic to the punctured disk $\mathbb{D}\smallsetminus\{0\}$. We have a holomorphic embedding $D\smallsetminus\{p\}\to\mathbb{D}\smallsetminus\{0\}$ and the Riemann extension theorem plus the maximum principle imply that this must extend to a holomorphic function $D\to\mathbb{D}$ which takes p to 0. This function is injective and so cannot have p as a multiple zero. This implies that it is in fact a holomorphic embedding on D. We conclude:

Proposition-definition 4.8 (Cusps on a hyperbolic surface). *The hyperbolic structure on (S°,J) makes of any $p\in P$ a* cusp, *by which we mean that a neighborhood D of p is isomorphic to a neighborhood of 0 in the unit disk $\mathbb{D}$ (with p mapping to 0) in such a manner that it becomes an isometry of $D\smallsetminus\{p\}$ onto $\mathbb{D}\smallsetminus\{0\}$, where the latter has been endowed with the hyperbolic metric it receives as the orbit space of $\mathbb{H}$ with respect to the group of integral translations.*

Pants decomposition in the presence of cusps

We have a notion of a pants decomposition of S° as before: it is given by closed compact submanifold of S which does not meet P and is given up to isotopy relative to P, and is such that each connected component of $S^\circ\smallsetminus A$ is diffeomorphic to the interior of a pair of pants (a thrice punctured sphere). A connected component of the boundary of $S^\circ\smallsetminus A$ is now a connected component of A or a singleton in P. We can still encode this in terms of a graph $G(S,P;A)$: the vertex set is $\pi_0(S^\circ\smallsetminus A)$ as before, but edges come in two types: those indexed by $\pi_0(A)$ (and called interior edges) and those indexed by P (the exterior edges). Edges of the last kind appear as 'leaves': they are connected with just one vertex.

Exercise 4.9. Prove that the graph $G(S,P;A)$ is trivalent, has $2g-2+n$ vertices and $3g-3+n$ interior edges.

If we are given a conformal structure J on S, then for the hyperbolic structure on S°, connected components of of $S^\circ \smallsetminus A$ are hyperbolic pairs of pants, except that some pants have become cusps. These are obtained as glueing a possibly degenerate hyperbolic hexagon (a $(6-n)$-gon with n improper points, where $n \leq 3$) onto its mirror image. By allowing this possibility, the arguments in the proof of Theorem 4.7 subsist and we find that this theorem continues to hold for $\mathscr{T}(S,P)$ as stated:

Theorem 4.10 (Fenchel-Nielsen parametrization in the punctured case)**.** *The action of $\mathbb{R}^{\pi_0(A)}$ on $\mathscr{T}(S,P)$ by geodesic shears makes $\ell : \mathscr{T}(S,P) \to \mathbb{R}^{\pi_0(A)}_{>0}$ a principal $\mathbb{R}^{\pi_0(A)}$-fibration. This bundle admits a section so that we have a homeomorphism of $(\mathbb{R}_{>0} \times \mathbb{R})^{\pi_0(A)}$ onto $\mathscr{T}(S)$. The Dehn twists around the connected components of A generate a free abelian subgroup of* $\mathrm{Mod}(S,P)$ *of rank $\pi_0(A)$.*

Remark 4.11. An embedded circle on S° which does not bound a disk that contains at most one point of P can always be made part of a pants decomposition. So this theorem implies the converse of what you proved in Exercise 1.1: a Dehn twist defined by such an embedded circle is never the identity.

The Fenchel-Nielsen parametrization makes $\mathscr{T}(S,P)$ homeomorphic to $(\mathbb{R}_{>0} \times \mathbb{R})^{3g-3+n}$. This homeomorphism gives it even the structure of a smooth manifold, although this a priori depends on the pants decomposition. But one can show that given this decomposition, then for any other isotopy class of embedded circles, the associated length function and shearing action is differentiable in terms of our $6g-6+2n$ Fenchel-Nielsen parameters. This implies that the manifold structure is independent of the pants decomposition. We shall see that $\mathscr{T}(S,P)$ has even a natural structure of a complex manifold of dimension $3g-3+n$.

5 Quadratic differentials

Local normal form

Let C be a Riemann surface. A *holomorphic quadratic differential* on C is a complex valued function on the total space of the tangent bundle TC which is quadratic, i.e., a function $\eta : TS \to \mathbb{C}$ such that for $v \in TC$ and $\lambda \in \mathbb{C}$ we have $\eta(\lambda v) = \lambda^2 \eta(v)$ and such that in terms of any local coordinate z, η takes the form $f dz^2$, with f holomorphic in z. (This is the same thing as a holomorphic section of the sheaf $\Omega_C^{\otimes 2} := \Omega_C \otimes_{\mathscr{O}_C} \Omega_C$.) We have a similar notion of a *meromorphic quadratic differential* (f is then meromorphic).

Lemma 5.1. *Let η be a meromorphic quadratic differential at $p \in C$ and let $k \in \mathbb{Z}$ be its order at p.*

If $k \geq -1$, then for some local coordinate z at p, $\eta = (k/2+1)^2 z^k dz^2$ (which we can write, at least formally, as $(d(z^{k/2+1}))^2$) and this coordinate is unique up to multiplication by a $(k+2)$th root of unity.

If $k = -2$, then there is a nonzero constant $c \in \mathbb{C}$ and a coordinate z at p such that $\eta = c(d\log z)^2 = cz^{-2}(dz)^2$ and c is unique (and called the squared residue *of η at p, denoted by* $\operatorname{Resq}_p \eta$*) and z is unique up to a scalar.*

Proof. We may as well assume that C is open in $\mathbb{C}$ and $p = 0$. So we have $\eta = f(z)dz^2$ with f meromorphic at $0 \in \mathbb{C}$ and of order k at 0. Suppose first $k \geq -1$. Then $\sqrt{f}$ is $z^{k/2}$ times a holomorphic function which is nonzero in 0. Since $k/2 > -1$, we can integrate $\sqrt{f}dz$ in the sense that there exists a g such that $g' = \sqrt{f}$ with g of the form $z^{k/2+1}u(z)$ for some holomorphic u with $u(0) \neq 0$. This implies that g can be written as the $(k/2+1)$th power of a local coordinate $\tilde{z}$. So then $\eta = (\sqrt{f}dz)^2 = (dg)^2 = (d(\tilde{z}^{k/2+1}))^2$. If $\tilde{w}$ is another local coordinate with this property, then $\tilde{w}^{k/2+1}$ and $\pm\tilde{z}^{k/2+1}$ differ by a constant, hence are equal: $\tilde{w}^{k/2+1} = \pm\tilde{z}^{k/2+1}$. It follows that $\tilde{w}/\tilde{z}$ is a $(k+2)$th root of unity.

Now assume $k = -2$. Then we may write $f(z) = z^{-2}u(z)$, where u is holomorphic with $u(0) \neq 0$. Let $c := u(0)$. Any other coordinate $\tilde{z}$ at 0 can be written as ze^{ϕ}, with ϕ holomorphic in 0. Then

$$c\left(\frac{d\tilde{z}}{\tilde{z}}\right)^2 = c(1+z\phi')^2\left(\frac{dz}{z}\right)^2.$$

So we must solve the equation $c(1+z\phi')^2 = u$ for ϕ. Now u/c takes the value 1 in 0 and so it has a square root v taking the value 1 in 0. If we write this square root as $1+zg$ with g holomorphic so that $u = c(1+zg)^2$, then we observe that the equation is solved by taking for ϕ an integral of g. You may check the uniqueness assertion yourself. □

Exercise 5.2. The preceding lemma does not extend as such to meromorphic quadratic differentials with a pole of even order ≥ 4: gives a meromorphic quadratic differential at $0 \in \mathbb{C}$ which has in 0 a pole of order 4, but can not be brought into the simple form $z^{-4}dz^2$. (Hint: use a notion of residue.)

Structure defined by a holomorphic quadratic differential

Let η be a meromorphic quadratic differential on C. Suppose first it has neither poles nor zeroes. According to Lemma 5.1 we can find at any point of C a local coordinate z, unique up to sign and an additive constant, such that $\eta = dz^2$. This means that $|\eta|$ defines on C a flat Euclidean metric. The conformal structure defined by this metric is clearly the given one. So if S is the surface underlying C, then the function $\eta : TS \to \mathbb{C}$ allows us to reconstruct the conformal structure.

Exercise 5.3. Prove that this is even true if η is meromorphic and not identically zero on any connected component of C. (Hint: show that a conformal structure on the unit disk which coincides with the usual one outside the origin, must be equal to the usual conformal structure.)

But η gives us more: we find an atlas whose transition maps are of the form $w \mapsto \pm w + \text{constant}$. In terms of a local coordinate z as above, the foliation defined by $\mathrm{Im}(z) = \text{constant}$ is therefore independent of that local coordinate and is hence globally defined. Alternatively, a tangent vector $v \in T_pC$ is tangent to this foliation precisely when $\eta_p(v)$ is real and ≥ 0 (we here think of η_p as a quadratic function $\eta_p : T_pC \to \mathbb{C}$). Similarly, the foliation defined by $\mathrm{Re}(z) = \text{constant}$ is globally defined and its tangent vectors are those on which η is real and ≤ 0. These two foliations are perpendicular with respect to the metric $|\eta|$. We call these the *horizontal* and the *vertical* foliations attached to η.

Now consider the case when η has order $k \geq -1$ at $p \in C$. Then the Euclidean structure and the two foliations are defined on a punctured neighborhood of p and we wish to understand what happens to that structure near p. According to Lemma 5.1 we can find a local coordinate z at p such that it acquires the standard form $\eta = (d(z^{k/2+1}))^2$. Let us first assume $k = -1$, so that $\eta = (d(z^{1/2}))^2$. The root $z^{1/2}$ is of course multivalued. It is best to understand this locally as a passage to the double cover defined by $w^2 = z$ so that the pull-back of η to this double cover becomes dw^2, i.e., the case we just analyzed. In particular, the horizontal and the vertical foliations pull back to the obvious ones. We recover geometrically the z-plane from the w-plane by taking the closed upper half plane in $\mathbb{C}_w$ defined by $\mathrm{Im}(w) \geq 0$ and identifying opposite points on its boundary (the real axis). This helps us to understand what happens to the two foliations. For instance, the horizontal foliation has as a leaf the positive real axis, the origin is a singular point, and its other leaves go almost around this half line. Similarly, the vertical foliation has as a leaf the negative real axis.

For general $k \geq -1$ we must take $k+2$ copies $(H_i)_{i \in \mathbb{Z}/(k+2)}$ of the closed upper half plane $\mathrm{Im}(w) \geq 0$ (with its two foliations and its Euclidean structure) and identify for $x \geq 0$, $x \in H_i$ with $-x \in H_{i-1}$. So the final result consists of $k+2$ sectors whose gluing rays are leaves of the horizontal foliation (except that when $k = 0$ the two rays join to form a single leaf). The same is true for vertical foliation (whose rays bisect the sectors). As an Euclidean metric $|\eta|$ is degenerate at such a point, but the notion of distance survives (so that we have a metric space in the usual sense). This is because for $a > 0$, $\int_0^a t^{k/2} dt$ converges and is finite for $k \geq -1$.

The case $k = -2$ is special. It then best to use the coordinate $w := \log z$ and to think of z as obtained from the strip $0 \leq \mathrm{Im}(w) \leq 2\pi$ by identifying its two boundary lines by means of a shift over $2\pi\sqrt{-1}$. Since the pull back of η is now $c dw^2$, this helps us understand the structure near p. In particular, we see that a punctured neighborhood of p is a metric product of an interval (a, ∞) and a circle so that a punctured neighborhood basis of p consists of the cylinders $\{(a', \infty) \times \text{circle}\}_{a' > a}$.

The case $c \in \mathbb{R}$ is of special interest. If $c > 0$, then in terms of the above metric product, the horizontal foliation is radial and the vertical foliation it circular: it has closed leaves only. For $c < 0$ the roles of horizontal and vertical foliation are interchanged. Notice that in either case the length of circular leaf is that of the circumference of the circle (measured by means of $|\eta|$, of course), which is here $2\pi\sqrt{|c|}$.

Quadratic differentials and the Teichmüller flow

Let S be a surface. Suppose we are given a conformal structure J on S and a holomorphic quadratic differential η for (S,J) without zeroes.

This generates a remarkable one-parameter family of pairs $(J_t, \eta_t)_{t\in\mathbb{R}}$, where J_t is a conformal structure on S and η_t is a holomorphic quadratic differential for J_t such that $(J_0, \eta_0) = (J, \eta)$ and η_t defines the same horizontal resp. vertical foliation as η. The definition is as follows: let $\sigma_t : \mathbb{C} \to \mathbb{C}$ be the $\mathbb{R}$-linear transformation defined by

$$\sigma_t(x + y\sqrt{-1}) = e^{-t/2}x + e^{t/2}y\sqrt{-1}.$$

Notice that $t \mapsto \sigma_t$ defines a group homomorphism $\mathbb{R} \to \mathrm{SL}(2,\mathbb{R})$. Now (J,η) defines an atlas of charts (U open in $\mathbb{C}, z : U \to \mathbb{C}$) in which η takes the form dz^2 and the horizontal resp. vertical foliations are given as the level sets of $\mathrm{Im}(z)$ resp. $\mathrm{Re}(z)$. For this atlas, any coordinate change is of the form $w \mapsto \pm w + c)$, where $c \in \mathbb{C}$. If we replace (U,z) by $(U, z_t := \sigma_t z)$, then the corresponding coordinate change will be $w \mapsto \pm w + \sigma_t(c)$ and hence will still be of this form. This shows that we have thus defined a new conformal structure J_t as well as a holomorphic quadratic differential η_t for (S, J_t) which in terms of the chart z_t is given as $(dz_t)^2$. The definition is canonical in the sense that if we start doing this for $(J_{t'}, \eta_{t'})$ instead of (J,η), then after time t we get $(J_{t+t'}, \eta_{t+t'})$. This has all the features of a geodesic in a Riemannian manifold (and indeed, it can be shown that its projection in $\mathscr{T}(S)$ is one relative to a Finsler metric).

The assumption that the quadratic differential be without zeroes or poles can be eliminated: all we need is that η is not identically zero on every connected component of S: the zeroes are then isolated. But then we must be willing to adapt the differentiable structure of S at such points, for the conformal structure (and the quadratic differential) need not extend for the given differentiable structure: the glueing of the $k+2$ copies of the upper half planes after applying σ_t to each copy will induce in the z-coordinate a homeomorphism that is no longer a diffeomorphism at its origin. Fortunately, there is a path of homeomorphisms connecting this homeomorphism with a diffeomorphism and the isotopy class of that diffeomorphism is unique. The details are not so hard, but tedious and so we omit them.

Exercise 5.4. Let $\sigma = \left(\begin{smallmatrix} a & b \\ c & d \end{smallmatrix}\right) \in \mathrm{SL}(2,\mathbb{R})$. Prove that for (U,z) as above, the charts $(U, {}^\sigma z := \sigma z)$ define a pair $({}^\sigma J, {}^\sigma \eta)$ consisting of a conformal structure plus a quadratic differential that is holomorphic for that structure. Prove that ${}^\sigma J = J$ if and only if $\sigma \in \mathrm{SO}(2)$. Bearing in mind that the stabilizer of $\sqrt{-1} \in \mathbb{H}$ in $\mathrm{SL}(2,\mathbb{R})$ is $\mathrm{SO}(2)$, conclude now that ${}^{\sigma}J$ only depends on $\sigma(\sqrt{-1}) \in \mathbb{H}$, so that we have a well-defined continuous map $\gamma : \mathbb{H} \to \mathscr{T}(S)$. Prove also that $\gamma(e^t\sqrt{-1}) = J_t$ as defined above.

We will see that if S is a connected surface and $P \subset S$ is finite such that S° has negative Euler characteristic, then for any conformal structure J on S, the cotangent space of $\mathscr{T}(S,P)$ at $[J]$ can be identified with the space of quadratic differentials on the Riemann surface (S,J) that have a pole of order ≤ 1 at the points of P and are holomorphic elsewhere. But we first show how certain meromorphic quadratic differentials can be used to almost triangulate $\mathscr{T}(S,P)$.

6 Ribbon graphs and ideal triangulation of Teichmüller space

Jenkins-Strebel differentials

Given a foliation on a surface, then a closed leaf of it either has a neighborhood consisting of closed leaves or it lies in the closure of an infinite number of (necessarily nonclosed) leaves. With this observation in mind we make the following definition.

Definition 6.1. A *Jenkins-Strebel differential* on a closed Riemann surface C is a meromorphic quadratic differential η on C which has everywhere order ≥ -2 and is such that all but a finite number of the leaves of its horizontal foliation are closed and encircle a point where η has order -2 with real negative squared residue.

By the above observation, the finitely many exceptional leaves can not be closed and will not have an accumulation point on a closed leaf. So they will join points where η has a zero or a pole of order one and are therefore the edges of an embedded graph $G(\eta) \subset C$ whose vertices are the aforementioned points. So if η has at $q \in C$ order $k \in \{-1,1,2,3,\dots\}$, then q is a vertex of order $k+2 \in \{1,3,4,5,\dots\}$ of $G(\eta)$. Each leaf has a length that can be measured be means of $|\eta|$: a closed leaf will encircle a point $p \in C$ where η has a pole of (exact) order 2, and then the preceding discussion shows that its length $\ell_p(\eta)$ equals $2\pi\sqrt{-\mathrm{Resq}_p\,\eta}$. A leaf that is an edge of the graph and has a simple pole of η as a vertex has still finite length because $\int_0^1 t^{-1/2}dt$ exists and is finite. Thus $G(\eta)$ is endowed with a metric. Notice that the orientation of C also determines a cyclic order of the leaves emanating from a given vertex.

Exercise 6.2. Verify that the quadratic differential $\eta^o := (z^2-z)^{-1}dz^2$ has no zeroes, has a simple pole in 0 and 1, and an order two pole at ∞ with squared residue -1. Prove that its horizontal foliation has the real interval $(0,1)$ as its only nonclosed leaf and that all the other leaves must be closed, encircle the real interval $[0,1]$ and fill up all of $\mathbb{C}-[0,1]$. Conclude that η^o is a Jenkins-Strebel differential on $\mathbb{P}^1$.

We now give a converse construction that starts out from a metrized graph such that the set of emanating edges of each vertex has been endowed with a cyclic order. This will lead in a surprising manner to an ideal triangulation of the Teichmüller space $\mathscr{T}(S,P)$.

Combinatorial description of a graph

We will only be concerned with finite graphs. We prefer to give a graph in combinatorial terms, similar to what we did for pants decompositions. That is, a graph consist of a finite set X (that we call the *set of oriented edges*, sometimes also called *half edges*), a fixed point free involution σ_1 of X and an equivalence relation $\sim_0$ on X. We denote by $x \in X \mapsto x_1 \in X_1$ the formation of the set of orbits of σ_1 (so this is a surjective 2-to-1 map) and by $x \in X \mapsto x_0 \in X_0$ the formation of the set of equivalence classes of $\sim_0$. We refer to X_0 resp. X_1 as the *vertex set* resp. *edge set* of the graph, in view of the fact that they are the thus named items of the geometric realization $G(X;\sim_0,\sigma_1)$ of the graph: this is by definition the quotient of a disjoint union of copies of $[-1,1]$ indexed by X: $\coprod_{x\in X}[-1,1]_x$, by the following two sets of relations: $t_x \sim -t_{\sigma_1 x}$ and $-1_x \sim -1_{x'}$ if $x \sim_0 x'$. We see here X_0 appear as the set of vertices (and realized as the image of $\{\pm 1_x\}_{x\in X}$) and X_1 as the set of midpoints of the edges (and realized as the image of $\{0_x\}_{x\in X}$). A *half edge* is the image of $[-1,0]_x$ for some $x \in X$. These cover the graph, for then $[0,1]_x = [-1,0]_{\sigma_1 x}$). Half edges are effectively labeled by X, with the members of an equivalence class of $\sim_0$ yielding the half edges (or equivalently, the oriented edges) coming out of the associated vertex. If we drop the condition that σ_1 be fixed point free, then this produces a graph 'with leaves' such as one which describes a pants decomposition in the presence of cusps.

Ribbon graphs

If we are also given a permutation σ_0 of X that induces the equivalence relation $\sim_0$ (this means that $x \sim_0 x'$ if and only if $x' = \sigma_0^r x$ for some r), then we call the triple $(X;\sigma_0,\sigma_1)$ (often abbreviated by $\sigma_\bullet$) a *set of ribbon data*. Notice that this amounts to giving a cyclic structure on the set of half edges emanating from a vertex: σ_0 then assigns to every such half edge a successor. We are going to show that then the above geometric realization canonically expands to an embedding of $G(\sigma_\bullet) = G(X;\sigma_0,\sigma_1)$ in a closed combinatorial surface $S(\sigma_\bullet)$ (oriented, as usual) in such a manner that the cyclic structure on the half edges emanating from a vertex comes from the orientation of $S(\sigma_\bullet)$. In fact, we will combinatorially reproduce almost everything that we attached to a Jenkins-Strebel differential.

To this end, we define another permutation σ_∞ of X by the property $\sigma_0\sigma_1\sigma_\infty = 1$. So if $x \in X$ represents an oriented edge, then $\sigma_\infty x = \sigma_1^{-1}\sigma_0 x$ represents the oriented edge that begins in the end point of x and precedes the oppositely oriented $\sigma_1 x$ for the cyclic ordering. We denote the formation of its orbit set by $x \in X \mapsto$

$x_\infty \in X_\infty$ and for reasons that become clear in a moment, we sometimes call a point of X_∞ a *boundary cycle*. Consider now the half strip $D \subset \mathbb{R}^2$ consisting of the $(t,u) \in \mathbb{R}^2$ with $|t| \leq 1$ and $u \geq 0$. Its one point compactification D^+ is the disjoint union of D and ∞ and for which a neighborhood basis of ∞ meets D in the subsets defined by $u > N$; we therefore parametrize it by the point $u = \infty$ and t undetermined. It is homeomorphic to a solid triangle: for example,

$$(t,u) \mapsto (a,b) := \frac{(t,u)}{u+1}, \qquad \infty \mapsto (0,1),$$

is a homeomorphism of D^+ onto the solid triangle defined by $|a| + b \leq 1$, $b \geq 0$. Consider the disjoint union copies of D^+ indexed by X, $\coprod_{x \in X} D_x^+$, and let $S(\sigma_\bullet)$ be the quotient of that union by the set of relations $(t,0)_x \sim (-t,0)_{\sigma_1 x}$ and $(1,u)_x \sim (-1,u)_{\sigma_\infty x}$ (including the case $u = \infty$). Let us write

$$\pi : \coprod_{x \in X} D_x^+ \to S(\sigma_\bullet)$$

for the quotient map. The π-images of the solid triangles D_x^+ cover $S(\sigma_\bullet)$ and define a triangulation of $G(\sigma_\bullet)$, except that in case $\sigma_1 x = \sigma_0 x$, the two vertical sides get identified via $(1,u)_x \sim (-1,u)_{\sigma_\infty x} = (-1,u)_x$. The quotient $S(\sigma_\bullet)$ is a closed combinatorial surface with an orientation inherited from the copies of D^+. (See Fig. 2) The sets $X_1 = \{(\pm 1, 0)_x\}_{x \in X}$, $X_1 = \{(0,0)_x\}_{x \in X}$ and $X_\infty = \{\infty_x\}_{x \in X}$ are effectively labeled by the orbit sets of σ_0, σ_1 and σ_∞ in X respectively. We sometimes refer to elements of X_∞ as *cusps*. Notice that in the presence of the first relation, the last one implies $(-1,0)_x \sim (-1,0)_{\sigma_1 x}$ and so we see $G(\sigma_\bullet)$ embedded in $S(\sigma_\bullet)$ as the image of $\cup_{x \in X}([-1,1] \times \{0\})_x$.

We have arranged things in such a manner that the cyclic ordering on the emanating edges of a vertex of $G(\sigma_\bullet)$ that comes from the orientation of $S(\sigma_\bullet)$ coincides with the one defined by σ_0. We also note that the projection $(t,u) \in D_x \mapsto t \in [-1,1]_x$ is a deformation retract, which descends to one of $S(\sigma_\bullet) \smallsetminus X_\infty$ onto $G(\sigma_\bullet)$. Therefore, $S(\sigma_\bullet) \smallsetminus X_\infty$ may be thought of as a thickened version of $G(\sigma_\bullet)$ which implements in a geometric (ribbonlike) manner the ribbon structure, whence the name. The obvious horizontal and vertical foliations on D determine (in a combinatorial sense) a horizontal and a vertical foliation of $S(\sigma_\bullet) \smallsetminus (X_0 \cup X_\infty)$. The horizontal foliation has all its leaves closed except the leaves that make up the edges of $G(\sigma_\bullet)$.

Notice that $S(\sigma_\bullet)$ is connected precisely if the group generated by σ_0 and σ_1 acts transitively on X. In that case we can compute the genus g by means of an Euler characteristic count: we have as many 2-cells as there are 1-cells having a point of X_∞ in their boundary, the remaining 1-cells are indexed by X_1 and the 0-cells are indexed by $X_0 \cup X_\infty$. It follows that

$$2 - 2g = |X_0| - |X_1| + |X_\infty|.$$

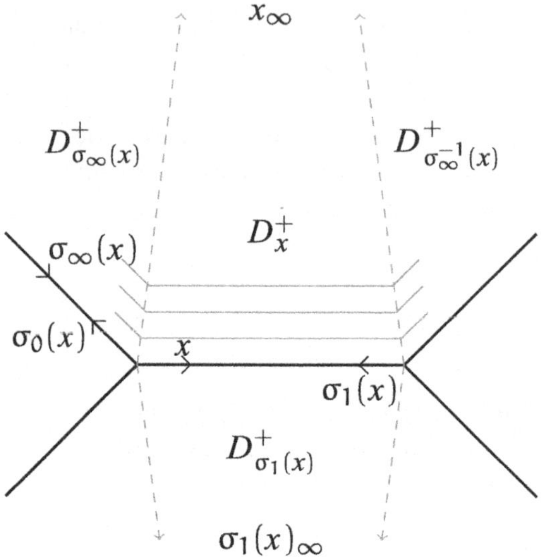

Fig. 2 $S(\sigma_\bullet)$ with some leaves of its horizontal foliation

We estimate the number of edges in terms of g:

Lemma 6.3. *Put* $n_\infty := |X_\infty|$ *and denote by* n_1 *resp.* n_2 *the number of vertices of degree* 1 *resp.* 2*. Then* $G(\sigma_\bullet)$ *has at most* $6g-6+3n_\infty+n_2+2n_1$ *edges and we have equality precisely when there are no vertices of degree* >3.

Proof. If n_r denotes the number of vertices of degree r (so that $|X_0| = \sum_r n_r$), then the cycle decomposition of σ_0 shows that $\sum_r rn_r = |X| = 2|X_1|$. On the other hand, the formula for the Euler characteristic yields $2g-2+n_\infty = |X_1| - |X_0|$. Hence $6g-6+3n_\infty = 3|X_1| - 3|X_0| = |X_1| - n_2 - 2n_1 + \sum_{r>3}(r-3)n_r$, from which the assertion follows. □

Exercise 6.4. The minimal (and degenerate) ribbon graph is defined by a singleton $X = \{*\}$ and so with $\sigma_0, \sigma_1, \sigma_\infty$ all trivial (we should think here of a graph that consists of a single half edge). Although σ_1 is no longer fixed point free, the above construction still makes sense. Prove that $S_o := S(1_\bullet)$ can be identified with $\mathbb{P}^1$ in such a manner that $G_o := G(1_\bullet)$ is mapped to $[0,1]$ and the three special points x_0, x_1, x_∞ are mapped to their namesakes $0, 1, \infty$. Try also to argue that this can even be done in such a way that this also respects the horizontal and vertical foliations associated to the quadratic differential η^o that we encountered in Exercise 6.2.

Exercise 6.5. Now construct for any set of ribbon data $(X; \sigma_0, \sigma_1)$ a natural (combinatorially defined) map $\pi : S(\sigma_\bullet) \to S_o$ with the property that $G(\sigma_\bullet) = \pi^{-1}G_o$ and the horizontal resp. vertical foliation is the preimage of that of S_o.

Remark 6.6. The map in the preceding exercise is a covering projection over the thrice punctured sphere $S_o \setminus \{x_0, x_1, x_\infty\}$. If we choose a base point on the interior of the half line

G_o, then the fiber over it is effectively indexed by X in such a manner that the monodromy around x_0 resp. x_1 is given by σ_0 resp. σ_1. This construction leads to a realization of the Galois group of $\mathbb{Q}$ as a permutation group of the collection of isomorphism types of ribbon graphs as follows: if $S(\sigma_\bullet) \to S_o$ is as in Exercise 6.5 and S_o has been identified with $\mathbb{P}^1$ as in Exercise 6.4, then there is a unique conformal structure on $S(\sigma_\bullet)$ for which the map $S(\sigma_\bullet) \to S_o \cong \mathbb{P}^1$ is holomorphic; the resulting Riemann surface $C(\sigma_\bullet)$ then naturally comes as a branched covering of $\mathbb{P}^1$ with branching in $\{0,1,\infty\}$ only, where over 1 we only have ramification of order 2. Such a Riemann surface is in fact an algebraic curve that is defined over a finite field extension of $\mathbb{Q}$: it can be given by homogeneous polynomial equations with coefficients in a finite extension K of $\mathbb{Q}$ contained in $\mathbb{C}$. If τ is a field automorphism of K, then applying τ to the coefficients of the equations that define $C(\sigma_\bullet)$ yields a new variety $C(\sigma_\bullet)^\tau$. Algebraic geometry tells us that this variety is again a smooth Riemann surface which still comes with a holomorphic map to $\mathbb{P}^1$ that is unramified over $\mathbb{P}^1 - \{0,1,\infty\}$ with over 1 only ramification of order 2. The preimage of $[0,1]$ in $C(\sigma_\bullet)^\tau$ yields a ribbon graph $(X,\sigma_\bullet)^\tau$ and $C(\sigma_\bullet)^\tau$ is the Riemann surface associated to the graph. We have thus defined an action of the Galois group $\mathrm{Gal}(\overline{\mathbb{Q}}/\mathbb{Q})$ of $\mathbb{Q}$ on the set of isomorphism types of ribbon graphs. The group $\mathrm{Gal}(\overline{\mathbb{Q}}/\mathbb{Q})$ is elusive and hard to understand as a group, whereas the set of isomorphism classes of ribbon graphs is rather concrete. Unfortunately no general method is known that can tell us whether two given ribbon graphs lie in the same Galois orbit.

The dual ribbon graph and the arc complex

The identity $\sigma_0\sigma_1\sigma_\infty = 1$ implies $\sigma_\infty^{-1}\sigma_1\sigma_0^{-1} = 1$ and so the triple $(\sigma_0^* := \sigma_\infty^{-1}, \sigma_1^* := \sigma_1, \sigma_\infty^* := \sigma_0^{-1})$ defines another ribbon graph $G(\sigma_\bullet^*)$, called the *dual* of $G(\sigma_\bullet)$. The associated surface $S(\sigma_\bullet^*)$ is homeomorphic with $S(\sigma_\bullet^*)$ under an orientation reserving homeomorphism that sends $X_0(\sigma_\bullet)$ to $X_\infty(\sigma_\bullet^*)$, $X_\infty(\sigma_\bullet)$ to $X_0(\sigma_\bullet^*)$ and $X_1(\sigma_\bullet)$ to $X_1(\sigma_\bullet^*)$. In particular, it sends the left half of $D_x^+ \subset S(\sigma_\bullet)$ with vertices $(-1,0)_x$, $(0,0)_x$, ∞_x to the right half of $(D_x^*)^+ \subset S(\sigma_\bullet^*)$ with vertices ∞_x^*, $(0,0)_x^*$, $(1,0)_x$ respectively. So under this homeomorphism, $G(\sigma_\bullet^*)$ is embedded in $S(\sigma_\bullet)$ as the union of the symmetry axes of the D_x which connect $(0,0)_x$ with ∞_x. We denote the latter by $G^*(\sigma_\bullet) \subset S(\sigma_\bullet)$. Notice that under this duality a vertex of degree one of $G(\sigma_\bullet)$ becomes the center of a loop of $G(\sigma_\bullet^*)$ and vice versa.

Definition 6.7. An *arc* of (S,P) is an isotopy class relative to P of an embedded interval which connects two points of P. We also allow the two points to coincide (so that we have a loop), but then require the loop not to bound a disk in S°. The *arc complex* $\mathscr{A}(S,P)$ is the simplicial complex whose vertex set is the set of arcs of (S,P) with a finite nonempty set of arcs representing a simplex precisely if that set can be simultaneously be represented by pairwise disjoint arcs. If we can do this in such a manner that their union $A \subset S$ has the property that every connected component of $S \smallsetminus A$ is homeomorphic to an open disk which meets P in at most one point, then we say that the simplex is *proper*.

Remark 6.8. The notion of an arc is such that if S° is endowed with a complete hyperbolic structure, then an arc is uniquely represented by an embedded complete geodesic connecting two cusps and two distinct arcs that have disjoint representatives yield disjoint geodesics (compare the discussion following Proposition of 4.2). So the arc complex is determined by its 1-skeleton: a finite set of arcs defines a simplex if and only if each pair of that set defines a 1-simplex.

Observe that $\mathrm{Mod}(S,P)$ acts on $\mathscr{A}(S,P)$.

Lemma 6.9. *The group* $\mathrm{Mod}(S,P)$ *has finitely many orbits in* $\mathscr{A}(S,P)$ *and so* $\mathbb{M}_{g,P} := \mathrm{Mod}(S,P)\backslash|\mathscr{A}(S,P)|$ *is compact.*

Proof. If we cut open S along an arc we obtain a surface of finite type with boundary. The resulting punctured surface has the property that the Euler characteristic of every connected component is smaller than that of S°. Now use the fact that there are only finitely many such surfaces up to diffeomorphism. □

If a simplex is improper, then so is every face of that simplex and so the improper simplices define a subcomplex $\mathscr{A}_\infty(S,P)$ of $\mathscr{A}(S,P)$. The point is that a pair $(S(\sigma_\bullet), X_\infty(\sigma_\bullet))$, where $\sigma_\bullet$ defines a ribbon graph, comes with a proper simplex of its arc complex, namely $G^*(\sigma_\bullet)$. Indeed, $G^*(\sigma_\bullet)$ has vertex set $X_\infty(\sigma_\bullet)$ and each connected component of $S(\sigma_\bullet) \setminus G^*(\sigma_\bullet)$ is an open cell with center a point of $X_0(\sigma_\bullet)$. This observation will be helpful in understanding the following definition:

Definition 6.10. An *(S,P)-marked ribbon graph* is a ribbon graph $(X,\sigma_\bullet)$ endowed with an orientation preserving piecewise-linear homeomorphism $h : S \to S(\sigma_\bullet)$ given up to an isotopy relative to P, with $h(P) \subset X_\infty \cup X_0$ and $h(P)$ containing X_∞ and all vertices of degree ≤ 2.

Notice that in terms of the notation of Lemma 6.3 we then have that $n := |P| \geq n_\infty + n_1 + n_2$. So the lemma in question shows that the number of edges $|X_1|$ is bounded by $3(2g-2+n)$, with equality holding if and only if the graph in question is trivalent (i.e., every vertex has degree 3).

Observe that the mapping class group $\mathrm{Mod}(S,P)$ acts on the class of (S,P)-marked ribbon graphs by letting $[g] \in \mathrm{Mod}(S,P)$ send $(X,\sigma_\bullet,[h])$ to the triple $(X,\sigma_\bullet,[hg^{-1}])$.

Given an (S,P)-marked ribbon graph $(\sigma_\bullet,[h])$, then h^{-1} maps an edge of $G^*(\sigma_\bullet)$ to an arc connecting two distinct points of P or to a loop at p when $h(p)$ is a vertex of degree one. We see that thus is defined a proper simplex $\Delta(\sigma_\bullet,[h])$ of $\mathscr{A}(S,P)$ which only depends on $(\sigma_\bullet,[h])$. Conversely, if we are given a proper simplex of $\mathscr{A}(S,P)$, then a representative set of pairwise disjoint arcs define a ribbon graph. If the dual of this ribbon graph is defined by $\sigma_\bullet$, then it is easily seen that we have naturally defined isotopy class of orientation reserving homeomorphisms relative to P, $S \cong S(\sigma_\bullet)$, which sends P to a subset of $X_0(\sigma_\bullet) \cup X_\infty(\sigma_\bullet)$ so that we have an (S,P)-marked ribbon graph.

Metrized ribbon graphs

Here is a beautiful construction that starts out with a set of ribbon data $(X,\sigma_\bullet)$ and a metric on $G(\sigma_\bullet)$, i.e., a function $e \in X_1 \mapsto l(e) \in \mathbb{R}_{>0}$, and leads to a Riemann surface plus a Jenkins-Strebel differential on that surface. Consider on D_x the complex coordinate $z_x := \frac{1}{2}l(x_1)dt + \sqrt{-1}du$. This identifies D_x with the half strip in $\mathbb{C}$ defined by $|\mathrm{Re}(z)| \le \frac{1}{2}l(x_1)$, $\mathrm{Im}(z) \ge 0$ and the associated holomorphic quadratic differential dz_x^2 yields the given horizontal and vertical foliation on D_x. The point is now that the holomorphic quadratic differentials $\eta_x := dz_x^2$ fit together nicely to form one on $S(\sigma_\bullet) \smallsetminus (X_0 \cup X_\infty)$, turning the latter into a Riemann surface. Indeed, $\pi(D_x)$ and $\pi(D_{\sigma_\infty x})$ are adjacent in the sense that have in common a vertical boundary piece and in the way they are fitted together, we see that their union has a natural conformal structure for which the differentials dz_x and $dz_{\sigma_\infty x}$ are restrictions of a holomorphic differential on the interior of their union (which is identified with an open half strip of width $l(x_1) + l((\sigma_\infty x)_1)$). The interior of $\pi(D_x \cup D_{\sigma_1 x})$ can be identified with the strip $|\mathrm{Re}(z)| < \frac{1}{2}l(x_1)$, at least when x and y lie not in the same σ_0-orbit (analyze that situation yourself), and we then see that z_x continues as a holomorphic function to $-z_{\sigma_1 x}$. So η_x then extends by $\eta_{\sigma_1 x}$. We thus end up with conformal structure $J(l)$ on $S(\sigma_\bullet) \smallsetminus (X_0 \cup X_\infty)$ plus a quadratic differential $\eta(l)$ on it. Both extend across $X_0 \cup X_\infty$: at a vertex in X_0 the quadratic differential will have a zero of order equal to the degree of that vertex less 2 (so a pole of order one if the vertex is an end vertex and no zero if it is of degree 2) and at any cusp $p \in X_\infty$, $\eta(l)$ will have a pole of order 2 with a real, negative squared residue.

We have set things up in such a manner that the following is true:

Observation 6.11. Let C be a closed connected Riemann surface endowed with a Jenkins-Strebel differential η. Then the nonclosed leaves of its horizontal foliation define an abstract graph and the orientation of C endows this graph with a ribbon structure so that we have a ribbon graph $G(\sigma_\bullet)$. If l is the metric on this graph defined by $|\eta|$, then there is a natural piecewise linear homeomorphism of $S(\sigma_\bullet)$ onto C such that $\eta(l)$ is the pull-back of η.

Let us return to the metrized ribbon graph $(S(\sigma_\bullet), l)$ above. We assume however that the metric is *unital*: the total length of the graph is 1. This means that l is in the relative interior $|\Delta(X_1)|^\circ$ of the simplex $|\Delta(X_1)|$ spanned by the edge set of σ.

Given $p \in X_\infty$, then as we know, the circumference of the closed leaves near p of the horizontal foliation is equal to $\ell_p := 2\pi\sqrt{-\mathrm{Resq}_p\,\eta}$. This is also the length of a boundary cycle:

$$\ell_p = \sum_{\{x | x_\infty = p\}} l(x_1).$$

Since $x \mapsto x_1$ is a 2-to-1 map, it follows that

$$\sum_{p\in X_\infty} \ell_p = \sum_{p\in X_\infty} \sum_{\{x|x_\infty = p\}} l(x_1) = \sum_{x\in X} l(x_1) = 2\sum_{e\in X_1} l(e) = 2.$$

Hence we have defined an affine-linear map

$$\ell/2 = (\ell_p/2)_{p\in P} : |\Delta(X_1)|^\circ \to |\Delta(X_\infty)|^\circ.$$

If we identify X_1 with X_1^*, then the conformal structures thus defined have a common piecewise linear structure on $S(\sigma_\bullet)$ and so this defines a continuous map

$$\Phi_{\sigma_\bullet} : |\Delta(X_1)|^\circ \to \mathscr{T}(S(\sigma_\bullet), X_\infty) \times |\Delta(X_\infty)|^\circ,$$

where the second component is $\ell/2$. In fact the construction does more: it endows each $S(\sigma_\bullet)$ with a Jenkins-Strebel differential. We may replace X_∞ by a subset P of $X_0 \cup X_\infty$ which contains X_∞ and then regard this as a map to $\mathscr{T}(S(\sigma_\bullet), P)) \times |\Delta(X_\infty)|^\circ \subset \mathscr{T}(S(\sigma_\bullet), P)) \times |\Delta(P)|$. So if we are given an (S,P)-marking $h : S \to S(\sigma_\bullet)$ of $(X, \sigma_\bullet)$, then we may regard this also as continuous map $\Phi_{\sigma_\bullet} : |\Delta(X_1)|^\circ \to \mathscr{T}(S,P) \times |\Delta(P)|$. It is then convenient to regard the (S,P)-marking simply as the given of a proper simplex of $\mathscr{A}(S,P)$. We thus showed that a proper simplex Δ of $\mathscr{A}(S,P)$ defines a continuous map

$$\Phi_\Delta : |\Delta|^\circ \to \mathscr{T}(S,P) \times |\Delta(P)|.$$

Let us now verify that when Δ' is a codimension one face of Δ that is still proper, then map $\Phi_{\Delta'}$ is simply the restriction of a continuous extension of Φ_Δ. It is clear that Δ' is obtained by removal of an arc that defines a vertex of Δ. In terms of the (S,P)-marked ribbon graph $(X, \sigma_\bullet)$ associated to Δ, this amounts to the contraction of an edge e so that we still have an (S,P)-marked ribbon graph. This means that e is not allowed to have its end vertices in (the image of) P, and that in case e is a loop, e must bound a disk containing a point of P: e can be represented by an $x \in X$ that is a fixed point of σ_∞, but the other representative $\sigma_1(x)$ may not have this property (and we still require that the vertex $\pi_0(x)$ of e is not an image of P). We can then indeed collapse e inside $G(\sigma_\bullet)$ and still get a ribbon graph with $X \smallsetminus \{x, \sigma_1 e\}$ as its set of oriented edges (in case e happens to be a loop, then the boundary cycle associated to x disappears). We call this an *edge contraction* and denote the resulting graph by $(X/e, (\sigma/e)_\bullet)$. A metric on the latter can be understood as a degenerate metric on $G(\sigma_\bullet)$, namely one which gives e zero length. We can also obtain the associated surface $S((\sigma/e)_\bullet)$ as a quotient of $S(\sigma_\bullet)$: the image of D_x^+ gets collapsed to an interval or to point: if e is not a loop, then identify two points in the image of D_x if they have the same imaginary part; otherwise contract the image of D_x^+ to a point. We do of course the same with $\sigma_1(x)$, but the images of D_x^+ and $D_{\sigma_1(x)}^+$ cannot both be contracted to a point. So the final result is the collapse of a topological disk in $S(\sigma_\bullet)$ onto an interval (Fig. 3). Although this is not a homeomorphism, such a collapse determines a natural isotopy class of homeomorphisms between them relative to any prescribed closed subset that is disjoint with the disk $\pi(D_x \cup D_y)$. The maps $\Phi_\Delta \cong \Phi_{\sigma_\bullet}$ and $\Phi_{\Delta'} \cong \Phi_{(\sigma/e)_\bullet}$ combine to define a continuous map defined on $\Delta \cup \Delta'$.

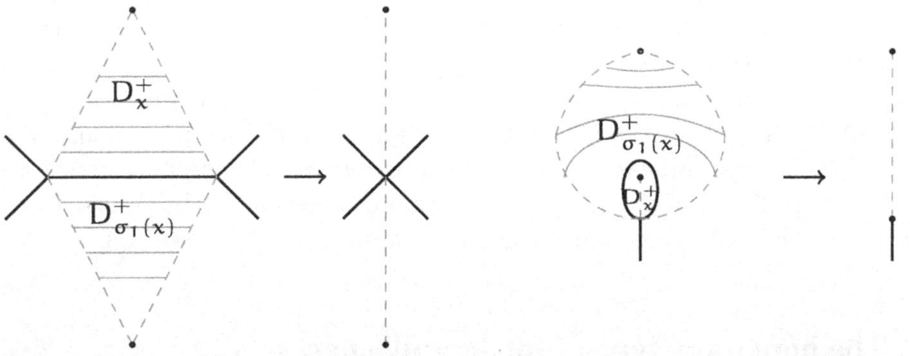

Fig. 3 The two kinds of collapses for trivalent ribbon graphs, producing a vertex of degree 4 resp. 1. The grey lines are leaf parts of the horizontal foliation and are fibers of the collapse; in the second case, the closure of D_x^+ is also a fiber

Ideal triangulation of thickened Teichmüller space

We may iterate this construction and the order in which we do so does not matter. So if we put $|\mathscr{A}(S,P)|^\circ := |\mathscr{A}_\infty(S,P)| \smallsetminus |\mathscr{A}_\infty(S,P)|$, then we have thus constructed a continuous map $\Phi_{(S,P)} : |\mathscr{A}(S,P)|^\circ \to \mathscr{T}(S,P) \times |\Delta(P)|$. One of the main results of the Jenkins-Strebel theory states:

Theorem 6.12. *The map $\Phi_{(S,P)} : |\mathscr{A}(S,P)|^\circ \to \mathscr{T}(S,P) \times |\Delta(P)|$ constructed above is a* $\mathrm{Mod}(S,P)$*-equivariant homeomorphism and (hence) drops to a homeomorphism of* $\mathbb{M}(g,P)^\circ := \mathrm{Mod}(S,P)\backslash|\mathscr{A}(S,P)|^\circ$ *onto* $\mathscr{M}_{g,P} \times |\Delta(P)|$*. In particular,* $\mathbb{M}_{g,P}$ *defines a compactification of* $\mathscr{M}_{g,P} \times |\Delta(P)|$.

This says essentially the following: if we are given a closed connected Riemann surface of genus g and a nonempty finite subset $P \subset C$ of n elements, then for every $\ell = \sum_{p \in P} \ell_p p \in |\Delta(P)|$ there is a unique Jenkins-Strebel differential η_ℓ on C whose polar set is contained in P and with $\mathrm{Resq}_p(\eta_\ell) = \ell_p^2/\pi$ for all $p \in P$. So $\ell_p = 0$ if and only if η_ℓ has a pole or order at most one at p. This is also equivalent to p not lying on one of the arcs make up the corresponding simplex of $\mathscr{A}(S,P)$ (S minus the union of these arcs has then a connected component that is an open disk which meets P in $\{p\}$).

We will not prove this theorem. One proof bears some formal similarity to the proof of the Fenchel-Nielsen parametrization and consists in showing that $\Phi_{(S,P)}$ is a covering projection; the simple connectivity of the range then implies that we must have a homeomorphism. Notice that the proper simplices are effectively indexed by the (S,P)-marked ribbon graphs.

Example 6.13. Let us do the case $(g,n) = (1,1)$. Following Lemma 6.3, a trivalent ribbon graph with just one boundary cycle that yields a genus 1 surface, has 3 edges. The Euler characteristic of the graph will be $2-2g-1 = -1$ and so we have two vertices. We then easily see that this ribbon graph is unique. We can contract one edge of this trivalent ribbon graph and still have a ribbon graph, but we cannot contract two. So we have 2-simplices and 1-simplices (each make up a single $\mathrm{Mod}_{1,1}$-orbit), but no 0-simplices: $\mathscr{A}_{1,1}$ is a simplicial complex of dimension two and $(\mathscr{A}_\infty)_{1,1}$ is simply the set of its vertices. We identified in Corollary 2.4 the pair $(\mathrm{Mod}_{1,1}, \mathscr{T}_{1,1})$ with $(\mathrm{SL}(2,\mathbb{Z}), \mathbb{H})$. In terms of this identification, $|\mathscr{A}_{1,1}|^\circ$ and its ideal triangulation is familiar from the theory of automorphic forms: $|\mathscr{A}_{1,1}|$ amounts to adding to $\mathbb{H}$ the set $\mathbb{Q}\cup\{\infty\}$ (or more canonically, $\mathbb{P}^1(\mathbb{Q})$) and a typical simplex is the ideal hyperbolic triangle with vertices $-1, 1, \infty \in \mathbb{P}^1(\mathbb{Q})$.

7 The homotopy type of the moduli spaces $\mathscr{M}_{g,n}$

In this section we give some applications of Theorem 6.12.

The homotopy type of $\mathscr{M}_{g,1}$

We here treat the case when P is a singleton: $P = \{*\}$. Theorem 6.12 then tells us that $|\mathscr{A}(S,*)|^\circ$ is equivariantly homeomorphic to the Teichmüller space $\mathscr{T}(S,*)$. The key point of this discussion is:

Lemma 7.1. *The dimension of $\mathscr{A}(S,*)$ is $6g-4$ and $\mathscr{A}_\infty(S,*)$ contains the $(2g-2)$-skeleton of $\mathscr{A}(S,*)$.*

Proof. A simplex Δ of $\mathscr{A}(S,*)$ not contained in $\mathscr{A}_\infty(S,*)$ defines a decomposition of S into polygonal cells with 0-skeleton $\{*\}$ and such that the vertices of Δ bijectively correspond to the 1-cells. So if c_k denotes the number of its k-cells, then $c_0 = 1$ and $\dim\Delta = c_1 - 1$. Since $c_0 - c_1 + c_2 = 2 - 2g$, we have $c_1 = 2g - 1 + c_2 \geq 2g$. The maximal value for c_2 is attained by a triangulation, in which case $3c_2 = 2c_1$: then $c_1 = 6g - 3$. So $2g - 1 \leq \dim\Delta \leq 6g - 4$. Since every improper simplex is a face of a proper one, the lemma follows. □

Corollary 7.2. *The complex $\mathrm{Mod}(S,*)$-manifold $\mathscr{T}_{g,*}$ has the equivariant homotopy type of a CW-complex with $\mathrm{Mod}(S,*)$-action of dimension $\leq 4g-3$. In particular, $\mathscr{M}_{g,1}$ has the homotopy type of a finite CW-complex of dimension $\leq 4g-3$.*

Before we give the proof, we recall a standard construction: if K is a simplicial complex, then the barycentric subdivision K' of K has as vertices canonically indexed by the nonempty simplices of K, with the vertex defined by the simplex σ of K here denoted b_σ. A finite set of those spans a simplex of K' if and only if its elements come from a strictly increasing sequence of simplices of K. We then identify $|K'|$ with $|K|$ in an obvious manner: the vertex b_σ of K' maps to the barycenter of σ.

Suppose now given a subcomplex $K_\infty \subset K$. Let $K_o \subset K'$ be the full subcomplex spanned by the vertices b_σ of K' defined by the simplices σ not in K_∞. This is called the *spine* of $|K| \smallsetminus |K_\infty|$, as there is a canonical deformation retraction from

$|K| \smallsetminus |K_\infty| = |K'| \smallsetminus |K'_\infty|$ onto $|K_o|$: any $x \in |K| \smallsetminus |K_\infty|$ lies in the relative interior of a simplex of K' not in K'_∞. This simplex can be represented by an increasing sequence of simplices in K, $\sigma_0 \subsetneq \sigma_1 \subsetneq \cdots \subsetneq \sigma_k$, with σ_k not in K_∞. Let i be the smallest index for which σ_i is not in K_∞ so that $\sigma_i \subsetneq \sigma_{i+1} \subsetneq \cdots \subsetneq \sigma_k$ defines a simplex of K_o. If we write $x = \sum_j x_j b_{\sigma_j}$, then $x_j \neq 0$ for some $j \geq i$, and hence $x' := \sum_{j \geq i} x_j b_{\sigma_j} / \sum_{j \geq i} x_j$ is defined and lies in $|K_o|$. The deformation retraction is then simply given by $(t,x) \mapsto (1-t)x + tx'$. Note that if K has finite dimension d and K_∞ contains the r-skeleton of K, then $\dim K_o \leq d - r - 1$ so that $|K| \smallsetminus |K_\infty|$ will have the homotopy type of a CW complex of dimension $\leq d - r - 1$.

Proof of Corollary 7.2. We apply the preceding discussion to $K = \mathscr{A}(S,*)$ and $K_\infty = \mathscr{A}_\infty(S,*)$. Invoking Lemma 7.1, we then find a canonical deformation retraction onto the geometric realization of a subcomplex of dimension $\leq (6g - 4) - (2g-2) - 1 = 4g - 3$. Since this deformation retraction is canonical, it is $\mathrm{Mod}(S,*)$-equivariant and the corollary follows, with the finiteness assertion being a consequence of the fact that $\mathrm{Mod}(S,*)$ has only finitely many orbits in the set of simplices of $\mathscr{A}(S,*)$. □

We use a similar argument to prove:

Corollary 7.3. *The mapping class group* $\mathrm{Mod}(S,P)$ *is finitely presented.*

Proof. We assume $P \neq \emptyset$ as the case $P = \emptyset$ can be deduced from this. We take $K = \mathscr{A}(S,P)$ and $K_\infty = \mathscr{A}_\infty(S,P)$. Then $|K_o|$ is contractible and hence simply connected. Moreover, $\mathrm{Mod}(S,P)$ acts on $|K_o|$ simplicially and with finite kernel and has a finite number of orbits in the set of its simplices. It is a general fact that this implies that $\mathrm{Mod}(S,P)$ is finitely presented. □

The moduli space of curves as a virtual classifying space

If Γ is an abstract group and M an abelian group on which Γ acts by automorphisms, then one has defined the cohomology groups $H^k(\Gamma;M)$. These can be defined directly as follows: regard $\mathbb{Z}$ as a (trivial) $\mathbb{Z}\Gamma$-module and choose a resolution $\cdots \to F_2 \to F_1 \to F_0 \to \mathbb{Z} \to 0$ by free $\mathbb{Z}\Gamma$-modules. Write $F_\bullet$ for the complex $\cdots \to F_2 \to F_1 \to F_0 \to 0$ and consider the (co)complex $\mathrm{Hom}_{\mathbb{Z}\Gamma}(F_\bullet, M)$. Then the kth cohomology group of this complex is independent of the resolution and is denoted $H^k(\Gamma, M)$: it is the kth cohomology group of Γ with values in M. The automorphism group $\mathrm{Aut}(\Gamma)$ of Γ acts on each of these groups, but as interior automorphisms are easily seen to act trivially, that is an action is via $\mathrm{Out}(\Gamma)$.

We can reproduce these groups topologically as the cohomology of a local system: let E be a contractible space on which Γ acts properly and freely (the space and the action must be decent, for instance, the space has admits the structure of a CW complex which is preserved by Γ) and denote by B the orbit space. Then the homotopy type of B only depends on Γ and B is called a *classifying space* of Γ. Moreover, the constant sheaf on E defined by M comes with a Γ-action and so determines a locally constant sheaf $\underline{M}$ on B; it has the property that $H^\bullet(B,\underline{M}) = H^\bullet(\Gamma,M)$. We say that Γ has *cohomological dimension* $\leq d$ (and write $\mathrm{cd}(\Gamma) \leq d$), if $H^k(\Gamma,M) = 0$ for every Γ-module and every $k > d$. So if B has the homotopy type of a CW-complex of dimension $\leq d$, then $\mathrm{cd}(\Gamma) \leq d$.

Example 7.4. Let F_n be a free group on n generators. Then the one point union of n circles may serve as a classifying space for F_n and so F_n has cohomological dimension 1.

There are corresponding 'virtual' notions: we say that a group Γ has *virtual cohomological dimension* $\leq d$ if it possesses a cofinite subgroup of cohomological dimension $\leq d$. This implies that it possesses the above vanishing properties for $\mathbb{Q}\Gamma$-modules. The geometric counterpart arises when we drop the freeness assumption on the action of Γ on E, but retain its properness and ask that a cofinite subgroup $\Gamma_1 \subset \Gamma$ acts freely. In that case, $\underline{M}$ need not be locally constant anymore (although it still so when we regard B as an orbispace; we then say that B is a *virtual classifying space* for Γ). In particular, $H^\bullet(B,\mathbb{Q}) \cong H^\bullet(\Gamma,\mathbb{Q})$.

This is precisely the situation that we have here for the action of the mapping class group $\mathrm{Mod}(S,P)$ on the Teichmüller space $\mathscr{T}(S,P)$. For we have defined a cellular decomposition of $\mathscr{T}(S,P) \times |\Delta(P)|$ that is invariant under the action of $\mathrm{Mod}(S,P)$. It has the property that the $\mathrm{Mod}(S,P)$-stabilizer of any cell is finite and that $\mathrm{Mod}(S,P)$ has only a finite number of orbits on the collection of cells. So there exists a cofinite subgroup $\Gamma \subset \mathrm{Mod}(S,P)$ which acts freely and properly on $\mathscr{T}(S,P) \times |\Delta(P)|$ and hence also on $\mathscr{T}(S,P)$. The Teichmüller space $\mathscr{T}(S,P)$ is contractible and so we have:

Proposition 7.5. *The orbifold $\mathscr{M}_{g,P}$ is a virtual classifying space for $\mathscr{M}_{g,P}$ so that the rational cohomology of $\mathscr{M}_{g,P}$ can be identified with the rational cohomology of* $\mathrm{Mod}_{g,P}$.

This proposition is supplemented by:

Theorem 7.6. *The virtual cohomological dimension of* $\mathrm{Mod}_{g,n}$ *can be bounded as follows:*

$$\mathrm{vcd}(\mathrm{Mod}_{g,n}) \leq \begin{cases} 4g-4+n, & \text{when both } g>0 \text{ and } n>0, \\ n-3, & \text{when } g=0 \text{ and } n\geq 3, \\ 4g-5, & \text{when } n=0 \text{ and } g \geq 2. \end{cases}$$

Proof. We first assume $g > 0$ and $n > 0$. Then the case $n = 1$ is covered by Corollary 7.2. We proceed with induction on n and assume the theorem verified for a given value of $n \geq 1$, i.e., there exists a cofinite subgroup $\Gamma \subset \mathrm{Mod}_{g,n}$ with $\mathrm{cd}(\Gamma) \leq 4g-4+n$. Now recall the Birman exact sequence

$$1 \to \pi_1(S^\circ, p_{n+1}) \to \mathrm{Mod}_{g,n+1} \to \mathrm{Mod}_{g,n} \to 1,$$

where $S^\circ := S \smallsetminus \{p_1,\dots,p_n\}$. Then its preimage $\tilde{\Gamma} \subset \mathrm{Mod}_{g,n+1}$ is cofinite in $\mathrm{Mod}_{g,n+1}$ and hence it suffices to see that $\mathrm{cd}(\tilde{\Gamma}) \le 1+\mathrm{cd}(\Gamma)$. If $\tilde{M}$ is any $\mathbb{Z}\tilde{\Gamma}$-module, then we have a spectral sequence

$$E_2^{p,q} = H^p(\Gamma, H^q(\pi_1(S^\circ, p_{n+1}), \tilde{M})) \Rightarrow H^{p+q}(\tilde{\Gamma}, \tilde{M}).$$

The left hand side is trivial for $p > \mathrm{cd}(\Gamma)$ or when $q > \mathrm{cd}(\pi_1(S^\circ, p_{n+1})) = 1$ (we here use that $\pi_1(S, p_{n+1})$ is free) and hence when $p+q > \mathrm{cd}(\Gamma)+1$. This yields the induction step: $H^r(\tilde{\Gamma}, \tilde{M}) = 0$ for $r > 1+\mathrm{cd}(\Gamma)$.

The same argument works $g=0$, albeit that the induction begins with $n=3$ (recall that $\mathrm{Mod}_{0,3}$ is trivial). Finally this argument is used in the opposite direction when $g \ge 2$, $n=0$: let $\Gamma \subset \mathrm{Mod}_g$ be a cofinite subgroup which acts freely on $\mathcal{T}_g$ and let M be a $\mathbb{Z}\Gamma$-module. Then $H^p(\Gamma, M) = 0$ for $p > \dim_{\mathbb{R}} \mathcal{T}_g$. If $H^\bullet(\Gamma, M)$ is nonzero, then let d be the largest degree of a nonzero summand. We must show that $d \le 4g-5$. From our spectral sequence $E_2^{p,q} = H^p(\Gamma, H^q(\pi_1(S,*), \tilde{M})) \Rightarrow H^{p+q}(\tilde{\Gamma}, \tilde{M})$, we then see that $E_2^{p,q}$ is zero for $p > d$ or $q > 2$. So $E_2^{d,2} = H^d(\Gamma, H^2(\pi_1(S,*), \tilde{M}))$ appears as a quotient of $H^{d+2}(\tilde{\Gamma}, \tilde{M})$. But $H^2(\pi_1(S,*) \cong H^2(S;\mathbb{Z})$ is a trivial $\mathbb{Z}\Gamma$-module, as Γ preserves orientation) and so $H^d(\Gamma, H^2(\pi_1(S,*), \tilde{M})) \cong H^d(\Gamma, M) \neq 0$. This implies that $d+2 \le \mathrm{cd}(\tilde{\Gamma}) = 4g-3$, that is, $d \le 4g-5$. □

In the above proof we could also have argued more geometrically by considering the map $\tilde{\Gamma}\backslash\mathcal{T}_{g,n+1} \to \Gamma\backslash\mathcal{T}_{g,n}$. This is a fiber bundle with typical fiber S° and the spectral sequence in question becomes one of Leray type.

Remark 7.7. Harer has shown in [5] that the the mapping class groups $\mathrm{Mod}_{g,n}$ are in fact (Bieri-Eckmann) virtual duality groups of the dimensions given by Theorem 7.6. This means that the bounds given there are optimal so that it in fact yields the virtual cohomological dimension.

8 A brief review of Dolbault cohomology and Serre duality for Riemann surfaces

With in mind applications to deformation theory, we briefly review Dolbault cohomology, Riemann-Roch and Serre duality for a Riemann surface.

Dolbault cohomology of a coherent sheaf

For a manifold M we denote by $\mathcal{E}_M^k$ the sheaf of its (C^∞) k-forms and by $\mathcal{E}_{M,\mathbb{C}}^k := \mathbb{C} \otimes_{\mathbb{R}} \mathcal{E}_M^k = \mathcal{E}_M^k + \sqrt{-1}\mathcal{E}_M^k$ its complexification. The exterior derivative extends complex-linearly to $d_{\mathbb{C}} : \mathcal{E}_{M,\mathbb{C}}^k \to \mathcal{E}_{M,\mathbb{C}}^{k+1}$.

We begin with a linear discussion. Let T be an oriented real vector space of dimension 2. If J is a conformal structure on T (understood here as a complex structure on T, i.e., as a transformation of T whose square is -1), then the complexification $T_{\mathbb{C}} := T \otimes_{\mathbb{R}} \mathbb{C}$ decomposes according to the eigen values of $J_{\mathbb{C}}$: the $\sqrt{-1}$-eigen space is $T^{1,0} := \{v - \sqrt{-1}Jv\}_{v\in T}$ and the $-\sqrt{-1}$-eigen space is $T^{0,1} := \{v + \sqrt{-1}Jv\}_{v\in T}$. Notice that these complex lines are complex conjugates of each other. Likewise, the 2-dimensional complex vector space $\mathrm{Hom}_{\mathbb{R}}(T,\mathbb{C})$ splits into two one-dimensional complex vector spaces: $(T^*)^{1,0}$ which consists of the maps ϕ that are $\mathbb{C}$-linear: $\phi(Jv) = \sqrt{-1}\phi(v)$, and its complex conjugate $(T^*)^{0,1}$, consisting of those that are anti-linear: $\phi(Jv) = -\sqrt{-1}\phi(v)$.

Let C be a Riemann surface. Then $\mathscr{E}^1_{C,\mathbb{C}}$ decomposes accordingly:

$$\mathscr{E}^1_{C,\mathbb{C}} = \mathscr{E}^{1,0}_C \oplus \mathscr{E}^{0,1}_C$$

and the two summands are interchanged by complex conjugation. We write ∂ for the first component of $d_{\mathbb{C}} : \mathscr{E}^0_{C,\mathbb{C}} \to \mathscr{E}^1_{C,\mathbb{C}} = \mathscr{E}^{1,0}_C \oplus \mathscr{E}^{0,1}_C$ so that $d_{\mathbb{C}} = \partial + \bar\partial$. If C happens to be an open subset U of $\mathbb{C}$, then $\bar\partial f = \frac{\partial f}{\partial \bar z} d\bar z$ and so this is the form that $\bar\partial$ will take in terms of a local coordinate. We thus see that complex valued function on C is in the kernel of $\bar\partial$ precisely, if in terms of any local coordinate its $\frac{\partial}{\partial \bar z}$-derivative is zero, and this is just saying that f is holomorphic.

This generalizes in a neat manner to holomorphic vector bundles over C. A holomorphic vector bundle over a Riemann surface is a complex vector bundle for which we are given a (maximal) atlas of charts for which the transition functions are holomorphic. But for our purpose a better way to introduce this notion is by singling out the class of its holomorphic local sections. If E/C is a complex vector bundle of rank r over a Riemann surface, then we denote by $\mathscr{E}_C(E)$ the sheaf of its smooth sections; this is a locally free $\mathscr{E}^0_{C,\mathbb{C}}$-module of rank r. But for the vector bundle $(T^*C)^{0,1} \otimes E$ we prefer to write $\mathscr{E}^{0,1}_C(E)$ instead of $\mathscr{E}_C((T^*C)^{0,1} \otimes E)$; if $s_1,\dots,s_r$ is a basis of sections of E over an open $U \subset C$, then a section $\omega \in \mathscr{E}^{0,1}(U,E)$ of $\mathscr{E}^{0,1}_C(E)$ over U has the form $\omega = \sum_{i=1}^r \omega_i \otimes s_i$, where each ω_i is a $(0,1)$-form on U (so if U happens to be the domain of a chart z, then $\omega_i = \phi_i d\bar z$ for some C^∞-function $\phi_i : U \to \mathbb{C}$).

Definition 8.1. A *holomorphic structure* on E/C amounts to giving a $\mathbb{C}$-linear sheaf map $\bar\partial_E : \mathscr{E}_C(E) \to \mathscr{E}^{0,1}_C(E)$ satisfying the following properties:

(i) if $s \in \mathscr{E}_C(E)$ and $f \in \mathscr{E}_{C,\mathbb{C}}$, then $\bar\partial_{E|U}(fs) = f\bar\partial_{E|U}(s) + \bar\partial f \otimes s$ and
(ii) $\mathscr{E}_C(E)$ is generated by $\mathrm{Ker}(\bar\partial_E)$.

If E/C has been equipped with such a structure, then we say that it is a *holomorphic vector bundle*. A local section $s \in \mathscr{E}_C(E)$ is then called *holomorphic* if $\bar\partial_E(s) = 0$.

Property (ii) says that we can cover C by open subsets U such that $E|U$ admits a basis of holomorphic sections. If $s_1,\dots,s_r$ is such a basis, then an arbitrary section

s of $E|U$ can be written $s = f_1 s_1 + \cdots f_r s_r$ and so $\bar{\partial}_{E|U}(s) = \sum_i \bar{\partial} f_i \otimes s_i$. We see that s is holomorphic if and only if the functions $f_1, \ldots, f_r$ are. So the local holomorphic sections of E form a sheaf of $\mathscr{O}_C$-modules locally free of the same rank as E/C. We denote that sheaf $\mathscr{O}(E)$. An integration argument (which we will not reproduce here) shows that $\bar{\partial}_E : \mathscr{E}_C(E) \to \mathscr{E}_C^{0,1}(E)$ is locally surjective, so that we have in fact a short exact sequence of abelian sheaves on C:

$$0 \to \mathscr{O}(E) \longrightarrow \mathscr{E}(C,E) \xrightarrow{\bar{\partial}_E} \mathscr{E}^{0,1}(C,E) \to 0.$$

Conversely, every sheaf of locally free $\mathscr{O}_C$-modules of finite constant rank comes from a vector bundle. This discussion also shows how we can pass from this definition to the atlas definition and vice versa.

Important examples are the holomorphic cotangent bundle of C, whose sheaf of sections is the sheaf of holomorphic differentials Ω_C, and its holomorphic dual, the sheaf of holomorphic vector fields θ_C. Notice that the standard ways of producing out of vector bundles a new one (like 'taking the dual' or 'tensor product') respect the category of holomorphic vector bundles.

Given a space X, then for every abelian sheaf $\mathscr{F}$ over X and every $k \in \mathbb{Z}$ is defined its cohomology group $H^k(X,\mathscr{F})$. This is zero for $k < 0$ and $H^0(X,\mathscr{F})$ is the space of its sections. These groups have among other things the property that a short exact sequence $0 \to \mathscr{F}' \to \mathscr{F} \to \mathscr{F}'' \to 0$ of such sheaves gives rise to a long exact sequence

$$\cdots \to H^k(X,\mathscr{F}') \to H^k(X,\mathscr{F}) \to H^k(X,\mathscr{F}'') \to H^{k+1}(C,\mathscr{F}') \to \cdots$$

which is canonical (relative to morphisms of such short exact sequences). Once one knows of a sufficiently large class of abelian sheaves that they do not have any cohomology in nonzero degree, then one is often able to compute these groups using the mentioned properties only (so without knowing the actual definition). An example of such a class consists of the sheaves that arise from differentiable vector bundles: the sheaf of C^∞-sections of a vector bundle over a manifold has no cohomology in nonzero degree. This can be used to show that the cohomology of the manifold M with coefficients in the constant sheaf $\mathbb{R}_M$ is the same as its De Rham cohomology.

A similar reasoning shows that for a holomorphic vector bundle E/C, $H^k(C, \mathscr{O}(E))$ is the cohomology of the *Dolbeault complex*

$$0 \to \mathscr{E}(C,E) \xrightarrow{\bar{\partial}_E} \mathscr{E}^{0,1}(C,E) \to 0,$$

meaning that

$$H^k(C,\mathscr{O}(E)) = \begin{cases} \mathrm{Ker}(\bar{\partial}_E) = H^0(C,\mathscr{O}(E)), & \text{if } k = 0, \\ \mathrm{Coker}(\bar{\partial}_E), & \text{if } k = 1, \\ 0, & \text{if } k \neq 0,1. \end{cases}$$

In case C is connected, but not compact, then a fundamental theorem asserts that $H^1(C,\mathcal{O}(E))=0$.

Riemann-Roch and Serre duality for a Riemann surface

In this subsection we assume that C is compact connected of genus g. We mention without proof:

Theorem 8.2. *The Dolbeault cohomology group $H^k(C,\mathcal{O}(E))$ and more generally, $H^k(C,\mathcal{F})$, where $\mathcal{F}$ is a coherent $\mathcal{O}_C$-module, is finite dimensional.*

The somewhat elusive $H^1(C,\mathcal{O}(E))$ can often be understood via Serre duality: denoting by $\mathcal{E}^{1,0}(C,E^*)$ resp. $\Omega(E^*)$ the sheaf of C^∞ resp. holomorphic sections of the holomorphic cotangent bundle of C tensored with E, then the duality pairing $E\otimes E^*\to\mathbb{C}_C$ determines a natural wedge product

$$\wedge:(\alpha,\beta)\in\mathcal{E}^{0,1}(C,E)\times\mathcal{E}^{1,0}(C,E^*)\to\alpha\wedge\beta\in\mathcal{E}^{1,1}(C)=\mathcal{E}^2(C)_\mathbb{C}.$$

Notice that if $\alpha=\bar{\partial}\alpha'$ with $\alpha'\in\mathcal{E}(C,E)$ and β is holomorphic (so $\beta\in H^0(C,\Omega(E^*))$), then $\alpha\wedge\beta=\bar{\partial}(\alpha'\wedge\beta)=d(\alpha'\wedge\beta)$. By Stokes, theorem this gets killed after integration over C. Hence $(\alpha,\beta)\mapsto\int_C\alpha\wedge\beta$ induces a pairing

$$H^1(C,\mathcal{O}(E))\times H^0(C,\Omega(E^*))\to\mathbb{C}.$$

The *Serre duality theorem* states that this pairing is nondegenerate.

Example 8.3. Serre duality identifies the dual of $H^1(C,\mathcal{O}_C)$ with $H^0(C,\Omega_C)$; in case C is connected, their common dimension is the genus of C (this is in fact the way to define the genus in Algebraic Geometry, where one cannot resort to topology). We also note the duality between $H^1(C,\Omega_C)$ and $H^0(C,\mathcal{O}_C)=\mathbb{C}$ and between $H^1(C,\theta_C)$ and $H^0(C,\Omega_C^{\otimes 2})$ (which is the space of holomorphic quadratic differentials).

For a coherent sheaf $\mathcal{F}$ of $\mathcal{O}_C$-modules, one defines its *Euler characteristic* $\chi(C,\mathcal{F})$ by $\chi(C,\mathcal{F}):=\dim H^0(C,\mathcal{F})-\dim H^1(C,\mathcal{F})$. If $0\to\mathcal{F}'\to\mathcal{F}\to\mathcal{F}''\to 0$ is a short exact sequence of coherent $\mathcal{O}_C$-modules, then its associated long exact sequence shows that $\chi(C,\mathcal{F}):=\chi(C,\mathcal{F}')+\chi(C,\mathcal{F}'')$. The Riemann-Roch theorem tells us that this Euler characteristic is of a topological nature. The formalism used here makes it easy to compute it from knowing $\chi(\mathcal{O}_C)$ only. In fact, if we put $\deg(\mathcal{F}):=\mathrm{rk}(\mathcal{F})(g-1)+\chi(\mathcal{F})$ so that

$$\chi(\mathcal{F})=\mathrm{rk}(\mathcal{F})(1-g)+\deg(\mathcal{F}),$$

then we show how to compute $\deg(\mathcal{F})$. With that recipe, the above definition gets mathematical content and is then called the *Riemann-Roch theorem*. We first observe that since χ and rk are additive for short exact sequences, deg is additive as well. The formula for the genus asserts that $\chi(\mathscr{O}_C) = 1 - g$ and so $\deg(\mathscr{O}_C) = 0$. If on the other hand $\mathscr{F}$ has finite support, then $H^1(C, \mathscr{F}) = 0$ and $\mathrm{rk}(\mathscr{F}) = 0$, so that $\deg(\mathscr{F}) = \dim H^0(C, \mathscr{F})$. Now let $D = \sum_{i=1}^k n_i(p_i)$ be a divisor on C, with $p_1, \dots, p_k$ distinct and with $n_i \geq 0$. Then we have a short exact sequence

$$0 \to \mathscr{O}_C \to \mathscr{O}_C(D) \to \mathscr{O}_C(D)/\mathscr{O}_C \to 0.$$

Notice that $\mathscr{O}_C(D)/\mathscr{O}_C$ has finite support and that $H^0(C, \mathscr{O}_C(D)/\mathscr{O}_C) \cong \oplus_{i=1}^k z^{-n_i}\mathbb{C}\{z\}/\mathbb{C}\{z\}$. It follows that the dimension of $H^0(C, \mathscr{O}_C(D)/\mathscr{O}_C)$ equals $\sum_{i=1}^k n_i$ and so $\deg(\mathscr{O}_C(D)/\mathscr{O}_C) = \dim H^0(C, \mathscr{O}_C(D)/\mathscr{O}_C) = \sum_{i=1}^k n_i$. Since $\deg(\mathscr{O}_C) = 0$, it then follows that $\deg(\mathscr{O}_C(D)) = \sum_{i=1}^k n_i$.

Exercise 8.4. Prove in a similar fashion that the same formula holds if more generally $n_i \in \mathbb{Z}$.

Exercise 8.5. Prove that $\deg(\Omega_C) = 2g - 2$.

9 Deformation theory of Riemann surfaces

Orbifolds

Let G be a Lie group which acts smoothly on a manifold M and denote by $\pi : M \to G\backslash M$ the formation of the quotient space. We assume that G acts properly (this means that $(g,p) \in G \times M \mapsto (gp, p) \in M \times M$ is proper) and that the G-stabilizer G_p of every $p \in M$ is finite.

Example 9.1. Let $k, l \in \mathbb{Z} \smallsetminus \{0\}$ and let the group $\mathbb{C}^\times$ act on $\mathbb{C}^2 \smallsetminus \{(0,0)\}$ by assigning to $\lambda \in \mathbb{C}^\times$ the diagonal matrix with entries (λ^k, λ^l). This action is proper. The stabilizer of (z_1, z_2) is a group of roots of unity of order $\gcd(k,l)$, k, l depending on whether $z_1 z_2 \neq 0$, $z_1 \neq 0 = z_2$ or $z_1 = 0 \neq z_2$.

Exercise 9.2. Prove that if k and l have the same sign and $d := \gcd(k,l)$, then $(z_1, z_2) \in \mathbb{C}^2 \smallsetminus \{(0,0)\} \mapsto [z_1^{l/d} : z_1^{k/d}] \in \mathbb{P}^1$ defines the orbit space. Show that if k and l have different sign, then the orbits of $(1,0)$ and $(0,1)$ cannot be separated in the orbit space.

Definition 9.3. A *slice* to the G-action on M is a submanifold $N \subset M$ such that the evaluation map $(g,q) \in G \times N \mapsto gq \in M$ sends a neighborhood of $\{1\} \times N$ in $G \times N$ diffeomorphically onto an open subset of M. We call it a *good slice* if in addition every $g \in G$ with $gN \cap N \neq \emptyset$ stabilizes N: $gN = N$.

Exercise 9.4. Prove that for any slice $N \subset M$, $\pi|N : N \to G\backslash M$ is open. Prove that if N is a good slice and G_N is the group of $g \in G$ with $gN = N$, then $\pi|N$ factors through a homeomorphism of $G_N\backslash N$ onto an open subset of $G\backslash M$.

Lemma 9.5. *Through every point $p \in M$ passes a good slice N whose G-stabilizer is the finite group G_p. If N' is another good slice through p, then after shrinking both N and N', there is a natural G_p-equivariant diffeomorphism $N \cong N'$ which commutes with π.*

Proof. Choose a Riemannian metric on M and make it G_p-invariant by replacing it by the sum of its G_p-transforms. The open unit ball B in the orthogonal complement of T_pGp in T_pM is clearly invariant under G_p. For $\varepsilon > 0$ small, the exponential map will map εB diffeomorphically onto a submanifold N_ε of M which meets every orbit transversally. Clearly, N_ε will be G_p-invariant.

The derivative of the map $(g,q) \in G \times N_\varepsilon \mapsto gq \in M$ in $(1,p)$ is a linear isomorphism and so by the inverse function theorem there exists a neighborhood U of $1 \in G$ and an $\varepsilon > 0$ such that the evaluation map $U \times N_\varepsilon \to M$ is a diffeomorphism onto an open subset of M.

Since G acts properly discontinuously on M we can take U and $\varepsilon > 0$ so small that $gN_\varepsilon \cap N_\varepsilon \neq \emptyset$ implies that $g \in UG_p$. So if $g \in G$ and $q \in N_\varepsilon$ are such that $gq \in N_\varepsilon$, then $g = g''g'$ with $g' \in G_p$ and $g'' \in U$. But then $g'q \in N_\varepsilon$ (for N_ε is G_p-invariant) and since $U \times N_\varepsilon \to M$ is injective, it follows that $g'' = 1$. Hence $N := N_\varepsilon$ is as desired.

If N' is another good slice, then after shrinking N, there is for every $q \in N$ a unique $g \in U$ with $gq \in N$. Then $q \in N \mapsto gq \in N'$ maps the good slice N diffeomorphically onto a good slice contained in N'. □

If N is a slice at p, then T_pN is a supplement in T_pM of the tangent space $T_p(Gp)$ of the G-orbit of p and so we may identify T_pN with the normal space $T_pM/T_p(Gp)$, or equivalently, with the cokernel of the map $Lie(G) \to T_pM$ obtained from the derivative of the G-action (where $Lie(G)$ denotes the Lie algebra of G).

Notice that if G_p is trivial, then a good slice maps homeomorphically onto an open subset of $G\backslash M$. This exhibits a C^∞-structure on that open subset. Lemma 9.5 shows that this structure is independent of the slice and that the tangent space of $G\backslash M$ at $\pi(p)$ may be identified with $T_pM/T_p(Gp)$.

When the stabilizers G_p are nontrivial one still feels that there is some structure on $G\backslash M$ worth remembering (and elucidating) and whose complexity does not go beyond that of the case when G is finite. A satisfactory way to do this is by regarding the good slices, or rather, their maps to open subsets of $G\backslash M$, as charts of $G\backslash M$, just as one may define a differentiable structure on a manifold. We will here loosely refer to this structure as an *orbifold*.

Remark 9.6. The actual definition of an orbifold does not assume that everything comes from a Lie group acting on a manifold. An *orbifold atlas* for a Hausdorff space X is a collection $\mathfrak{N}$ of triples (N, G_N, π_N), where N is a smooth manifold, G_N a discrete group acting properly discontinuously on N and $\pi_N : N \to X$ a continuous open map which is constant on the G_N-orbits and which induces a homeomorphism of $G_N\backslash N$ onto an open subset of X. In addition one assumes that there are the analogues of inclusions between open subsets that turn $\mathfrak{N}$ into a category: a morphism $(N', G_{N'}, \pi_{N'}) \to (N, G_N, \pi_N)$ is given by a group monomorphism $\phi : G_{N'} \to G_N$ and an open $G_{N'}$-equivariant embedding $f : N' \to N$ over X (i.e., such that $\pi_N f = \pi_{N'}$ and $fg = \phi(g)f$ for all $g \in G_{N'}$) which identifies $G_{N'}$ with the G-stabilizer of $f(N')$. This is set up in such a manner that the morphisms with fixed target (N, G_N, π_N) render the same service for $G_N\backslash N$ as the whole structure is supposed to provide for the original context (i.e., $G\backslash M$): we demand that any open covering of N can be refined as a covering by the images of $\mathfrak{N}$-morphisms.

An *orbifold structure* on X is simply an orbifold atlas that is not contained in a strictly bigger one.

An example to keep in mind is $M = \mathrm{Conf}(TN/N)$ and $G = \mathrm{Diff}^+(N,P)$. Its sheer infinite dimensionality makes this fall outside the scope of the discussion above. Yet the associated orbifold exists and is here representable by a single triple, namely $(\mathscr{T}(N,P), \mathrm{Mod}(N,P), \mathscr{T}(N,P) \to \mathscr{M}_{g,P})$. The notion of a good slice will here only serve as a useful heuristic that helps us to justify the notion of universal deformation that we will introduce in order to put more structure on $\mathscr{M}_{g,P}$.

Towards the complex tangent bundle of Teichmüller space

Here S is a closed connected surface of genus g and $P \subset S$ is an n-element subset. Let J be a conformal structure on S, so that we have a Riemann surface C. The stabilizer of J in $\mathrm{Diff}^+(S,P)$ is the group of automorphisms of C that are the identity on P. In the non-hyperbolic cases $(g,n) = (0,0), (0,1), (0,2), (1,0)$, this group is infinite, but otherwise it acts on C° as a group of isometries and (hence) is finite. We therefore assume that $2g-2+n > 0$. Let J be a conformal structure on S, so that we have a Riemann surface C. We want to compute the tangent space of $\mathscr{T}(S,P)$ at the point defined by $[J]$. For this we take a naive approach: we first determine the space of first order deformations of J in $\mathrm{Conf}(S)$ and divide out by the space of first order deformations that come from $\mathrm{Diff}^o(S,P)$ (of which we think as vector fields on S that are zero on P).

Let T be an oriented real vector space of dimension 2 as before. If J is a conformal structure on T, then we have defined a decomposition $T_{\mathbb{C}} = T_J^{1,0} \oplus T_J^{0,1}$ and the composite map $T \subset T_{\mathbb{C}} \to T_{\mathbb{C}}/T_J^{0,1}$ has the property that this yields a $\mathbb{C}$-linear isomorphism of (T,J) onto $T_{\mathbb{C}}/T_J^{0,1}$. We see from this description that J is completely determined by $T_J^{0,1}$. Conversely, if $L \subset T_{\mathbb{C}}$ is such that the $\mathbb{R}$-linear map $T \subset T_{\mathbb{C}} \to T_{\mathbb{C}}/L$ is an orientation preserving isomorphism of real vector spaces (where we give $T_{\mathbb{C}}/L$ the orientation defined by its complex structure), then we can transfer the complex structure of $T_{\mathbb{C}}/L$ to T and get an element of $\mathrm{Conf}(T)$. If J is given as above, then we cannot have $L = T_J^{1,0}$ (for $T \subset T_{\mathbb{C}} \to T_{\mathbb{C}}/T_J^{1,0}$ is orientation reversing) and so L is the graph of a $\mathbb{C}$-linear map $\phi : T_J^{0,1} \to T_J^{1,0}$. Which ϕ occur this way? If $e \in T$ is nonzero, then $e - \sqrt{-1}Je \in T_J^{1,0}$ resp. $e + \sqrt{-1}Je \in T_J^{0,1}$

are generators and so if $\phi(e+\sqrt{-1}Je)=\lambda(e-\sqrt{-1}Je)$, with $\lambda\in\mathbb{C}$, then

$$\begin{aligned} e+\sqrt{-1}Je+\lambda(e-\sqrt{-1}Je) &= (1+\lambda)e+\sqrt{-1}(1-\lambda)Je \\ &= (1+\lambda)\Big(e+\sqrt{-1}\frac{1-\lambda}{1+\lambda}Je\Big) \end{aligned}$$

is a generator of L. Since L cannot be real, it follows that $\mathrm{Re}(\frac{1-\lambda}{1+\lambda})\neq 0$, or equivalently, $|\lambda|\neq 1$. We need to be in the part where the complex structure is compatible with the given orientation and this amounts to the condition $|\lambda|<1$. We thus find:

Lemma 9.7. *The space of conformal structures on T, $\mathrm{Conf}(T)$, can be identified with the open subset of the complex projective line $\mathbb{P}(T_\mathbb{C})$ parametrizing the complex lines $L\subset T_\mathbb{C}$ for which $T\subset T_\mathbb{C}\to T_\mathbb{C}/L$ is an orientation preserving $\mathbb{R}$-linear isomorphism, hence has a natural complex structure (it is in fact a copy of the complex unit disk). Its tangent space $T_{[J]}\mathrm{Conf}(T)$ is as a complex vector space naturally isomorphic $\mathrm{Hom}_\mathbb{C}(T_J^{0,1},T_J^{1,0})=(T^*)_J^{0,1}\otimes_\mathbb{C}T_J^{1,0}$.*

We return to $C=(S,J)$. Lemma 9.7 shows that the tangent space of $\mathrm{Conf}(TS/S)$ at $[J]$ is the space of sections of the complex vector bundle $\mathrm{Hom}(T^{0,1}C,T^{1,0}C)$. But this is just $\mathscr{E}^{0,1}(T^{1,0}C)$.

Suppose we have a curve $t\in(-\varepsilon,\varepsilon)\mapsto h_t\in\mathrm{Diff}^0(S)$ with $h_0=1$. Then this defines a vector field $X:=\frac{d}{dt}\big|_{t=0}h_t$ on S, which we consider as a section of the holomorphic tangent bundle $T^{1,0}C$. We thus regard $\mathscr{E}(C,T^{1,0}C)$ as the Lie algebra of $\mathrm{Diff}^0(S)$. The tangent vector of $\mathrm{Conf}(TS/S)$ defined by the curve $t\in(-\varepsilon,\varepsilon)\mapsto h_t^*J$ is given as $\frac{d}{dt}\big|_{t=0}h_t^*J\in\mathrm{Hom}(T^{0,1}C,T^{1,0}C)$ and we want to understand this as an element of $\mathscr{E}^{0,1}(T^{1,0}C)$. If (U,z) is a coordinate chart for C, then

$$h_t^*dz=\frac{\partial h_t}{\partial z}dz+\frac{\partial h_t}{\partial\bar z}d\bar z.$$

So the family of conformal structures h_t^*J is in terms of our chart given by

$$t\mapsto -\frac{\frac{\partial h_t}{\partial\bar z}}{\frac{\partial h_t}{\partial z}}\frac{\partial}{\partial z}\otimes d\bar z.$$

Since $h_t^*dz\equiv dz \pmod t$ (for h_0 is the identity), its derivative at $t=0$ is

$$-\frac{\partial}{\partial t}\Big|_{t=0}\frac{\partial h_t}{\partial\bar z}d\bar z=\frac{\partial}{\partial\bar z}\Big(\frac{\partial h_t}{\partial t}\Big|_{t=0}\Big)d\bar z=-\bar\partial X.$$

So the tangent vector of $\mathrm{Conf}(TS/\underline{S})$ is on U given by $-\bar\partial X\in\mathscr{E}^{0,1}(C,T^{1,0}C)$. This makes it reasonable to regard $\bar\partial\mathscr{E}(C,T^{1,0}C)\subset\mathscr{E}^{0,1}(C,T^{1,0}C)$ as the tangent space at J at its $\mathrm{Diff}^0(S)$-orbit so that the tangent space $T_{[J]}\mathscr{T}(S)$ of $\mathscr{T}(S)$ at $[J]$ can be identified as the Dolbeault cohomology space $H^1(C,\theta_C)$. Note this tangent space then comes a natural complex structure. By Serre duality its complex dual (the cotangent space $T^*_{[J]}\mathscr{T}(S)$) should be identified with the space of quadratic differentials $H^0(C,\Omega_C^{\otimes 2})$.

A similar heuristic reasoning shows that the tangent space $T_{[J]}\mathscr{T}(S,P)$ of $\mathscr{T}(S)$ at $[J]$ can be identified with $H^1(C,\theta_C(-P))$, where $\theta_C(-P)$ is the sheaf of holomorphic vector fields which vanish in P.

Deformation theory

The Ehresmann fibration theorem says that a proper submersion $f:\mathscr{M}\to B$ between manifolds is C^∞-locally trivial. So for any $o\in B$ there exist an open neighborhood U of o in B and a differentiable retraction $r:\mathscr{M}_U\to M_o$ such that $(f_U,r):\mathscr{M}_U\to U\times M_o$ is a diffeomorphism. Notice that then for every $b\in U$, r restricts to a diffeomorphism $r_b:M_b\cong M_o$. This diffeomorphism varies smoothly with b and since r is a retract, r_o is the identity. If U is path connected, then the isotopy class of r_b is independent of the choice of r, for if r' is another differentiable retraction $r:M_U\to M_o$ as above, then choose a path $\gamma:[0,1]\to U$ from o to b. The map $t\in[0,1]\mapsto h_t:=r'_{\gamma(t)}r^{-1}_{\gamma(t)}\in\mathrm{Diff}(M_o)$ is a path with h_0 the identity and so $h_t r_o$ connects $h_0 r_b=r_b$ with $h_1 r_b=r'_b r_b^{-1} r_b=r'_b$. In other words, we have a natural identification of M_b with M_o up to isotopy. The Ehresmann fibration theorem has also relative version: if in the situation above, $\mathscr{N}\subset\mathscr{M}$ is a closed submanifold such that $f|\mathscr{N}$ is also a submersion, then we may choose U and r in such a manner that (f_U,r) maps $\mathscr{N}_U$ onto $U\times N_o$. We find that for U path connected, we have for every $b\in U$ a well-defined isotopy class of diffeomorphisms of (M_b,N_b) onto (M_o,N_o). This makes the Ehresmann fibration theorem a convenient tool for the study of complex structures up to isotopy. In particular, it suggests:

Definition 9.8 (A family of pointed Riemann surfaces)**.** Let B be a complex manifold. A *family of P-pointed Riemann surfaces of genus g* consists a proper holomorphic submersion of complex manifolds $f:\mathscr{C}\to B$ endowed with a set x_P pairwise disjoint holomorphic sections $(x_p:B\to\mathscr{C})_{p\in P}$ such that every fiber $C_b:=f^{-1}(b)$ is a connected Riemann surface of genus g. We call the domain and range of f the *total space* and the base respectively.

So if $U\subset B$ is a path connected open subset such that f is C^∞-locally trivial over U relative to the sections indexed by P with fiber (S,P), then we have define a map from to the Teichmüller space $\mathscr{T}(S,P)$. We will also need a local version:

Definition 9.9 (Deformation of a pointed Riemann surface)**.** Let C be a closed Riemann surface and $P\subset C$ a finite subset. A *deformation* of (C,P) consists of a holomorphic submersion of complex manifolds $f:\mathscr{C}\to B$, a set x_P of holomorphic sections $(x_p:B\to\mathscr{C})_{p\in P}$, a point $o\in B$ and an isomorphism of complex manifolds $\iota:C\cong C_o$ such that $\iota(p)=x_p(o)$ for all p. It is here understood that for any neighborhood U of o in B, the system $(f_U:\mathscr{C}_U\to U,x_P|U,o,\iota)$ defines the same deformation.

A deformation is therefore given as a system $\mathscr{D} = (f, x_P, \iota)$ where $f : (\mathscr{C}, C_o) \to (B, o)$ is a *germ* of a submersion, $\iota : C \cong C_o$ an isomorphism, and $x_p : (B, o) \to (\mathscr{C}, C_o)$ a germ of a section with $\iota(p) = x_p(o)$. We need not require that the sections are pairwise disjoint: since their values in o are distinct, so will be their values in a neighborhood of o. The preceding shows that if S is the surface underlying C, then for B connected but so small that the sections x_p are pairwise disjoint, then such a deformation yields a continuous map $B \to \mathscr{T}(S, P)$. We call this the *classifying map* of the deformation (although strictly speaking this is only defined as map-germ $(B, o) \to \mathscr{T}(S, P)$). Note also that f then defines a deformation of (C_b, P) for every $b \in B$.

The deformations are the objects of a category $\mathfrak{Def}_{C,P}$ for which a morphism $\mathscr{D}' \to \mathscr{D}$ is given by a Cartesian diagram

$$\begin{array}{ccc} (\mathscr{C}', C'_o) & \xrightarrow{\Phi} & (\mathscr{C}, C_o) \\ {\scriptstyle f'}\downarrow & & \downarrow{\scriptstyle f} \\ (B', o') & \xrightarrow{\phi} & (B, o) \end{array}$$

of holomorphic maps such that $\iota = \Phi\iota'$ and $\Phi x'_p = x_p\phi$ for all $p \in P$. We recall that the Cartesian property means that the diagram is commutative and that the resulting map from $(\mathscr{C}', C'_o)$ to the fiber product $(B', o') \times_{(B,o)} (\mathscr{C}, C_o)$ is an isomorphism, where

$$B' \times_B \mathscr{C} = \{(b', q) \in B' \times \mathscr{C} \mid \phi(b') = f(q)\}.$$

The Cartesian property implies that Φ maps $C'_{b'}$ isomorphically onto $C_{\phi(b')}$, taking $x'(p)$ to $x(p)$. So the mere existence of a morphism $\mathscr{D}' \to \mathscr{D}$ implies that $\mathscr{D}$ is richer than $\mathscr{D}'$ in the sense that every deformed complex structure appearing in the family $\mathscr{D}'$ appears in $\mathscr{D}$.

Remark 9.10. One actually wants to allow the base germ (B, o) to be singular and ask that f be local-analytically trivial on $\mathscr{C}$: at every point $x \in C_o$, f is the last component of an isomorphism of germs $(\mathscr{C}, x) \cong (C_o, x) \times (B, o)$ (this is usually phrased by asking that f be *smooth*), but the more restrictive definition above is good enough for our purposes.

Definition 9.11. We say that a deformation $\mathscr{D}$ is *universal* if it is a final object of the deformation category $\mathfrak{Def}_{C,P}$.

We recall that a universal object of a category must be unique up to unique isomorphism. The notion of a universal deformation will serve as a workable substitute for our previously defined notion of a good slice. Let us show for instance, that if $\mathscr{D} = (f, x_P, \iota)$ is a universal deformation, then the stabilizer of the point of $\mathscr{T}(S, P)$ defined by the conformal structure defined by J acts on it. To see this, note that if $h \in \mathrm{Aut}(C, P)$, then $\mathscr{D}_h = (f, x_P, \iota h^{-1})$ is also a deformation (C, P) and so there is a unique morphism $(\Phi_h, \phi_h) : \mathscr{D}_h \to \mathscr{D}$. This means that the triple

(h,Φ,ϕ_h) defines an endomorphism of the diagram $C\xrightarrow{\iota}(\mathscr{C},C_o)\xrightarrow{f}(B,o)$ in such a way that it commutes with the sections $(x_p)_{p\in P}$. If $h'\in \mathrm{Aut}(C,P)$ is another automorphism, then the uniqueness property of the universal deformation implies that $\Phi_h\Phi_{h'}=\Phi_{hh'}$ and $\phi_h\phi_{h'}=\phi_{hh'}$. It then follows that $h\mapsto(h,\Phi_h,\phi_h)$ defines an action of $\mathrm{Aut}(C,P)$ on the universal deformation with $(h^{-1},\Phi_{h^{-1}},\phi_{h^{-1}})$ giving the two-sided inverse of (h,Φ,ϕ_h).

The Kodaira Spencer map

We define for a (not necessarily universal) deformation $\mathscr{D}=(f,x_P,\iota)$ a linear map $T_oB\to H^1(C,\theta_C)$ that can be understood as the derivative of the associated map-germ $(B,o)\to\mathscr{T}(S,P)$. Since f is a submersion and ι an isomorphism onto its fiber over o, we have for every $q\in C$ an exact sequence

$$0\to T_qC\xrightarrow{D_q\iota}T_{\iota(q)}\mathscr{C}\xrightarrow{D_{\iota(q)}f}T_oB\to 0.$$

This leads to the following exact sequence of vector bundles

$$0\to\theta_C\to\iota^*\theta_{\mathscr{C}}\to\mathscr{O}_C\otimes T_oB\to 0.$$

Let $\theta_{\mathscr{C}}(\log x_P)\subset\theta_{\mathscr{C}}$ denote the sheaf of vector fields on $\mathscr{C}$ that are tangent to the sections $(x_p)_{p\in P}$. So $\theta_{\mathscr{C}}(\log x_P)$ equals $\theta_{\mathscr{C}}$ away from $\cup_{p\in P}x_p(B)$. At a point $x_p(b)$ we can describe it as follows: choose local coordinates $(w_1,\dots,w_k)$ on a neighborhood V of $b\in B$ and a holomorphic function z_0 on a neighborhood U of $x_p(b)$ in $f^{-1}V$ which is zero on $x_p(B)$ and is such that $(z_0,z_1:=f^*w_1,\dots,z_k:=f^*w_k)$ is a coordinate system on U. Then the sheaf $\theta_{\mathscr{C},x_P}|U$ has as an $\mathscr{O}_U$-module the basis $z_0\frac{\partial}{\partial z_0},\frac{\partial}{\partial z_1},\dots,\frac{\partial}{\partial z_k}$. It follows that $\iota^*\theta_{\mathscr{C}}(\log x_P)$ is equal to $\iota^*\theta_{\mathscr{C}}$ away from P and if $p\in P$, then for a coordinate system as above with $b=o$ (so that $x_p(o)=\iota(p)$), $\iota^*\theta_{\mathscr{C}}|U_o$ has the $\mathscr{O}_{U_o}$-basis $z_0\frac{\partial}{\partial z_0},\frac{\partial}{\partial z_1},\dots,\frac{\partial}{\partial z_k}$, where $U_o=\iota^{-1}U$. We now see that the above sequence becomes the exact sequence

$$0\to\theta_C(-P)\to\iota^*\theta_{\mathscr{C}}(\log x_P)\to\mathscr{O}_C\otimes T_oB\to 0.$$

Its long exact cohomology sequence begins as

$$0\to H^0(C,\theta_C(-P))\to H^0(C,\iota^*\theta_{\mathscr{C}}(\log x_P))\to T_oB\xrightarrow{\delta}H^1(C,\theta_C(-P))\to\dots$$

and the map $\delta: T_oB\to H^1(C,\theta_C(-P))$ is the one are after. It is called the *Kodaira Spencer* map.

Lemma 9.12. *We have $H^0(C,\theta_C(-P))=0$ if and only if we are in the hyperbolic case ($2g-2+n>0$), and then $H^1(C,\theta_C(-P))$ is of dimension $3g-3+n$.*

Proof. Observe that $H^0(C,\theta_C(-P))$ is the space of holomorphic vector fields on C that are zero in P. Any nontrivial vector field on C has as many zeroes (counted with multiplicity) as its Euler characteristic $2-2g$. So for $g>1$, there is no such vector field, for $g=1$ it is without zeroes, and for $g=0$ it has two zeroes. These exceptions correspond to the non-hyperbolic cases.

Assuming that $H^0(C,\theta_C(-P))=0$, then the Riemann-Roch formula implies that $\dim_{\mathbb{C}} H^1(C,\theta_C(-P)) = g-1-\deg(\theta_C(-P))$. We have $\deg(\theta_C) = -\deg(\Omega_C)=2-2g$ and hence $\deg(\theta_C(-P))=2-2g-n$. The lemma follows. □

Here is the central result:

Theorem 9.13. *Assume that we are in the hyperbolic case: S° has negative Euler characteristic. Then a deformation (C,P) is universal if and only if its Kodaira-Spencer map is an isomorphism. If these two equivalent conditions are satisfied, then the deformation can be represented by a family which defines a universal deformation of each of its fibers and whose classifying map is homeomorphism onto an open subset of $\mathscr{T}(S,P)$.*

The proof, which we omit, involves a fair amount of work.

The classifying maps of universal deformations as above now provide an atlas for $\mathscr{T}(S,P)$. Since a universal deformation is unique up to unique (holomorphic) isomorphism, we conclude:

Corollary 9.14. *Suppose S° has negative Euler characteristic. Then $\mathscr{T}(S,P)$ has the structure of a complex manifold that is characterized by the property that the classifying maps of universal deformations are local isomorphisms. In fact, $\mathscr{T}(S,P)$ supports the* universal Teichmüller family*: it is the base of a P-pointed family $(\mathscr{C}(S,P)\to\mathscr{T}(S,P),x_P)$ of genus g Riemann surfaces in such a manner that for every $b\in\mathscr{T}(S,P)$, the germ of this family at b is a universal deformation of the fiber (C_b,P). The complex tangent space of $\mathscr{T}(S,P)$ at b is naturally identified with $H^1(C_b,\theta_{C_b}(P))$.*

The mapping class group $\mathrm{Mod}(S,P)$ *acts on this family as a group of biholomorphic transformations. The* $\mathrm{Mod}(S,P)$*-stabilizer of any $b\in\mathscr{T}(S,P)$ is just the group of automorphisms of C_b that leave every $x_p(b)$ fixed.*

So the group $\mathrm{Mod}(S,P)$ acts faithfully on $\mathscr{C}(S,P)$. It need not be so on $\mathscr{T}(S,P)$ as the two examples below will show.

Example 9.15. The case $(g,n)=(1,1)$. We know that $(\mathrm{Mod}_{1,1},\mathscr{T}_{1,1})$ can be identified with $(\mathrm{SL}(2,\mathbb{Z}),\mathbb{H})$ and so $\mathrm{SL}(2,\mathbb{Z})$ acts on $\mathbb{H}$ with kernel its center $\{\pm1\}$. The universal family pointed family $(f:\mathscr{C}\to\mathbb{H},x_0)$ is then obtained as the quotient of $\mathbb{H}\times\mathbb{C}$ by the lattice of translations acting by $(k,l)\in\mathbb{Z}^2:(\tau,z)\mapsto(\tau,z+k+l\tau)$ and $x_0:\mathbb{H}\to\mathscr{C}$ is given by $\tau\mapsto(\tau,0)$. The mapping class group action on this pointed family is then defined by letting $\left(\begin{smallmatrix}a&b\\c&d\end{smallmatrix}\right)\in\mathrm{SL}(2,\mathbb{Z})$ act on $\mathscr{C}$ as

$$(\tau,z+\mathbb{Z}+\tau\mathbb{Z})\in\mathscr{C}\mapsto\left(\frac{a\tau+b}{c\tau+d},\frac{z}{c\tau+d}+\mathbb{Z}+\frac{a\tau+b}{c\tau+d}\mathbb{Z}\right)\in\mathscr{C}.$$

This is best seen by regarding $\mathbb{H}$ as the space of oriented $\mathbb{R}$-bases of $\mathbb{C}$ up to multiplication by a complex nonzero scalar: then every basis is uniquely represented by one of the form $(1,\tau)$ with $\tau \in \mathbb{H}$. The group $\mathrm{SL}(2,\mathbb{Z})$ acts by letting $\left(\begin{smallmatrix} a & b \\ c & d \end{smallmatrix}\right)$ send the basis $(1,\tau)$ to the basis $(c\tau+d, a\tau+b)$, which is a basis of the same lattice $\mathbb{Z}+\tau\mathbb{Z}$. But that basis is represented by $(1,\frac{a\tau+b}{c\tau+d})$. Thus $(1,\tau;z+\mathbb{Z}+\tau\mathbb{Z}) \in \mathscr{C}$ goes to $(c\tau+d, a\tau+b; z+\mathbb{Z}+\tau\mathbb{Z})$, and this is represented by $(1,\frac{a\tau+b}{c\tau+d};\frac{z}{c\tau+d}+\mathbb{Z}+\frac{a\tau+b}{c\tau+d}\mathbb{Z}) \in \mathscr{C}$.

The element $-1 \in \mathrm{SL}(2,\mathbb{Z})$ leaves every fiber C_τ of $\mathscr{C} \to \mathbb{H}$ fixed and acts there as minus the identity: $z+\mathbb{Z}+\tau\mathbb{Z} \mapsto -z+\mathbb{Z}+\tau\mathbb{Z}$. That action has in C_τ as its fixed points the subgroup of elements of order ≤ 2: the image of $(\frac{1}{2}\mathbb{Z})^2/\mathbb{Z}^2$ in $C_\tau = \mathbb{C}/\mathbb{Z}+\tau\mathbb{Z}$. The orbit space $\{\pm 1\}\backslash C_\tau$ has the structure of a compact Riemann surface. Since -1 has 4 fixed points, we have the following identity for the Euler characteristic $2\chi(\{\pm 1\}\backslash C_\tau) - 4 = 2\chi(C_\tau) = 0$, and so $\chi(\{\pm 1\}\backslash C_\tau) = 2$. This implies that $\{\pm 1\}\backslash C_\tau \cong \mathbb{P}^1$. So if we divide out $\mathscr{C} \to \mathbb{H}$ by $\{\pm 1\}$, we find a family $\overline{\mathscr{C}} := \{\pm 1\}\backslash \mathscr{C} \to \mathbb{H}$ whose fibers are isomorphic to $\mathbb{P}^1$. It has four pairwise disjoint sections indexed by $P := \mathbb{Z}/2 \times \mathbb{Z}/2$, three of which can be used to trivialize this bundle.

In fact, we can say a bit more. Since $\overline{\mathscr{C}} \to \mathbb{H}$ with its four sections indexed by $P = (\frac{1}{2}\mathbb{Z})^2/\mathbb{Z}^2 \cong (\mathbb{Z}/2)^2$ is a family of P-pointed genus zero curves, we have a classifying map $\mathbb{H} = \mathscr{T}_{1,1} \to \mathscr{T}_{0,P}$ covered by a map $\overline{\mathscr{C}} \to \mathscr{C}_{0,P}$. Both are actually isomorphisms, for an inverse is defined as follows: given an embedding $i : P \hookrightarrow \mathbb{P}^1$, then we can make a double cover $C(i) \to \mathbb{P}^1$ of $\mathbb{P}^1$ ramified in the image of P and this produces a genus one Riemann surface for which the covering transformation acts as minus the identity relative to one of the ramification points (recall that an automorphism of a genus one surface can be linearized). If $g \in \mathrm{PSL}(2,\mathbb{C})$, then g lifts to an isomorphism $\tilde{g} : C(i) \cong C(gi)$ and $\tilde{g}$ is unique up to sign. This defines the inverses $\mathscr{T}_{0,P} \to \mathbb{H}$ and $\mathscr{C}_{0,P} \to \overline{\mathscr{C}}$.

Thus $\overline{\mathscr{C}} \to \mathbb{H}$ acquires the interpretation of $\mathscr{C}_{0,P} \to \mathscr{T}_{0,P}$. Under this identification, the mapping class group $\mathrm{Mod}_{0,P}$ becomes a subgroup of $\mathrm{PSL}(2,\mathbb{Z})$: $\mathrm{SL}(2,\mathbb{Z})$ acts on $\mathbb{Z}^2$ hence on the three element set $(\mathbb{Z}/2)^2 - \{(0,))\} = P - \{(0,0)\}$ (which we might think of as the projective line over $\mathbb{Z}/2$) and that last action is via $\mathrm{PSL}(2,\mathbb{Z})$. The image is the full symmetric group of these three points. The subgroup of $\mathrm{PSL}(2,\mathbb{Z})$ which leaves each of these fixed, i.e., the kernel of $\mathrm{PSL}(2,\mathbb{Z}) \to \mathrm{PSL}(2,\mathbb{Z}/2)$, appears here as $\mathrm{Mod}_{0,P}$.

This discussion shows that $\mathscr{T}_{1,1}$ and $\mathscr{T}_{0,P} \cong \mathscr{T}_{0,4}$ are isomorphic as complex manifolds, but not as orbifolds: this only happens after we have divided out the orbifold $\mathscr{T}_{1,1}$ by the trivially acting group $\{\pm 1\}$; we therefore better write this as $\overline{\mathscr{T}}_{1,1} \cong \mathscr{T}_{0,4}$.

Example 9.16. The case $(g,n) = (2,0)$ is similar to the preceding one. It is known that a closed connected Riemann surface C of genus 2 is hyperelliptic: it has an involution ι_C with the property that the orbit space $\iota_C\backslash C$ of ι_C is isomorphic to the Riemann sphere $\mathbb{P}^1$. This involution is unique and has 6 fixed points. Conversely, given six distinct points in $\mathbb{P}^1$, then the double cover of $\mathbb{P}^1$ ramified in these six points is of genus two. We have Mod_2 acting on $\mathscr{T}_2$ with a kernel $\{1,\iota\}$. The quotient $\overline{\mathrm{Mod}_2} := \mathrm{Mod}_2/\{1,\iota\}$ acts on $\mathscr{T}_{0,6}$, but permutes the six points and its image is in fact the full permutation group of these points. So we get an identification of $\overline{\mathscr{T}}_2 := \{1,\iota\}\backslash\mathscr{T}_2$ with $\mathscr{T}_{0,6}$ and an identification of $\mathrm{Mod}_{0,6}$ with the kernel of a surjection $\overline{\mathrm{Mod}_2} \to \mathfrak{S}_6$.

At this point it is clear that if divide out the universal Teichmüller family by $\mathrm{Mod}(S,P)$, then we do not get a family $\mathscr{C}_{g,P} \to \mathscr{M}_{g,P}$ of P-pointed Riemann surfaces over $\mathscr{M}_{g,P}$ when orbifold type structure is ignored. For a fibre is always the quotient of a Riemann surface by its group of automorphisms which leave P pointwise fixed. But if we take into account the orbifold structure, then it becomes legitimate to think of this as universal family. The notation $\mathscr{C}_{g,P} \to \mathscr{M}_{g,P}$ is then understood accordingly.

Exercise 9.17. The elliptic curve defined by the Gauß lattice $\mathbb{Z}+\sqrt{-1}\mathbb{Z}$, i.e., $C := \mathbb{C}/(\mathbb{Z}+\sqrt{-1}\mathbb{Z})$ (with the image of 0 as origin) has automorphism group of order 4. Prove this and analyze how the group acts on the universal deformation of C. Try to determine what kind of singularities appear on the total space if we divide out the universal deformation by this group.

The Weil-Petersson metric

Recall that if $b \in \mathscr{T}(S,P)$, then the complex tangent space $T_b\mathscr{T}(S,P)$ can be identified with $H^1(C_b, \theta_{C_b}(-P))$. Let us write C for C_b. By Serre duality, the complex dual $T_b\mathscr{T}(S,P)$ is then identified with $H^0(C,\Omega_C^2(P))$. We also have a complete hyperbolic metric on $C^\circ := C \smallsetminus P$. We show how this leads to a Hermitian inner product on $H^0(C,\Omega_C^2(P))$. Such an inner product determines one on its dual and so $T_b\mathscr{T}(S,P)$ receives an inner product as well. This is what is called the *Weil-Petersson metric* g_{WP} on $\mathscr{T}(S,P)$.

Let η_1,η_2 be sections of $\Omega_C^2(P)$. In terms of a local coordinate z on C° they are written $\eta_i = f_i dz^2$ and so $\eta_1 \otimes \overline{\eta_2}$ then takes the form $f_1\overline{f_2}|dz|^4$. If we write the hyperbolic metric g as $\phi|dz|^2$ (with ϕ real valued and positive), then $\eta_1 \otimes \overline{\eta_2}/g$ is given by $\phi^{-1}f_1\overline{f_2}|dz|^2$. We may interpret this as a 2-form by replacing $|dz|^2 = dx^2+dy^2$ by $dx\wedge dy = \frac{1}{2}\sqrt{-1}dz\wedge d\bar{z}$. We denote this 2-form $\eta_1 \star_{WP} \eta_2$ and then define the *Weil-Petersson inner product* by

$$\langle \eta_1, \eta_2\rangle_{WP} := \int_{C^\circ} \eta_1 \star_{WP} \eta_2.$$

It is however not yet clear that this integral converges, for C° is not compact. Let us see what the situation is near a point $p \in P$. Then f_i may have a pole of order one. On the other hand, by Proposition 4.8 we have a local coordinate z at p such that for $0 < |z| < \varepsilon$ such that if we write $z = \exp(2\pi\sqrt{-1}\tau)$, then the metric pulls back to $(\operatorname{Im}\tau)^{-2}|d\tau|^2$. We have $z^{-1}dz^2 = -4\pi^2\exp(2\pi\sqrt{-1}\tau)d\tau^2$ and so if we write $\tau = s+\sqrt{-1}t$, and $f_i = z^{-1}\phi_i$ with ϕ_i holomorphic and bounded by $C_i > 0$, say, then

$$|f_1 dz^2|.|f_2 dz^2|/g \leq C_1 C_2.|z^{-1}dz^2|^2/g = 16\pi^4 C_1 C_2 t^2 \exp(-4\pi t) ds \wedge dt,$$

and this is clearly absolutely integrable over the strip $0 \leq s \leq 1, t \geq \frac{-1}{2\pi}\log\varepsilon$. Hence $\int_{C^\circ} \eta_1 \star_{WP} \eta_2$ converges absolutely.

The Weil-Petersson metric identifies $T_b\mathscr{T}(S,P)$ and $H^0(C,\Omega_C^2(P))$ as real vector spaces, but under this identification their complex structures are opposite: we really have an identification of $T_b^{0,1}\mathscr{T}(S,P) \cong H^0(C,\Omega_C^2(P))$ as complex vector spaces. Via this identification we may understand the Teichmüller flow defined in Section 6 as one which is generated by a tangent vector of $\mathscr{T}(S,P)$. It turns out that the Weil-Petersson metric is not complete.

Let T be a real vector space endowed with complex structure J. If $H : T \times T \to \mathbb{C}$ is a Hermitian form on (T,J) then its imaginary part $A := \operatorname{Im} h : T \times T \to \mathbb{R}$ is antisymmetric (for $\operatorname{Im} H(v',v) = \operatorname{Im} \overline{H(v,v')} = -\operatorname{Im} H(v,v')$) and invariant under the complex structure in the sense that $H(Jv,Jv') = \operatorname{Im} H(v,v')$. The identity $H(v,v') = A(Jv,v') + \sqrt{-1}A(v,v')$ shows that H is determined by A and J and that H is nondegenerate if and only if A is. So a Hermitian metric h on a complex manifold M has its imaginary part $\operatorname{Im}(h)$ a nondegenerate 2-form. If that 2-form is closed, we say that h is a *Kähler* metric. Then the underlying C^∞-manifold endowed with the symplectic form $\operatorname{Im}(h)$ is a symplectic manifold.

Ahlfors proved that the Weil-Petersson metric is a Kähler metric. Wolpert found a nice relation with the shearing action that we mention without proof:

Theorem 9.18. *Let α be an isotopy class of embedded circles in S° which do not bound an open disk or an open cylinder and let $\ell_\alpha : \mathscr{T}(S,P) \to \mathbb{R}_{>0}$ be the corresponding length function. Then half the associated Hamiltonian action on $\mathscr{T}(S,P)$ is the shearing action defined by α on $\mathscr{T}(S,P)$: if ∂_α is the vector field on $\mathscr{T}(S,P)$ defined by the derivative of that action, then $d\ell_\alpha = \frac{1}{2}\operatorname{Im} h(-,\partial_\alpha)$.*

If we let α run over the $3g-3+n$ connected components of a pants decomposition A of S° , then the corresponding shearing actions commute and since $3g-3+n = \frac{1}{2}\dim_{\mathbb{R}} \mathscr{T}(S,P)$, we thus get a *completely integrable system.* This can be simpler stated in terms of the diffeomorphism $(\ell_\alpha, t_\alpha)_{\alpha\in\pi_0(A)} : \mathscr{T}(S,P) \to (\mathbb{R} \times \mathbb{R}_{>0})^{\pi_0(A)}$ found in Theorem 4.10:

$$\operatorname{Im} g_{WP} = \sum_{\alpha\in\pi_0(A)} \tfrac{1}{2} d\ell_\alpha \wedge dt_\alpha.$$

The quasi-projective structure on the universal family

The construction of the universal orbifold family $\mathscr{C}_{g,P} \to \mathscr{M}_{g,P}$ can also be carried out in the context of algebraic geometry. In fact, this can be done over any algebraically closed field and it is then constructed as a *Deligne-Mumford stack,* which is the algebro-geometric analogue of an orbifold. As this requires a substantial background in algebraic geometry, we only sketch this approach and use freely the language of algebraic geometry. Assume $2g-2+n > 0$ and let K be an algebraically closed field. We construct a vector space V, a representation W of $\mathrm{SL}(V)$ and Zariski closed $\mathrm{SL}(V)$-invariant subsets $Z \subset \mathbb{P}(W)$ and $\mathscr{C}_Z \subset \mathbb{P}(V) \times Z \subset \mathbb{P}(V) \times \mathbb{P}(W)$ with $\mathrm{SL}(V)$-equivariant sections $(x_p : Z \to \mathscr{C}_Z)_{p \in P}$ of the projection $f_Z : \mathscr{C}_Z \to Z$ (all defined over K) such that the system $(f_Z, (x_p)_{p\in P})$ is a family of P-pointed nonsingular connected curves of genus g on which $\mathrm{GL}(V)$ acts properly with finite stabilizers. It will have the property that the associated orbifold, or rather its algebro-geometric counterpart, is a K-model of $\mathscr{C}_{g,P} \to \mathscr{M}_{g,P}$.

We proceed as follows. Let (C,P) be a P-pointed nonsingular connected projective curve over K. The Riemann-Roch theorem implies that a line bundle on C of degree $d \geq 2g+1$ is very ample and that its linear system embeds C in a projective space of dimension $d-g$. The degree of $\Omega_C(P)$ is $2g-2+n$ and so for every $r \geq 3$, $\Omega_C(P)^{\otimes r}$ is very ample and embeds C in $\mathbb{P}(H^0(C,\Omega_C(P)^{\otimes r})^*)$. The image of C is then the common intersection of all the degree d hypersurfaces passing through C, at least for d large enough (and depending only on r, g and n). Their defining equations make up a subspace of $\mathrm{Sym}^d\left(H^0(C,\Omega_C(P)^{\otimes r})\right)$ that can be identified with the kernel of the evident linear map

$$\mathrm{Sym}^d\left(H^0(C,\Omega_C(P)^{\otimes r})\right) \to H^0(C,\Omega_C(P)^{\otimes rd}).$$

With the help of Riemann-Roch one can show that this map is also surjective. We put $N_r := \chi(\Omega_C(P)^{\otimes r}) = r(2g-2+n)-g+1$ so that $\dim H^0(C,\Omega_C(P)^{\otimes r}) = N_r$ and the above kernel has codimension N_{rd}. Now let V be a fixed K-vector space of dimension N_r, e.g., K^{N_r}. Choose an isomorphism of V^* onto $H^0(C,\Omega_C(P)^{\otimes r})$. Then the above kernel determines a linear subspace $E \subset \mathrm{Sym}^d V^*$ of codimension N_{rd} and if we think of this subspace as a linear space of homogeneous polynomials of degree d on V, then the projective variety it defines in $\mathbb{P}(V)$ is a copy C_E of C. The subspace $E \subset \mathrm{Sym}^d V^*$ of codimension N_{rd} corresponds to a subspace $E^\perp \subset \mathrm{Sym}^d V$ of dimension N_{rd}. Notice that $E^\perp$ is completely given by its determinant line $\wedge^{N_{rd}} E^\perp \subset \wedge^{N_{rd}} \mathrm{Sym}^d V$, for it is the set of vectors $u \in \mathrm{Sym}^d V$ with $u \wedge (\wedge^{N_{rd}} E^\perp) = 0$. Each $p \in P$ also defines a line $L_p \in V$ and so we have a line $\wedge^{N_{rd}} E^\perp \otimes \bigotimes_{p\in P} L_p$ in the vector space $W := \wedge^{N_{rd}} \mathrm{Sym}^d V \otimes \bigotimes_{p\in P} V_p$, or equivalently a point $z := [\wedge^{N_{rd}} E^\perp \otimes \bigotimes_{p\in P} L_p] \in \mathbb{P}(W)$. We claim that z determines C_E as well as a P-pointing of C_E, so that that (C_E,P) is isomorphic to (C,P). Observe first that from z we can completely recover E and the L_p's: any generator v of the line defined by z can be written $u_E \otimes \bigotimes_{p\in P} v_p$ with $u_E \in \wedge^{N_{rd}} E^\perp$, $v_p \in V_p$ all nonzero and these vectors are unique up to scalar. We thus recover each L_p and $E^\perp$. Hence we recover E and C_E. The point defined by L_p lies on C_E and defines a P-pointing of C_E as asserted.

Notice that W can be viewed as a representation of $\mathrm{GL}(V)$. Another choice of isomorphism may determine another such line in W but always lies in the $\mathrm{GL}(V)$-orbit of L. In $\mathbb{P}(W)$ this amounts to a $\mathrm{PSL}(V)$-orbit, or, as we prefer, a $\mathrm{SL}(V)$-orbit. So we have found a $\mathrm{SL}(V)$-invariant subset $Z \subset \mathbb{P}(W)$ whose $\mathrm{SL}(V)$-orbits are in bijective correspondence with the isomorphism types of pairs (C,P). It can be shown that this subset is in fact a (not necessarily closed) subvariety of $\mathbb{P}(W)$ and that the group $\mathrm{SL}(V)$ acts properly on W with finite stabilizers: the map $(g,z) \in \mathrm{SL}(W) \times Z \mapsto (g(z),z) \in Z \times Z$ is a finite morphism. Since $z \in Z$ determines a P-pointed curve in $\mathbb{P}(V)$, we see that we have a family of P-pointed curves $(f_Z : \mathscr{C}_Z \subset \mathbb{P}(V) \times Z \to Z, x_P)$. It has the property that the $\mathrm{SL}(V)$-orbits in Z are in bijective correspondence with the isomorphism types of P-pointed nonsingular connected genus g curves over K. Dividing out this family by the $\mathrm{SL}(V)$-action yields the desired K-model of $\mathscr{C}_{g,P} \to \mathscr{M}_{g,P}$.

We will see that this construction also leads to projective compactification of $\mathscr{M}_{g,P}$ via geometric invariant theory.

10 Harvey bordification and Deligne-Mumford compactification

In this section S is a closed connected surface of genus g and $P \subset S$ a finite subset of n elements with $2g-2+n>0$. We retain our convention to abbreviate $S \smallsetminus P$ by S°.

Our goal is to describe two related compactifications of $\mathscr{M}_{g,P}$, the *Harvey bordification* $\mathscr{M}^+_{g,P}$, and the *Deligne-Mumford compactification* $\overline{\mathscr{M}}_{g,P}$. Both are geometrically meaningful (they have a modular interpretation, one might say) in the sense that the boundary parameterizes degenerate objects. In the former case, we allow the conformal structure (or rather the hyperbolic metric) to degenerate on a fixed surface of genus g. The boundary will then have real codimension one and $\mathscr{M}^+_{g,P}$ becomes an orbifold with corners. On the other hand, the Deligne-Mumford compactification allows the topology of S to degenerate so that it becomes a singular surface and this can be carried out in the complex-analytic (even algebro-geometric) category so that $\overline{\mathscr{M}}_{g,P}$ is a complex-analytic orbifold. The boundary is then a normal crossing divisor. The two compactifications are related through a natural map $\mathscr{M}^+_{g,P} \to \overline{\mathscr{M}}_{g,P}$, which can be understood as a real oriented blowup. Both compactifications can be obtained as $\mathrm{Mod}(S,P)$-quotients of extensions $\mathscr{T}^+_{g,P}$ resp. $\overline{\mathscr{T}}_{g,P}$ of $\mathscr{T}_{g,P}$.

The curve complex

Consider the collection $\mathscr{C}(S^\circ)$ of isotopy classes of embedded circles in S° that do not bound a disk in S which meets P in at most one point. We make this collection the vertex set of a simplicial complex, called the *curve complex*, as follows: we agree that a nonempty finite subset $\mathbf{a} \subset \mathscr{C}(S^\circ)$ spans a simplex if and only if its elements can be represented by pairwise disjointly embedded circles. So the simplex $\mathbf{a}$ of $\mathscr{C}(S^\circ)$ can be thought of as an isotopy class of closed 1-dimensional submanifolds $A \subset S^\circ$ with the property that every connected component of $S^\circ \smallsetminus A$ has negative Euler characteristic. Observe that $\mathrm{Mod}(S,P)$ acts on this complex.

Exercise 10.1. Prove that a pants decomposition of S° defines a maximal simplex of $\mathscr{C}(S^\circ)$ and that every maximal simplex is of this form. Conclude that the geometric realization $|\mathscr{C}(S^\circ)|$ of $\mathscr{C}(S^\circ)$ is everywhere of dimension $3g-4+n$.

Example 10.2 (The case of a punctured torus)**.** We illustrate the preceding with the case $(g,n)=(1,1)$. Let us first observe that the curve complex of the once pointed torus $(S := (\mathbb{R}^2/\mathbb{Z}^2, P=\{o\})$ is discrete (there are no 1-simplices), where o is the origin. If $\alpha \subset S^\circ$ is an embedded circle which does not bound a disk in S, then cutting S open along α is diffeomorphic to a cylinder that has the image of o a distinguished point in its interior. Any two such pointed cylinders are diffeomorphic to each other and from this we can easily deduce that there the group $\mathrm{Diff}^+(S,P)$ acts transitively the set of such embedded circles. So $\mathrm{Mod}_{1,1}$ acts transitively on $\mathscr{C}(S^\circ)$.

We claim the identification of $(\mathscr{T}_{1,1}, \mathrm{Mod}_{1,1})$ with $(\mathbb{H}, \mathrm{SL}(2,\mathbb{Z}))$ identifies of the $\mathrm{Mod}_{1,1}$-set $\mathscr{C}(S^\circ)$ with the projective line $\mathbb{P}^1(\mathbb{Q})$ with its evident $\mathrm{SL}(2,\mathbb{Z})$-action. The correspondence goes like this: Let $q \in \mathbb{P}^1(\mathbb{Q})$ and represent q by a pair $(a,b) \in \mathbb{Z}^2$ with a,b relatively prime. This representation is unique up to sign. Then the map $t \in \mathbb{R} \mapsto (a,b) \pmod{\mathbb{Z}^2}$ is periodic modulo $\mathbb{Z}$ and its image is an embedded circle in S which passes through p. The embedded circles parallel to this circle not passing through o are parametrized by an open interval and lie in a single $\mathrm{Diff}^0(S,P)$-orbit. If we orient any such circle by its parametrization, then it defines a homology class in $H_1(S^\circ)$ and that class is under the obvious isomorphism $H_1(S^\circ) \cong \mathbb{Z}^2$ equal to (a,b) (the opposite orientation yields $(-a,-b)$). This shows that q determines an element $a_q \in \mathscr{C}(S,P)$ and that $q \in \mathbb{P}^1(\mathbb{Q}) \mapsto \mathscr{C}(S^\circ)$ is injective. This map is also equivariant if we identify $\mathrm{SL}(2,\mathbb{Z})$ with $\mathrm{Mod}_{1,1}$. Since the latter is transitive on $\mathscr{C}(S^\circ)$, it follows that the map $\mathbb{P}^1(\mathbb{Q}) \to \mathscr{C}(S^\circ)$ is also surjective.

Let $A \subset S^\circ$ represent a simplex $\mathbf{a}$ of $\mathscr{C}(S^\circ)$, i.e., a closed one-dimensional submanifold of S° such that every connected component of $S^\circ \smallsetminus A$ has negative Euler characteristic, with the connected components of A effectively indexed by the finite set $\mathbf{a}$. So if S' is such a connected component, then it is of the same type as S°: we may make it compact by adding a finite set, namely $S' \cap P$ and a point for every oriented connected component of A which appears as an oriented boundary component of S'. If we do this for every connected component of $S^\circ \smallsetminus A$ we get a (possibly disconnected) surface S^A and a finite subset $P^A \subset S^A$ with $S^A \smallsetminus P^A$ identifiable with $S^\circ \smallsetminus A$. Notice that $|P^A| = n + 2|\mathbf{a}|$.

Another way of arriving at the pair (S^A, P^A) is by first collapsing in S each connected component of A to a point. The resulting quotient space S_A of S is singular, for at the image p_α of a connected component α of A, S_A has two branches: a neighborhood is homeomorphic to the union two disks having their center in common. We then obtain S^A from S_A as its topological normalization: there is natural map $S^A \to S_A$ which simply separates the two branches at each p_α.

The smooth part of S_A can be identified with $S \smallsetminus A$. Since the latter has the same Euler characteristic (namely $2-2g$) as S, it follows that g can be read off from S_A. Following established terminology, we call g the *arithmetic genus* of S_A. (The *geometric genus* of S_A is the sum of the genera of the connected components of its normalization S^A.)

Note that we have not specified a differentiable structure on S^A near the points lying over the p_α's. We can always choose one, but it will not be unique. It will be so however up to an isotopy relative to P^A that is allowed to be non-differentiable at the points over the p_α's.

Harvey's bordification

Recall that the punctured disk $0<|z|<1$ is universally covered by the upper half plane $\mathbb{H}$ by the map $\tau \in \mathbb{H} \mapsto z = \exp(2\pi\sqrt{-1}\tau)$. Since $z^{-1}dz = 2\pi\sqrt{-1}d\tau$ and $|z| = \exp(-2\pi \operatorname{Im}\tau)$, we find that the corresponding hyperbolic metric (often called *Poincaré metric*) is given by $(-|z|\log|z|)^{-2}|dz|^2$. Now consider the map ϕ : $(-1,1)\times\mathbb{C}_1 \to \mathbb{C}$ defined by $\phi(t,u) := tu$. Since both the restrictions $\phi|(-1,0)\times\mathbb{C}_1$ and $\phi|(0,1)\times\mathbb{C}_1$ parametrize the punctured disk, we get a hyperbolic metric on $((-1,1)-\{0\})\times\mathbb{C}_1$. We think of this as a degenerate metric on $(-1,1)\times\mathbb{C}_1$ and will refer to it as the *standard degenerate hyperbolic metric* on $(-1,1)\times\mathbb{C}_1$.

Definition 10.3. Let $A\subset S^\circ$ represent a simplex $\mathbf{a}$ of $\mathscr{C}(S^\circ)$ or be empty. We say that a complete hyperbolic metric on $S^\circ\smallsetminus A$ is *permissible with respect to the embedding* $S^\circ\smallsetminus A\subset S$ if every connected component of A admits a tubular neighborhood parametrized by the cylindrical coordinates: $f:(-r,r)\times\mathbb{C}_1\to S^\circ$ for some $0<r<1$ (with the central circle $t=0$ mapping to A) in which the hyperbolic metric takes the standard form above. We shall call a complete hyperbolic metric of this form (so that it is only defined on an open subset of the form $S^\circ\smallsetminus A$ for some A as above) a *hyperbolic quasi-metric* on S° (since we allow A to be empty, this includes the case of a complete hyperbolic metric on S°).

Exercise 10.4. Prove that for a metric on $S^\circ\smallsetminus A$ permissible with respect to $S^\circ - A\subset S$ the embedding and a connected component α of A, a parametrization of a tubular neighborhood of α, $f:(-r,r)\times\mathbb{C}_1\hookrightarrow S$, as in the definition is unique up a rotation in the second coordinate and the inversion $(t,u)\mapsto(-t,u^{-1})$.

A hyperbolic metric on $S^\circ\smallsetminus A$ permissible with respect to $S^\circ\smallsetminus A\subset S$ defines the structure of a Riemann surface on $S\smallsetminus A$ or rather, we get on S^A the structure of a Riemann surface C^A. Similarly, S_A can be given an analytic structure C_A for which its only singularities are *nodes*, that is, have as their local-analytic model, the union of the two coordinate lines $(z_1z_2=0)$ in a neighborhood of the origin of $\mathbb{C}^2$. Then the evident map $C^A\to C_A$ is complex-analytic and can be understood as the normalization of C_A. The space of permissible hyperbolic metrics $S^\circ\smallsetminus A$ up to $\operatorname{Diff}^0(S,P\cup A)$-isotopy only depends on $\mathbf{a}$ and so we feel justified in denoting it by

$\mathscr{T}^{\mathbf{a}}(S,P)$. For a similar reason the Teichmüller space $\mathscr{T}(S^A,P^A)$ (which is actually a product of Teichmüller spaces, with one factor for every connected component of S^A) may also be denoted $\mathscr{T}(S^{\mathbf{a}},P^{\mathbf{a}})$. It is clear that this construction defines a map from $\mathscr{T}^{\mathbf{a}}(S,P)$ to the Teichmüller space $\mathscr{T}(S^{\mathbf{a}},P^{\mathbf{a}})$. We also have a shearing action of $\mathbb{R}^{\mathbf{a}}$ on $\mathscr{T}^{\mathbf{a}}(S,P)$: if we are given a permissible hyperbolic metric on $S^\circ \setminus A$ and $f:(-r,r)\times\mathbb{C}_1 \hookrightarrow S$ parametrizes in an orientation preserving manner a tubular neighborhood of the connected component α of A as in the definition, then we can do the shearing in terms of this parametrization: for every $t_\alpha \in \mathbb{R}$ we have defined the sheared surface $S(t_\alpha)$ obtained from the union of $S \setminus \alpha$ and $(-r,r)\times\mathbb{C}_1$ by identifying (t,u) with $f(t,u)$ for $t<0$ and with $f(t,e^{-\sqrt{-1}t_\alpha}u)$ for $t>0$. It has the property that $C(t_\alpha)^A = C^A$. Notice that thus is defined an action of $\mathbb{R}^{\mathbf{a}}$ on $\mathscr{T}^{\mathbf{a}}(S,P)$ which preserves the fibers of $\mathscr{T}^{\mathbf{a}}(S,P) \to \mathscr{T}(S^{\mathbf{a}},P^{\mathbf{a}})$. In fact, it is not hard to see that this defines a principal bundle:

Lemma 10.5. *The shearing action of $\mathbb{R}^{\mathbf{a}}$ on $\mathscr{T}^{\mathbf{a}}(S,P)$ turns the map $\mathscr{T}^{\mathbf{a}}(S,P) \to \mathscr{T}(S^{\mathbf{a}},P^{\mathbf{a}})$ into a principal $\mathbb{R}^{\mathbf{a}}$-bundle in the differentiable category.*

Corollary 10.6. *The manifold $\mathscr{T}^{\mathbf{a}}(S,P)$ is diffeomorphic to an open cell of dimension $6g-6+2n-|\mathbf{a}|$.*

A hyperbolic quasi-metric arises as a limit of hyperbolic metrics on S° if we let the length of the geodesics in the isotopy classes of the connected components of A tend to zero (so that some of the Fenchel-Nielsen coordinates ℓ_α, $\alpha\in\mathbf{a}$, assume the value zero). Let us make this explicit with the help of Lemma 4.1. Scalar multiplication by $\exp(2\pi r)$ acts on $\mathbb{H}$ and leaves the first and second quadrant, denoted here by $\mathbb{H}^+$ and $\mathbb{H}^-$ respectively, invariant. The common boundary of $\mathbb{H}^+$ and $\mathbb{H}^-$ in $\mathbb{H}$ is the positive imaginary axis and is a geodesic. The function $\exp(\sqrt{-1}/r.(\log z)$ identifies the orbit space C_r of $\mathbb{H}$ by this transformation with the annulus $\exp(-\pi/r) < |w| < 1$ with metric

$$\frac{r^2|dw|^2}{|w|^2\sin^2(r\log|w|)}.$$

The image of $\mathbb{H}^\pm$ is a open half annulus $C_r^\pm \subset C_r$ whose boundary is the unique closed geodesic α_r of C_r defined by $|w| = \exp(-\pi/2r)$. The length of that geodesic is $2\pi r$. This models the general situation, for according to Lemma 4.1 a neighborhood of a closed geodesic of length $2\pi r$ on any hyperbolic surface is isometric to a neighborhood of α_r in C_r. This justifies that we focus on this case. We want to produce a sensible geometric limit if this length r goes to zero. Now note that for $r\downarrow 0$, the outer annulus $\exp(-\pi/2r) < |w| < 1$ tends to the punctured unit disk $0<|w|<1$ with the metric tending to the Poincaré metric $(|w|\log|w|)^{-2}|dw|^2$. This makes the whole inner annulus go to the origin. In order to preserve that annulus in the limit, we invert with respect to the central geodesic: if we write w^+ for w and define w^- by $w^+w^- = \exp(-\pi/r)$, then the pair (w^+,w^-) maps C_r isomorphically onto the curve in (w^+,w^-)-space defined by $|w^\pm| < 1$ and

$w^+w^- = \exp(-\pi/r)$. If we let $r \downarrow 0$, then the limit is the union of the two unit disks in the coordinate axes, each endowed with the Poincaré metric. Notice that the closed geodesic α_r on C_r is given on C_r as the locus where $|w^+| = |w^-|$ and that it tends to the origin of $\mathbb{C}^2$ as $t \downarrow 0$ (algebraic geometers recognize this as a *vanishing cycle* of the function w^+w^-).

Let $\mathbf{a}$ be a simplex of $\mathscr{C}(S^\circ)$ that defines a pants decomposition of S° (according to Exercise 10.1 this means that $\mathbf{a}$ is a maximal simplex). We then have an associated Fenchel-Nielsen parametrization of the Teichmüller space $\mathscr{T}(S,P)$ of hyperbolic metrics on S° given up to isotopy relative to P by the product of open half planes $(\mathbb{R}_{>0} \times \mathbb{R})^{\mathbf{a}}$. If we now allow the metrics to be permissible with respect to $S^\circ \smallsetminus A \subset A$, where A is a closed submanifold which represents $\mathbf{a}$, then we see that all the coordinates (ℓ_α, t_α) retain their meaning, except that the ℓ_α's can take the value zero. The resulting map from the space of A-permissible hyperbolic metrics to the product of closed half planes $([0,\infty) \times \mathbb{R})^{\mathbf{a}}$ is surjective and constant on the $\mathrm{Diff}^0(S,P)$-orbits. It is not hard to see that these $\mathrm{Diff}^0(S,P)$-orbits are just the fibers of this map. In other words, $([0,\infty) \times \mathbb{R})^{\mathbf{a}}$ parametrizes the $\mathrm{Diff}^0(S,P)$-orbits of such quasi-metrics. We then see that the $\mathrm{Diff}^0(S,P)$-orbit space of quasi-metrics has the structure of a manifold with corners:

Proposition-definition 10.7. *The* Harvey bordification $\mathscr{T}^+(S,P)$ *of* $\mathscr{T}(S,P)$ *is the* $\mathrm{Diff}^0(S,P)$*-orbit space of hyperbolic quasi-metrics on* S° *(with respect to the inclusion* $S^\circ \subset S$*). It is in a natural manner a manifold with corners with* $\mathrm{Mod}(S,P)$*-action. The corner strata of its boundary* $\partial\mathscr{T}^+(S,P)$ *are the* $\mathscr{T}^{\mathbf{a}}(S,P)$*, where* $\mathbf{a}$ *runs over the simplices of the curve complex* $\mathscr{C}(S^\circ)$ *and this indexing is order reversing:* $\mathscr{T}^{\mathbf{a}}(S,P)$ *is in the closure of* $\mathscr{T}^{\mathbf{b}}(S,P)$ *if and only if* $\mathbf{a} \supset \mathbf{b}$.

Moreover, any Fenchel-Nielsen parametrization $(\mathbb{R}_{>0} \times \mathbb{R})^{\mathbf{a}} \cong \mathscr{T}(S,P)$ *extends to a diffeomorphism of* $([0,\infty) \times \mathbb{R})^{\mathbf{a}}$ *onto an open subset of* $\mathscr{T}^+(S,P)$.

Example 10.8 (The case of a punctured torus continued)**.** We take up again the case $(\mathscr{T}_{1,1}, \mathrm{Mod}_{1,1})$. Recall that in Example 10.2 the triple $(\mathscr{T}_{1,1}, \mathscr{A}(S^\circ), \mathrm{Mod}_{1,1})$ was identified with $(\mathbb{H}, \mathbb{P}^1(\mathbb{Q}), \mathrm{SL}(2,\mathbb{Z}))$. The Harvey bordification corresponds under this identification to what it known as the *Borel-Serre compactification* $\mathbb{H}^{BS}$ of $\mathbb{H}$, which in this case is a manifold with boundary (there are here no corners) whose connected components are effectively indexed by $\mathbb{P}^1(\mathbb{Q})$. In fact, the boundary component indexed by $q \in \mathbb{P}^1(\mathbb{Q})$ can be identified with the set of complete *oriented* geodesics in $\mathbb{H}$ which tend to q. As such a geodesic is given by its 'source', a point of $\mathbb{P}^1(\mathbb{R})$ distinct from q, we see that as a topological space, this boundary component can be identified with the real affine line $A_q := \mathbb{P}^1(\mathbb{R}) - \{q\}$. For instance, $A_\infty = \mathbb{R}$. The topology on the disjoint union of $\mathbb{H}$ and the A_q's is essentially specified if require that it be $\mathrm{SL}(2,\mathbb{Z})$-invariant and that A_∞ is glued onto $\mathbb{H}$ as to form in an evident manner $\mathbb{R} \times (0,\infty]$ (so that A_∞ corresponds to $y = \infty$).

The group $\mathrm{SL}(2,\mathbb{Z})$ acts properly discontinuously on $\mathbb{H}^{BS}$, the stabilizer of A_q being the $\mathrm{SL}(2,\mathbb{Z})$-stabilizer of q. The latter is infinite cyclic and acts faithfully as a translation group in A_q. A fundamental domain of the $\mathrm{SL}(2,\mathbb{Z})$-action on $\mathbb{H}^{BS}$ is the closure D of its standard fundamental domain D in $\mathbb{H}$ defined by $|\mathrm{Re}(\tau)| \leq \frac{1}{2}$, $|\tau| \geq 1$ in $\mathbb{H}^{BS}$: the set of added points is the interval in A_∞ defined by $|x| \leq \frac{1}{2}$. This makes D compact. This helps us to understand the $\mathrm{SL}(2,\mathbb{Z})$-orbit space $\mathrm{Mod}_{1,1}\backslash\mathbb{H}^{BS}$, for we then find it to be a quotient of D where any identification takes place on the boundary. It adds to the J-line a circular boundary (namely the quotient of A_∞ by its stabilizer in $\mathrm{SL}(2,\mathbb{Z})$, i.e., the group of integral translations) so that the result is homeomorphic to a closed disk.

This example also illustrates a more general fact:

Stable pointed curves

Let C be a compact *nodal curve*, i.e., a complex-analytic curve whose only singularities are nodes. The normalization $\hat{C} \to C$ is then the smooth curve (Riemann surface) obtained by separating the branches of the nodes. Every node defines a pair in $\hat{C}$ and we can reconstruct C from $\hat{C}$ and this collection of pairs.

We recall that its arithmetic genus g is the Euler characteristic of its smooth part C_{reg}: $\chi(C_{reg}) = 2-2g$. We want to deal with the pointed situation as well. We stipulate that this means here that we are given a finite subset P *of the smooth part* C_{reg} of C. So P should not contain a node.

Definition 10.9. Let (C,P) be a compact nodal curve. We say that (C,P) is *stable* (more precisely, *Deligne-Mumford stable*) if every connected component of $C^\circ_{reg} := C_{reg} \smallsetminus P$ has negative Euler characteristic.

As we know, this implies that the conformal structure on C°_{reg} comes from a complete hyperbolic metric, that is in fact unique. We actually have encountered the stable pointed curves already via the converse construction: if (S,P) is a hyperbolic pair and $A \subset S^\circ$ is a closed submanifold of dimension 1 which represents a simplex of the curve complex, then for any A-permissible hyperbolic metric on $S^\circ \smallsetminus A$ we get a stable curve pointed curve (C_A,P). Thus the Harvey bordification $\mathscr{T}^+(S,P)$ parametrizes stable P-pointed curves of genus g.

Exercise 10.10. Prove that (C,P) is stable precisely when the group $\mathrm{Aut}(C,P)$ of automorphisms of C which preserve P pointwise is finite.

Stable pointed curves appear to have a deformation theory which is almost as good as the one for smooth pointed curves. If $\hat{P} \subset \hat{C}$ denotes the preimage of $P \cup C_{sg}$, then one way to deform (C,P) is by deforming each connected component of the pair $(\hat{C},\hat{P})$ and then identifying the point pairs in $\hat{P}$. This will produce stable pointed curves homeomorphic to (C,P).

Exercise 10.11. Prove that the Teichmüller space $\mathscr{T}(\hat{C},\hat{P})$ (which is a product of Teichmüller spaces with one factor for every connected component of $\hat{C}$) has complex dimension $3g-3+n-|C_{sg}|$.

However, we also want to include deformations which make the nodes disappear. If we look for a deformation space of the hoped for dimension $3g-3+n$, then the above exercise suggests that the smoothing of each node needs one extra dimension.

Deformations of nodes

Let C be an abstract complex-analytic curve with a nodal singularity at $p \in C$. This means that there exists a neighborhood C_0 of p in C and an isomorphism of (C_0, p) onto the locus in $U \subset \mathbb{C}^2$ defined by $z_1 z_2 = 0$, where U is defined by $|z_i| < \varepsilon$, $i = 1,2$. Thus is defined closed embedding ι of C_0 in U. If we let $f : U \to \mathbb{C}$ be defined by $f(z_1, z_2) = z_1 z_2$, then we may regard the pair (f, ι) as defining a *deformation* of C_0 in the sense that for small $|t|$ (e.g., $|t| < \varepsilon^2$), the fiber $U_t := f^{-1}(t)$ may be regarded as a 'deformed C_0'. Notice that in contrast to the deformation of a complex structure considered earlier, what is being deformed here is not just a complex structure, but also the underlying topological space: U_0 is a union of two disks with their center in common, whereas for $t \neq 0$ small, U_t is a nonsingular curve isomorphic to an annulus. On the other hand, this is very much a local affair: we can take C_0 and U as small as we like. Therefore the notion of a deformation here is really that one of a germ: what we are deforming is the curve germ (C, p) and the deformation is then the pair of analytic map-germs $(f : (\mathbb{C}^2, 0) \to (\mathbb{C}, 0), \iota : (C, o) \hookrightarrow (\mathbb{C}^2, 0))$.

We are here free to replace $(\mathbb{C}^2, 0)$ and $(\mathbb{C}, 0)$ by complex manifold germs $(\mathscr{C}, p)$ and (B, o) of dimension 2 and 1 respectively as long we can find coordinates (z_1, z_2) at $(\mathscr{C}, p)$ and a coordinate w at (B, o) such that wf takes the form $z_1 z_2$. There is in this case also a Kodaira-Spencer map (or rather its inverse) which grasps the first order content of the deformation: the derivative of $D_p f : T_p\mathscr{C} \to T_o B$ at p is zero (for p is a singular point), but then the Hessian of f is intrinsically defined as a *quadratic* map $H_p f : T_p\mathscr{C} \to T_o B$. If f is given in terms of coordinates as $f^* w = z_1 z_2$, then $H_p f$ is just given as the product of the linear forms $dz_1|_0$ and $dz_2|_0$. So it sends $a_1 \frac{\partial}{\partial z_1}|_0 + a_2 \frac{\partial}{\partial z_2}|_0$ to $a_1 a_2 \frac{\partial}{\partial w}$. We can also express this as follows: let $C^{\pm}$ be the two branches of C_0 (these are just names: there is no natural way of ordering them). Then the images of $T_p C^+$ and $T_p C^-$ in $T_p\mathscr{C}$ decompose the latter into two lines and the Hessian $H_p f$ defines a linear isomorphism $T_p C^+ \otimes T_p C^- \xrightarrow{\cong} T_o B$. Its inverse, $T_o B \xrightarrow{\cong} T_p C^+ \otimes T_p C^-$, is the analogue of the Kodaira-Spencer map: it goes from $T_o B$ to a space which only involves the object to be deformed (namely (C, p)).

A decent deformation theory should however not be restricted to bases of dimension one, nor should it require the total space germ $\mathscr{C}$ to be nonsingular. For instance, if (B, o) is an arbitrary smooth manifold germ, then the projection $(C, p) \times (B, o) \to (B, o)$ is perhaps uninteresting as a deformation (nothing is being deformed, after all), but should be allowed. These considerations lead to the following definition.

Definition 10.12. Let B be a complex manifold and $o \in B$. A *deformation* of the nodal curve germ (C, p) over (B, o) is a pair of analytic map-germs

$$(f : (\mathscr{C}, q) \to (B, o), \iota : (C, p) \hookrightarrow (\mathscr{C}, q)).$$

We demand that f factors through an isomorphism j of $(\mathscr{C},q)$ onto an analytic hypersurface germ in $(\mathbb{C}^2,0)\times(B,o)$ whose defining equation F has the form $z_1z_2=u$ for some holomorphic $u:(B,o)\to(\mathbb{C},0)$ and require that ι be an isomorphism of (C,q) onto the central fiber $(f^{-1}(o),q)$.

We call the deformation a *smoothing*, if the union of the singular fibers is nowhere dense in $\mathscr{C}$.

This covers the case considered at the beginning, where $f:(\mathbb{C}^2,0)\to(\mathbb{C},0)$ is given by $f(z_1,z_2)=z_1z_2$, for then we may take $u(t)=t$ so that hypersurface in $(\mathbb{C}^2,0)\times(\mathbb{C},0)$ is the graph of f, defined by $z_1z_2-t=0$. Notice that this is in fact a smoothing. As we have already seen, the general fiber of a smoothing is homeomorphic a cylinder $(-1,1)\times\mathbb{C}_1$ and the central fiber is obtained from it by collapsing its midcircle $\{0\}\times\mathbb{C}_1$ to a point.

Exercise 10.13. Prove that if in the above definition $\mathscr{C}$ is nonsingular, then f is a smoothing.

Remark 10.14. It is possible to set up a local deformation category in much the same way as $\mathfrak{Def}_{C,P}$. We shall not do that, but we observe that we have something like a universal object: for f,ι,j,u as in the definition, let $\iota':(C,p)\hookrightarrow(\mathbb{C}^2,0)$ be the restriction of j (so that $j(z)=(\iota'(z),b))$ and let $f':(\mathbb{C}^2,0)\to(\mathbb{C},0)$ be given by $f'(z_1,z_2)=z_1z_2$. Then (f',ι') defines a deformation of (C,p) and a morphism from (f,ι) to (f',ι') is then defined by the pair (Φ,ϕ), where $\Phi:(\mathscr{C},q)\to(\mathbb{C}^2,0)$ is the first component of j and and $\phi=u:(B,o)\to\mathbb{C}$. But the morphism (Φ,ϕ) is not unique and this is why (f',ι') is called a *semi-universal* deformation. It is so in first order though, for we may still define Kodaira-Spencer map $T_oB\to T_pC^+\otimes T_pC^-$ as the composite of $D_o\phi=D_ou:T_oB\to T_0\mathbb{C}$ with the inverse of the isomorphism $T_pC^+\otimes T_pC^-\cong T_0\mathbb{C}$ defined by the Hessian of f'. You may check that this definition is coordinate-independent (i.e., independent of j and u).

Deformations of nodal curves

We are going to combine the two types of deformations.

Definition 10.15. Let B be a complex manifold. A *family of stable P-pointed curves over B* consists a proper analytic map of complex-analytic varieties $f:\mathscr{C}\to B$ endowed with a set x_P of pairwise disjoint holomorphic sections $(x_p:B\to C)_{p\in P}$ such that every fiber $C_b:=f^{-1}(b)$ forms with $x_P(b)$ a stable P-pointed curve of genus g. We require that if $z\in\mathscr{C}$, then the germ of f at z is either a submersion (this presupposes that $\mathscr{C}$ is nonsingular at z) or that z is a node on the fiber $C_{f(z)}$ and the germ of f at z defines a deformation of this node.

We also have a version relative to a fixed stable pointed curve:

Definition 10.16. Given a stable P-pointed curve (C,P), a complex manifold B and a point $o \in B$, then a *deformation* of C over (B,o) is a system of analytic map-germs

$$\big(f:(\mathscr{C},C_o)\to(B,o), x_P=(x_p:(B,o)\to(\mathscr{C},C_o))_{p\in P}, \iota: C\hookrightarrow\mathscr{C}\big)$$

with ι an isomorphism onto $C_o := f^{-1}(o)$, $x_p(o)=\iota(p)$ for all $p\in P$, and for which some representative of (f,x_P) is a family of P-pointed stable curves.

These definitions clearly extend the ones we gave for nonsingular curves (= compact Riemann surfaces). Since the section $x_p : B\to\mathscr{C}$ of f takes its values in the part where f is a submersion, we have defined a line bundle $\mathscr{L}_{f,p}$ on B whose fiber in b is the cotangent space $T^*_{x_p(b)}C_b$. More precisely, $\mathscr{L}_{f,p}$ is the pull-back under x_p of the relative cotangent bundle of $\mathscr{C}/B$.

The deformations of a stable pointed curve (C,P) are the objects of a deformation category $\mathfrak{Def}_{C,P}$ (the definition of a morphism being the obvious one) and so we have a notion of a universal deformation. As before, it is formal consequence of the universal property that the (finite) automorphism group of (C,P) acts naturally on its universal deformation.

Remark 10.17. Let (C,P) be a connected stable curve. There is a first order deformation space which serves as the generalization of the one in the smooth case (i.e., $H^1(C,\theta_C)$) and is for a deformation the target of a Kodaira-Spencer map. We briefly explain. Suppose (C,P) appears as the central fiber in a deformation $(f:(\mathscr{C},C_o)\to(B,o),x_P,\iota:C\cong C_o)$ with $\mathscr{C}$ nonsingular. Then the derivative of f along C_o defines a map of $\mathscr{O}_C$-modules:

$$Df_{C,P}:\iota^*\theta_{\mathscr{C}}(\log x_P)\to\mathscr{O}_C\otimes T_oB,$$

where we recall that $\theta_{\mathscr{C}}(\log x_P)$ stands for the sheaf of holomorphic vector fields on $\mathscr{C}$ that are tangent to the sections x_p, $p\in P$. We will be concerned with the cohomology of the *mapping cone* of $Df_{C,P}$.

We digress for a moment to discuss that notion in its proper setting. We first mention that the notion of cohomology for sheaves extends to complexes of abelian sheaves that are zero for sufficiently small (negative) degree in such a manner that essentially all the fundamental properties subsist (it is then often referred to as *hypercohomology*). For instance, there is a long exact sequence attached to a short exact sequence $0\to\mathscr{A}^\bullet\to\mathscr{B}^\bullet\to\mathscr{C}^\bullet\to 0$ of sheaf complexes on a space X. This is also true if for every k the row $0\to\mathscr{A}_k\to\mathscr{B}_k\to\mathscr{C}_k\to 0$ is a sheaf complex (the composite is zero) and the sequence of cohomology sheaves $0\to\mathscr{H}^k(\mathscr{A}^\bullet)\to\mathscr{H}^k(\mathscr{B}^\bullet)\to\mathscr{H}^k(\mathscr{C}^\bullet)\to 0$ is exact.

A particular case of interest is when we only have a homomorphism $\phi:\mathscr{F}\to\mathscr{G}$ of abelian sheaves, for this then enables us to put the induced homomorphisms $H^k(X,\phi):H^k(X,\mathscr{F})\to H^k(X,\mathscr{G})$ into a long exact sequence defined as follows.

First regard ϕ as defining a complex $\mathscr{C}_\phi^\bullet$ by putting $\mathscr{C}_\phi^\bullet = 0$ for $k \neq -1, 0$ and $\mathscr{C}_\phi^{-1} \to \mathscr{C}_\phi^0$ be given by ϕ. This complex is called the *mapping cone* of ϕ. We then have a short exact sequence of complexes of sheaves

$$0 \longrightarrow \mathscr{G} \longrightarrow \mathscr{C}_\phi^\bullet \longrightarrow \mathscr{F}[1] \longrightarrow 0,$$

where $\mathscr{G}$ and $\mathscr{F}$ are now regarded as one term complexes placed in degree 0 and -1 respectively. (The convention is that if $C^\bullet$ is a complex, then $C^\bullet[n]$ denotes the complex whose term in degree k is C^{k+n}.) This yields the exact sequence

$$\cdots \to H^k(X, \mathscr{F}) \xrightarrow{H^k(X,\phi)} H^k(X, \mathscr{G}) \to H^k(X, \mathscr{C}_\phi^\bullet) \to H^{k+1}(X, \mathscr{F}) \xrightarrow{H^{k+1}(X,\phi)} \cdots$$

We may however also form

$$0 \longrightarrow \mathrm{Ker}(\phi)[1] \longrightarrow \mathscr{C}_\phi^\bullet \longrightarrow \mathrm{Coker}(\phi) \longrightarrow 0.$$

This is exact on the cohomology sheaves, for $\mathscr{H}^k(\mathscr{C}_\phi^\bullet)$ equals $\mathrm{Ker}(\phi)$ for $k = -1$, $\mathrm{Coker}(\phi)$ for $k = 0$, and is zero otherwise. So this gives the long exact sequence

$$\cdots \to H^{k-1}(X, \mathrm{Coker}(\phi)) \to H^{k+1}(X, \mathrm{Ker}(\phi)) \to H^k(X, \mathscr{C}_\phi^\bullet) \to H^k(X, \mathrm{Coker}(\phi)) \to \cdots$$

Returning to the case at hand, we take for ϕ the homomorphism $Df_{C,P}$ as defining a complex of $\mathscr{O}_C$-modules and put $\mathscr{T}^\bullet := \mathscr{C}^\bullet_{Df_{C,P}}[-1]$. Then the above discussion gives the exact sequence

$$\cdots \to H^0(C, \iota^*\theta_{\mathscr{C}}(\log x_P)) \to T_0B \xrightarrow{KS} H^1(C, \mathscr{T}^\bullet_{C,P}) \to H^1(C, \iota^*\theta_{\mathscr{C}}(\log x_P)) \to \cdots$$

We will see that the other long exact sequence helps us to understand $H^1(C, \mathscr{T}^\bullet_{C,P})$ as the space of first order deformations of the stable pointed curve (C, P) with KS playing the role of the Kodaira-Spencer map. For this we need to work out the kernel and cokernel of $Df_{C,P}$. The kernel of $Df_{C,P}$ can be identified with the sheaf of holomorphic fields tangent to C which vanish in P. At a node q this sheaf is in terms of local coordinates the $\mathscr{O}_{C,q}$-module generated by $z_1\frac{\partial}{\partial z_1}$ and $z_2\frac{\partial}{\partial z_2}$ and so that kernel is just the direct image of $\theta_{\hat{C},\hat{P}}$ under the normalization map $\pi : \hat{C} \to C$: $\mathrm{Ker}(Df_{C,P}) \cong \pi_*\theta_{\hat{C}}(-\hat{P})$. The cohomology of an abelian sheaf is the same as its direct image under a finite map and since π is such a map, we find that $H^k(C, \mathrm{Ker}(Df_{C,P})) \cong H^k(\hat{C}, \theta_{\hat{C}}(-\hat{P}))$. The cokernel is a skyscraper sheaf with support in the nodes: the stalk at a node y is the tensor product of the tangent spaces along the two branches $T_yC_y^+ \otimes T_yC_y^-$ and its cohomology is concentrated in degree zero. This yields the exact sequence

$$0 \to H^1(\hat{C}, \theta_{\hat{C}}(-\hat{P})) \to H^1(C, \mathscr{T}^\bullet_{C,P}) \to \oplus_{y \in C_{sg}} T_yC_y^+ \otimes T_yC_y^- \to 0.$$

The subspace $H^1(\hat{C}, \theta_{\hat{C}}(-\hat{P}))$ then corresponds to the space of first order deformations which preserve the nodes, and the quotient of these yields the first order deformations of the nodes.

The Deligne-Mumford compactification

We are now ready to state an extension to the central Theorem 9.13 in the stable case.

Theorem 10.18. *A deformation of the stable curve (C,P) is universal if and only if its Kodaira-Spencer map is an isomorphism. If these equivalent conditions are satisfied, then the deformation can be represented by a family $(f:\mathscr{C}\to B, x_P)$ endowed with an isomorphism $\iota : C \cong C_o$ which defines a universal deformation of each of its fibers and is such that*

(i) *B is contractible and admits a partial system of coordinates at $o\in B$ indexed by C_{sg}, $(t_q)_{q\in C_{sg}}$, such that $t_q=0$ defines the locus where $q\in C_{sg}$ subsists as a singular point in the fiber,*
(ii) *the locus $B_o\subset B$ where all the t_q's vanish parametrizes the fibers homeomorphic to (C,P) and (simultaneous) normalization of these fibers yields a universal deformation of the normalization $(\hat{C},P)$ of (C,P) over B_o as in Theorem 9.13,*
(iii) *the group* Aut(C,P) *of automorphisms of (C,P) extends its tautological action on (C,P) naturally to the family $(f:\mathscr{C}\to B, x_P)$.*

Notice that this theorem implies that the locus D_f parametrizing singular fibers is the normal crossing locus defined by $\prod_q t_q = 0$.

The theorem differs from Theorem 9.13 in that it does not involve Teichmüller space. Yet we may glue these together to form a family:

$$\left(\overline{\mathscr{C}}_{g,P}\to\overline{\mathscr{M}}_{g,P}, x_P=(x_p:\overline{\mathscr{M}}_{g,P}\to\overline{\mathscr{C}}_{g,P})_{p\in P}\right)$$

as an *orbifold* object in the holomorphic category: a family $(f:\mathscr{C}\to B, x_P)$ as in Theorem 10.18 defines a chart with Aut(C,P) as associated group. If we ignore the orbifold structures, then we get a map of complex-analytic spaces $\overline{C}_{g,P}\to\overline{M}_{g,P}$ with sections indexed by P for which the local model on $\overline{M}_{g,P}$ is given as by the Aut(C,P)-quotient of a family as in Theorem 10.18: $\mathrm{Aut}(C,P)\backslash\mathscr{C}\to\mathrm{Aut}(C,P)\backslash B$.

The difference $D_{g,P}:=\overline{\mathscr{M}}_{g,P}-\mathscr{M}_{g,P}$ is a normal crossing divisor (again in the orbifold sense). Such a normal crossing divisor is naturally partitioned into *strata*: these are the connected components of loci where this divisor has a fixed number of branches. We also see that the strata are effectively indexed by Mod(S,P)-orbits of simplices of the curve complex $\mathscr{C}(S,P)$.

We might say that we have a holomorphic line bundle $\mathscr{L}_p$ over $\overline{\mathscr{M}}_{g,P}$ $(p\in P)$ in the orbifold sense. If we ignore the orbifold structure, then $\mathscr{L}_p$ is still defines a sheaf L_p on $\overline{M}_{g,P}$: in terms of a chart as above, the sections of L_p over $\mathrm{Aut}(C,P)\backslash B\hookrightarrow\overline{M}_{g,P}$ are the Aut(C,P)-invariant sections of $\mathscr{L}_p$ on B.

Remark 10.19 (Smooth Galois covers of $\overline{\mathscr{M}}_{g,P}$)**.** The technicalities that usually accompany an abstract orbifold setting can here be avoided, because one can show that there exist finite quotients $\phi : \mathrm{Mod}(S,P) \twoheadrightarrow G$ of the mapping class group satisfying the following two properties:

(i) $\mathrm{Ker}(\phi)$ acts freely on $\mathscr{T}(S,P)$ so that $\mathscr{M}[\phi] := \mathrm{Ker}(\phi)\backslash\mathscr{T}(S,P)$ is a complex manifold and

(ii) the natural extension of the formation of the G-orbit variety $\mathscr{M}[\phi] \to \mathscr{M}_{g,P}$ to a G-covering over $\overline{\mathscr{M}}_{g,P}$, $\overline{\mathscr{M}[\phi]} \to \overline{\mathscr{M}}_{g,P}$ (in the language of algebraic geometry, this is the normalization of the map $\mathscr{M}[\phi] \to \overline{\mathscr{M}}_{g,P}$) has nonsingular total space. So $\overline{\mathscr{M}}_{g,P}$ then appears as the G-orbit space of a nonsingular variety. The orbifold line bundle $\mathscr{L}_p$ is obtained as a G-quotient of a genuine line bundle $\mathscr{L}_p[\phi]$ on $\overline{\mathscr{M}[\phi]}$.

Exercise 10.20. Prove that all but one of the irreducible components of $D_{g,P}$ are effectively indexed by unordered pairs $\{(g',P'),(g'',P'')\}$ with $g'+g''=g$, $P=P'\sqcup P''$ and such that $g'+|P'|\geq 2$, $g''+|P''|\geq 2$.

Let $\mathbf{a}$ be a simplex of the curve complex. If $A\subset S^\circ$ represents $\mathbf{a}$, then the choice of a permissible hyperbolic metric g on $S^\circ - A$ yields a stable pointed curve (C^A,P). We thus have defined a map $\tilde{\pi} : \mathscr{T}^+_{S,P} \to \overline{\mathscr{M}}_{g,P}$. This map is clearly constant on $\mathrm{Mod}(S,P)$-orbits and so it factors through a map $\pi : \mathscr{M}^+_{g,P} \to \overline{\mathscr{M}}_{g,P}$.

Let us denote by $\tilde{b} \in \mathscr{T}^+_{S,P}$ the point defined by $(S,P;A,g)$, by $b\in \mathscr{M}^+_{g,P}$ its image and by $c\in \overline{\mathscr{M}}_{g,P}$ the point defined by (C^A,P). Suppose we have likewise $\mathbf{a}'$, A' and g' with the property that the resulting stable pointed curve $(C^{A'},P)$ is isomorphic to (C^A,P). Then after some shearing, there is a element of $\mathrm{Diff}^+(S,P)$ which carries (A',g') onto (A,g). In other words, the element $\tilde{b}' \in \mathscr{T}^+_{S,P}$ defined by (A',g') lies in $\mathrm{Mod}(S,P).\mathbb{R}^{\mathbf{a}}.\ \tilde{b}$. The intersection $\mathrm{Mod}(S,P)\cap\mathbb{R}^{\mathbf{a}}$ consists of the Dehn twists around the vertices of $\mathbf{a}$ and so we see that on the fiber $\pi^{-1}(c)$ we have an principal action of the torus $(S^1)^{\mathbf{a}}$ and that the image of $\tilde{b}'$ in $\mathscr{M}^+_{g,P}$ is in the orbit of that torus. This shows that $\pi^{-1}(c)$ is an orbit of $(S^1)^{\mathbf{a}}$. We must be careful though, for it is possible that some mapping classes permute the elements of $\mathbf{a}$. So this action only lives in an orbifold setting. We thus find:

Proposition 10.21. *We have a natural map of orbifolds $\pi : \mathscr{M}^+_{g,P} \to \overline{\mathscr{M}}_{g,P}$. This map is proper and over the stratum of $\overline{\mathscr{M}}_{g,P}$ defined by the* $\mathrm{Mod}(S,P)$*-orbit of the simplex* $\mathbf{a}$ *of the curve complex it has the structure of a principal bundle of the torus $(S^1)^{C_{sg}}$ in the orbifold sense.*

The idea behind Geometric Invariant Theory

Before we discuss the projectivity of the moduli space of stable pointed curves, we need to review some basic facts from Geometric Invariant Theory. We assume here our base field K is algebraically closed of characteristic zero. If G is a reductive group (like $\mathrm{GL}(n,K)$ or $\mathrm{SL}(n,K)$) and W is a representation of G, then the algebra of G-invariant polynomial functions on W, $K[W]^G$, defines the *separated orbit space* of W: it is a finitely generated algebra and is able to separate any two orbits whose closures are disjoint. The geometric content of this statement is that $K[W]^G$

defines a variety (denoted $G\backslash\backslash W$) whose closed points are in bijection with the closed orbits in W. If we pass to the G-action on $\mathbb{P}(W)$, then we need to discard in W the closed orbit defined by the origin and hence with it also all orbits in W having the origin in their closure. The remaining open subset of W is called the *semi-stable locus* and is denoted W^{ss}. The closed orbits in $\mathbb{P}(W^{ss}) \subset \mathbb{P}(W)$ are then in bijective correspondence with the closed points of the projective variety $\mathrm{Proj}(K[W]^G)$ for which $K[W]^G$ is the homogeneous coordinate ring (we therefore denote this variety also by $G\backslash\backslash\mathbb{P}(W^{ss})$). A point of W is called *G-stable* if its G-orbit is closed and its G-stabilizer is finite. Such a point clearly lies in W^{ss} and defines an open (possibly empty) subset $W^{st} \subset W^{ss}$. It defines subsets of $G\backslash\backslash W$ and $G\backslash\backslash\mathbb{P}(W^{ss})$ that can be identified with $G\backslash W^{st}$ and $G\backslash\mathbb{P}(W^{st})$ respectively.

In practice one is only interested in a G-invariant subvariety $Z \subset \mathbb{P}(W)$. If $Z^{ss} := Z \cap \mathbb{P}(W^{ss})$ is closed in $\mathbb{P}(W^{ss})$, then so is its image in $\mathrm{Proj}(K[W]^G)$, making it a projective variety. If it so happens that $Z^{ss} \subset \mathbb{P}(W^{st})$, then we see that the G-orbit space $G\backslash Z^{ss}$ of Z^{ss} is realized as a closed subset in $\mathrm{Proj}(K[W]^G)$. So $G\backslash Z^{ss}$ has then the structure of a projective variety.

Projectivity of the moduli space of stable pointed curves

We now return to the discussion *the quasi-projective structure on the universal family* of Section 9. We there constructed a vector space V, a representation W of $\mathrm{SL}(V)$ and a $\mathrm{SL}(V)$-invariant a subvariety $Z \subset \mathbb{P}(W)$ on which $\mathrm{SL}(V)$ acts properly with finite stabilizers in such a manner that the orbit space is naturally identified with $M_{g,P}$. What was left unsaid is that if we take in *loc. cit.* the parameters r and d large enough, then $\overline{Z} \cap \mathbb{P}(W^{ss})$ is closed in $\mathbb{P}(W^{ss})$ and contained in $\mathbb{P}(W^{st})$: any $z \in \overline{Z} \cap \mathbb{P}(W^{ss})$ defines a projective stable P-pointed curve (C_z, P) of genus g and the image of the natural action of the stabilizer $\mathrm{SL}(V)_z$ on (C_z, P) yields all of $\mathrm{Aut}(C_z, P)$.

The above construction then leads to a projective compactification of $M_{g,P}$ which turns out to be just our $\overline{M}_{g,P}$. This implies in particular that if our base field is $\mathbb{C}$, then $M_{g,P}$ is compact in the Hausdorff topology. It then follows from Proposition 10.21 that the same is true for $M^+_{g,P}$.

Getting Harvey's bordification from the D-M compactification

We first need to discuss the notion of a real oriented blowup. If M is a C^∞-manifold and $N \subset M$ a closed submanifold, then the *real oriented blowup* of N in M is a C^∞-manifold $\tilde{M}_N$ with boundary which comes with a differentiable map $\pi : M_N \to M$ with the property that π is over M_N is a diffeomorphism and is over N is a the normal sphere bundle: a point over $p \in N$ is given by a ray in the normal space T_pM/T_pN. This sphere bundle is also the boundary of $\tilde{M}_N$. The differentiable structure on M_N can be given by local coordinates: if $(x', x'') = (x^1, \dots, x^n, x^{n+1}, \dots, x^m)$ is a coordinate system on $U \subset M$ such that $N \cap U$ is given by $x'' = 0$, then on $\pi^{-1}U$

we have a parametrization by an open subset of $\mathbb{R}^n \times [0,\infty) \times S^{m-n-1}$ such that π is given over U by $(x', r, \xi) \mapsto (x', r\xi)$.

In case M is a complex manifold and N is a complex submanifold of codimension one, then the boundary of M_N is in a natural way a principal homogeneous bundle of the circle group $U(1)$ (for the rays in $\mathbb{C}$ emanating from $0 \in \mathbb{C}$ make up a principal homogeneous space for that group). If $N' \subset M$ has another complex submanifold of codimension one which meets N transversally (so that $N \cap N'$ is a complex submanifold of codimension two), then the preimage of N' in M_N can be identified with manifold with boundary $N'_N \cap N'$ and the normal bundle of $N'_{N\cap N'}$ in M_N is the pull-back of the normal bundle of $N' \subset M$. So if we perform another real oriented blowup for the inclusion $N'_N \subset M_N$, then the resulting space $M_{N,N'}$ is a manifold with a boundary over $N \cup N'$ which has a corner over $N \cap N'$. Notice that a point in $M_{N,N'}$ over $p \in N \cap N'$ is a pair of rays, one in T_pM/T_pN' and another in T_pM/T_pN'. In fact, the obvious map $M_{N,N'} \to M$ is over $N \cap N'$ a principal $U(1) \times U(1)$-bundle and the order of blowing up is irrelevant: $M_{N,N'}$ may be identified with $M_{N',N}$. This generalizes in a straightforward manner to the situation of a normal crossing divisor: if D is a normal crossing divisor in M, then we have a real oriented blowup $M_D \to M$, which over the locus where D has precisely k local irreducible components has locally the structure of a principal $U(1)^k$-bundle; if $p \in M$, then a point over M_D is given by a ray in T_pM/T_pD_i for every local irreducible component (D_i, p) of (D, p).

We apply this in an orbifold setting: we take $M = \overline{\mathcal{M}}_{g,P}$ and we let D be the Deligne-Mumford boundary.

Proposition 10.22. *The Harvey bordification $\mathcal{M}^+_{g,P}$ is naturally identified with the real oriented blowup of the Deligne-Mumford boundary $D_{g,P}$ in $\overline{\mathcal{M}}_{g,P}$.*

Proof. We construct a map from this real oriented blowup to $\mathcal{M}^+_{g,P}$. A point $c \in M$ is given by a stable P-pointed curve (C, P). The local irreducible components of (D, c) correspond to the nodes of C and the normal space to the irreducible component indexed by the node q is naturally identified with $T_qC^+_q \otimes T_qC^-_q$. So a point over c is specified by giving for every node $q \in C_{sg}$ a ray in $T_qC^+_q \otimes T_qC^-_q$. We use these data to construct a surface with a permissible metric. Consider the normalization $\pi : \hat{C} \to C$ and denote by $N \subset \hat{C}$ the preimage of C_{sg}. This comes as a finite set of pairs: for each node $q \in C$ we get an unordered pair $\{q^+, q^-\}$ and we assume given a ray r_q in $T_q\hat{C}_{q^+} \otimes T_q\hat{C}_{q^-}$. Now consider the real-oriented blowup $\hat{C}_N \to C$. This is a surface with boundary. The boundary component defined by $q^{\pm}$ is the space of rays in $T_q\hat{C}_{q^\pm}$. We use the ray r_q to identify these boundary components: we identify the boundary point defined by the ray $r^+ \subset T_q\hat{C}_{q^+}$ the boundary point defined by $r^- \subset T_q\hat{C}_{q^-}$ if $r^+ \otimes r^- = r_q$. This produces in first instance a topological surface S over C that is with over each q an embedded circle A_q. But if we equip C°_{reg} with its hyperbolic metric, then S has a unique differentiable structure for which the hyperbolic metric is that is given on $S^\circ - \cup_{q \in C_{sg}} A_q \cong C^\circ_{reg}$ is permissible with respect to its embedding in S. The resulting object defines an element of the Harvey compactification $\mathcal{M}^+_{g,P}$.

This defines our map. It is not hard to check that it is continuous. It is also has a continuous two-sided inverse, which is somewhat easier to define and whose definition is clearly suggested by the above map: if $A \subset S^\circ$ defines a simplex of the curve complex, then a permissible metric on $S^\circ - A$ relative to its embedding in S defines a conformal structure on $S \smallsetminus A$ giving us a nodal curve C_A. We have (according to Exercise 10.4) at each connected component A_q of A a standard coordinate system that is unique up to a rotation and inversion. So an orientation of A_q determines on A_q the structure of a principal $U(1)$-space (with opposite orientations defining opposite $U(1)$-actions). This 'opposition' of $U(1)$-structures then determines a ray in the deformation space of the node $q \in C_A$. □

This implies an important compactness property for the Harvey bordification.

Corollary 10.23. *The mapping class group* $\mathrm{Mod}(S,P)$ *acts properly discontinuously on* $\mathcal{T}(S,P)^+$ *and the orbit space* $\mathcal{M}^+_{g,P}$ *and is a compact orbifold with corners and with interior* $\mathcal{M}_{g,P}$.

Proof. Since the Deligne-Mumford is compact, it follows from Proposition 10.22 that $\mathcal{M}^+_{g,P}$ is compact as well.

We show how the first assertion can also be deduced from a property of the Deligne-Mumford compactification. A point $z \in \overline{\mathcal{M}}_{g,P}$ is represented by a (C,P) be a stable P-pointed curve (C,P) and an orbifold atlas at that point is given by a universal deformation $(f : \mathcal{C} \to B, x_P)$ of that curve as in Theorem 10.18, the associated (finite) group being $\mathrm{Aut}(C,P)$. The discriminant of f is a normal crossing divisor $D \subset B$. The real oriented blowup $B_D \to B$ of D will have the homotopy type of its central fiber, a torus of dimension $|C_{sg}|$. Its universal cover $\tilde{B}_D \to B_D$ is then contractible and its covering group can be identified with $\mathbb{Z}^{C_{sg}}$, the generator indexed by $q \in C_{sg}$ being a Dehn twist over that point. Then we have an exact sequence of automorphisms of $\tilde{B}_D$:

$$0 \to \mathbb{Z}^{C_{sg}} \to \widetilde{\mathrm{Aut}(C,P)} \to \mathrm{Aut}(C,P) \to 1.$$

This group clearly acts properly discontinuously on $\tilde{B}_D$. From our assumptions on the universal deformation it follows that $\tilde{B}_D$ parametrizes an open subset $U \subset \mathcal{T}(S,P)^+$ (more precisely, a $\mathrm{Mod}(S,P)$-orbit of those) and that any mapping class $g \in \mathrm{Mod}(S,P)$ with the property that $gU \cap U \neq \emptyset$ must come from $\widetilde{\mathrm{Aut}(C,P)}$. Since the open sets thus obtained cover $\mathcal{M}^+_{g,P}$, it follows that $\mathrm{Mod}(S,P)$ acts properly discontinuously on $\mathcal{T}(S,P)^+$. □

Corollary 10.24. *For every* $\varepsilon>0$, *the locus* $\mathcal{T}(S,P)^\varepsilon \subset \mathcal{T}(S,P)$ *where all the geodesic length functions are* $\geq \varepsilon$ *is a* $\mathrm{Mod}(S,P)$*-invariant subset on which* $\mathrm{Mod}(S,P)$ *acts with compact fundamental domain. For small enough* ε, $\mathcal{T}(S,P)^\varepsilon$ *is a* $\mathrm{Mod}(S,P)$*-equivariant deformation retract of* $\mathcal{T}(S,P)^+$ *(and hence of* $\mathcal{T}(S,P)$*).*

Proof. Clearly, $\mathscr{T}(S,P)^{\varepsilon}$ is closed in $\mathscr{T}(S,P)$ and $\mathrm{Mod}(S,P)$-invariant. Hence its image $\mathscr{M}^{\varepsilon}_{g,P}$ in $\mathscr{M}^{+}_{g,P}$ is also closed. A closed subset of a compact set is compact and so $\mathscr{M}^{\varepsilon}_{g,P}$ is compact. With the help of Fenchel-Nielsen charts, we see the For $\varepsilon > 0$ small enough, $\mathscr{M}^{\varepsilon}_{g,P}$ is a deformation retract of $\mathscr{M}^{+}_{g,P}$. Such a deformation retraction will lift to a $\mathrm{Mod}(S,P)$-equivariant one of $\mathscr{T}^{+}_{g,P}$ onto $\mathscr{T}^{\varepsilon}_{g,P}$. □

11 Cohomological properties of $\mathscr{M}_{g,P}$

Harer's stability theorem

Let (S,P) be a P-pointed connected closed oriented surface of genus g such that $S^{\circ} = S \smallsetminus P$ has negative Euler characteristic. Let $o \in S \smallsetminus P$ and put $S^{\diamond} := S \smallsetminus \{o\}$ and $\tilde{P} := P \cup \{o\}$. Consider the group $\mathrm{Diff}_c(S^{\diamond},P)$ of diffeomorphisms of S that fix P and are the identity on an unspecified neighborhood of o (the subscript stands for *compact support*). It is a subgroup of the group $\mathrm{Diff}_{\vec{o}}(S^{\diamond},P)$ of diffeomorphisms of S that fix P and T_oC pointwise. It is not hard to verify that the inclusion $\mathrm{Diff}_c(S^{\diamond},P) \subset \mathrm{Diff}_{\vec{o}}(S^{\diamond},P)$ induces an isomorphism on connected components: $\mathrm{Mod}_c(S^{\diamond},P) \cong \mathrm{Mod}_{\vec{o}}(S,P)$ (it is in fact a weak homotopy equivalence). Let us fix an oriented isomorphism $\alpha : T_oS \cong \mathbb{C}$. Then we have a Teichmüller space $\mathscr{T}_{\vec{o}}(S,P)$ associated to $\mathrm{Mod}_{\vec{o}}(S,P)$, namely the space of conformal structures S that extend α given up to isotopy in the identity component of $\mathrm{Diff}_{\vec{o}}(S^{\diamond},P)$. If we divide out by $\mathrm{Mod}_{\vec{o}}(S,P)$, then we get the moduli space $\mathscr{M}_{g,P,\vec{o}}$ of triples $(C,\tilde{P},v)$, where the first two items define a $\tilde{P}$-pointed curve and $v \in T_oC$ is a nonzero tangent vector (the vector which under α maps to 1). The structure becomes clearer if we consider the forgetful map $\mathscr{T}_{\vec{o}}(S,P) \to \mathscr{T}(S,\tilde{P})$. The tangent space T_oS defines the line bundle over $\mathscr{T}(S,\tilde{P})$ and $\mathscr{T}(S,P,\vec{o})$ may be identified with the universal cover of the complement of the zero section of that line bundle (this has the structure of an affine line bundle over $\mathscr{T}(S,\tilde{P})$). The group of covering transformations is generated by the class $D_o \in \mathrm{Mod}_c(S^{\diamond},P)$ of a Dehn twist along a small circle centered at o, so that dividing out by $D_o^{\mathbb{Z}}$ yields the complement of the zero section of the line bundle defined above. We have a central extension

$$1 \to D_o^{\mathbb{Z}} \to \mathrm{Mod}_c(S^{\diamond},P) \to \mathrm{Mod}(S,\tilde{P}) \to 1$$

that can be understood as an exact sequence of orbifold fundamental groups of the associated orbifold $\mathbb{C}^{\times}$-bundle $\mathscr{M}_{g,P,\vec{o}} / \mathscr{M}_{g,\tilde{P}}$.

Suppose we are given an embedding $i : S^{\diamond} \to S'$ of $S^{\diamond}$ in a connected surface S' (necessarily of genus $\geq g$; it could be $S' = S$ with i the inclusion, but it is also possible for $i(S^{\diamond})$ to appear as the connected component of S' minus an embedded circle). Then 'extension by the identity' defines an embedding $\mathrm{Diff}_c(S^{\diamond},P) \hookrightarrow \mathrm{Diff}_c^{+}(S',P)$. This induces a group homomorphism $\mathrm{Mod}_c(S^{\diamond},P) \to \mathrm{Mod}_c(S',P)$. The stability theorem reads as follows.

Theorem 11.1 (Harer). *The homomorphism* $\mathrm{Mod}_c(S^\circ,P) \to \mathrm{Mod}_c(S',P)$ *induces an isomorphism on group (co)homology in degrees* $< \frac{2}{3}g$.

For a concise proof we refer to [20]. So if we take $S' = S$, then we see that $\mathrm{Mod}_c(S^\circ,P) \to \mathrm{Mod}_c(S,P) = \mathrm{Mod}(S,P) \cong \mathrm{Mod}_{g,P}$ has this property and if we take S' to be closed of genus $g+1$, then we see that $\mathrm{Mod}_c(S^\circ,P) \to \mathrm{Mod}_c(S',P) = \mathrm{Mod}(S',P) \cong \mathrm{Mod}_{g+1,P}$ has that property, too. The comparison maps

$$\mathrm{Mod}_{g,P} \cong \mathrm{Mod}(S,P) \leftarrow \mathrm{Mod}_c(S^\circ,P) \to \mathrm{Mod}(S',P) \cong \mathrm{Mod}_{g+1,P}$$

involve choices. But the ambiguity is always a conjugacy in the target group and since interior automorphisms act trivially on the cohomology of a group, the induced maps on cohomology are independent of these choices. So for a fixed k, $H^k(\mathrm{Mod}_{g,P})$ is independent of g if g is large enough. This so called *stable cohomology* is in fact the cohomology of the group $\mathrm{Mod}_{\infty,P} := \mathrm{Mod}_c(S_\infty,P)$ where S_∞ is connected orientable and of infinite genus.

Corollary 11.2. *For* $g > \frac{3}{2}k$, *we have a natural identification* $H^k(\mathcal{M}_{g,P};\mathbb{Q}) \cong H^k(\mathrm{Mod}_{\infty,P};\mathbb{Q})$.

Hopf algebra structure on stable cohomology

Let S_i be a closed connected surface of genus g_i and let $o_i \in S_i$, $i = 1,2,\dots$ and let S be another closed connected surface of genus g and $o \in S$. Suppose that for $i = 1,2$, we are given embeddings $j_i : S_i^\circ \to S^\circ$ with disjoint images. This induces a homomorphism of groups

$$\mathrm{Mod}_c(S_1^\circ) \times \mathrm{Mod}_c(S_2^\circ) \to \mathrm{Mod}_c(S^\circ).$$

If D_i is a small open disk on S_i centered at o_i, then $S_i \smallsetminus D_i$ is a compact surface with boundary ∂D_i and then the complement of the images of $S_1 \smallsetminus D_1$ and $S_2 \smallsetminus D_2$ in S° is a connected punctured surface of genus $g - g_1 - g_2$ with two boundary components. Such a surface is unique up to diffeomorphism. This is easily seen to imply that all embeddings of $(S_1 \smallsetminus D_1) \sqcup (S_2 \smallsetminus D_2)$ in S° lie in a single $\mathrm{Diff}_c(S^\circ)$-orbit. So the choice of another pair (j_1, j_2) may alter the homomorphism $\mathrm{Mod}_c(S_1^\circ) \times \mathrm{Mod}_c(S_2^\circ) \to \mathrm{Mod}_c(S^\circ)$ by conjugation with an element of $\mathrm{Mod}(S^\circ)$ only. Since an inner automorphism of a group acts trivially on its cohomology, the induced ring homomorphism

$$\begin{aligned}\Delta_{1,2} : H^\bullet(\mathrm{Mod}_c(S^\circ)) &\to H^\bullet(\mathrm{Mod}_c(S_1^\circ) \times \mathrm{Mod}_c(S_2^\circ)) \to \\ &\to H^\bullet(\mathrm{Mod}_c(S_1^\circ)) \otimes H^\bullet(\mathrm{Mod}_c(S_2^\circ))\end{aligned}$$

is independent of choices. The inclusion of the trivial group $\{1\}$ in $\mathrm{Mod}_c(S_2^\circ)$ induces the *co-augmentation* $\varepsilon_2 : H^\bullet(\mathrm{Mod}_c(S_2^\circ)) \to H^\bullet(\{1\}) = \mathbb{Z}$ and so if we compose $\Delta_{1,2}$ with $1 \otimes \varepsilon_2$, then the resulting map $H^\bullet(\mathrm{Mod}(S^\circ)) \to H^\bullet(\mathrm{Mod}(S_1^\circ))$ is just the map that we used to state the stability theorem.

If it so happens that $(S_1, D_1, o_1) = (S_2, D_2, o_2)$, then precomposing the embedding of $(S_1 \smallsetminus D_1) \sqcup (S_2 \smallsetminus D_2)$ in $S^\diamond$ with the exchange map alters the embedding by an element of $\mathrm{Diff}_c(S^\diamond)$. This implies that $\Delta_{1,2}$ does not change if we compose it with the exchange map (the graded involution which is on $H^k(\mathrm{Mod}_\infty) \otimes H^l(\mathrm{Mod}_\infty)$ is given by $u \otimes v \mapsto (-1)^{kl} v \otimes u$).

A similar argument shows that if we are given pairwise disjoint embeddings $j_i : S_i^\diamond \to S^\diamond$ for $i = 1,2,3$, such that j_1 and j_2 factor through an embedding $S_{12}^\diamond \to S^\diamond$ and j_2 and j_3 factor through an embedding $S_{23}^\diamond \to S^\diamond$, then the diagram below commutes.

$$\begin{array}{ccc}
H^\bullet(\mathrm{Mod}_c(S^\diamond)) & \xrightarrow{\Delta_{12,3}} & H^\bullet(\mathrm{Mod}_c(S_{12}^\diamond)) \otimes H^\bullet(\mathrm{Mod}_c(S_3^\diamond)) \\
\downarrow \Delta_{1,23} & & \downarrow \Delta_{1,2} \otimes 1 \\
H^\bullet(\mathrm{Mod}_c(S_1^\diamond)) \otimes H^\bullet(\mathrm{Mod}_c(S_{23}^\diamond)) & \xrightarrow{1 \otimes \Delta_{2,3}} & H^\bullet(\mathrm{Mod}_c(S_1^\diamond)) \otimes H^\bullet(\mathrm{Mod}_c(S_2^\diamond)) \otimes \\
& & H^\bullet(\mathrm{Mod}_c(S_3^\diamond)).
\end{array}$$

So if we invoke Harer's stability theorem, then we find a graded ring homomorphism $\Delta : H^\bullet(\mathrm{Mod}_\infty) \to H^\bullet(\mathrm{Mod}_\infty) \otimes H^\bullet(\mathrm{Mod}_\infty)$ with the property that if $\varepsilon : H^\bullet(\mathrm{Mod}_\infty) \to \mathbb{Z}$ denotes the co-augmentation map, then

co-unit: $(\varepsilon \otimes 1)\Delta$ and $(1 \otimes \varepsilon)\Delta$ are the identity,
co-associativity: $(\Delta \otimes 1) \otimes \Delta = (1 \otimes \Delta) \otimes \Delta$,
cocommutativity: $\Delta = T\Delta$, where T is the graded involution which on $H^k(\mathrm{Mod}_\infty) \otimes H^l(\mathrm{Mod}_\infty)$ is given by $u \otimes v \mapsto (-1)^{kl} v \otimes u$,

the last assertion being rather trivial. In other words, this makes $H^\bullet(\mathrm{Mod}_\infty)$ a Hopf algebra that is both graded-commutative and graded-cocommutative. If we tensor with $\mathbb{Q}$, so that we are dealing with $H^\bullet(\mathrm{Mod}_\infty; \mathbb{Q})$, then according to a theorem of Milnor-Moore [15] such an algebra has a very simple structure. Let us abbreviate $H^\bullet(\mathrm{Mod}_\infty; \mathbb{Q})$ by $H^\bullet$. The *primitive part* $H^\bullet_{\mathrm{pr}}$ is defined as the space of $u \in H^\bullet$ with $\Delta(u) = u \otimes 1 + 1 \otimes u$. This is a graded subspace and the inclusion of $H^\bullet_{\mathrm{pr}}$ in $H^\bullet$ extends of course to a $\mathbb{Q}$-algebra homomorphism

$$\mathrm{Sym}^\bullet(H_{\mathrm{pr}}^{\mathrm{even}}) \otimes_{\mathbb{Q}} \wedge^\bullet(H_{\mathrm{pr}}^{\mathrm{odd}}) \to H^\bullet.$$

The theorem in question asserts that this is in fact an isomorphism. Since Δ is a ring homomorphism, this description also determines the comultiplication on the left hand side. We may think of $\mathrm{Sym}^\bullet(H_{\mathrm{pr}}^{even})$ as the coordinate ring of the dual H^{pr}_{even} of H_{pr}^{even} (the *stable primitive homology*). The coproduct on $\mathrm{Sym}^\bullet(H_{\mathrm{pr}}^{even})$ simply describes the additive structure on H^{pr}_{even}.

12 Tautological algebras

Duality on orbifolds

Suppose X is a space and G is a finite group acting on X. Then the quotient map $X \to G\backslash X$ maps the singular cochain complex of $G\backslash X$ to the G-invariant part of the cochain complex of X and under mild conditions (e.g., X admits the structure of a G-invariant CW complex) this identifies $H^\bullet(G\backslash X;\mathbb{Q})$ with $H^\bullet(X;\mathbb{Q})^G$.

This is used to see that the space underlying an orbifold M of dimension m is a rational homology manifold of that same dimension. By the latter is meant that for every $p\in M$, $H_\bullet(M,M\smallsetminus\{p\};\mathbb{Q})$ is isomorphic to $H_\bullet(\mathbb{R}^m,\mathbb{R}^m\smallsetminus\{0\};\mathbb{Q})$ (in other words is of dimension one and concentrated in degree m). It is said to be *oriented* if for every $p\in M$ we have specified a generator μ_p for $H_m(M,M\smallsetminus\{p\};\mathbb{Q})$ which depends 'continuously' on p: if $U\subset M$ is the interior of an embedded disk, then for $p,q\in U$ the isomorphism $H_m(M,M\smallsetminus\{p\};\mathbb{Q})\xrightarrow{\cong}H_m(M,M\smallsetminus U;\mathbb{Q})\xleftarrow{\cong}H_m(M,M\smallsetminus\{q\};\mathbb{Q})$ takes μ_p to μ_q. In that case M satisfies $\mathbb{Q}$-Poincaré duality in dimension m: we have a natural isomorphism $H^k(M;\mathbb{Q})\cong H^{lf}_{m-k}(M;\mathbb{Q})$, where $H^{lf}_\bullet$ stands for the homology of the complex of possibly infinite, but *locally finite* chains (if M is open in a compact set K, such that $M-K$ is not too wild, e.g., a compact manifold with boundary having M as its interior, then $H^{lf}_\bullet(M)\cong H_\bullet(K,M\smallsetminus K)$). Note that a cochain with compact support can be evaluated on a locally finite chain.

The preceding applies to the space underlying a complex-analytic orbifolds such as $\mathcal{M}_{g,P}$ and $\overline{\mathcal{M}}_{g,P}$ and so they satisfy $\mathbb{Q}$-Poincaré duality.

Review of the Gysin map

Let $f:N\to M$ be a proper map between manifolds of dimension n and m respectively which are endowed with orientations μ_M and μ_N. Then we can use duality on both N and M to define the *Gysin map*: if $d:=m-n$ denotes the *formal codimension* of f, then put

$$f_!:H^k(N)\overset{\mu_N}{\cong}H^{lf}_{n-k}(N)\xrightarrow{f_*}H^{lf}_{n-k}(M)\overset{\mu_M}{\cong}H^{d+k}(M).$$

The ring homomorphism $f^*:H^\bullet(M)\to H^\bullet(N)$ turns $H^\bullet(N)$ into a $H^\bullet(M)$-module and with respect to that structure $f_!$ is a homomorphism of $H^\bullet(M)$-modules: $f_!(f^*y\cup x)=y\cup f_!(x)$. We further note that if $g:M\to P$ is another proper map with P an oriented manifold, then gf is proper and $(gf)_!=g_!f_!$. The following two cases are of special interest.

If f is a proper embedding of codimension d, then $f_!:H^\bullet(N)\to H^{d+\bullet}(M)$ can be obtained as the composite of the Thom isomorphism

$$H^\bullet(N)\cong H^{d+\bullet}(U,U\smallsetminus N))\cong H^{d+\bullet}(M,M\smallsetminus N)$$

(where U is a tubular neighborhood of N) and the natural map $H^{d+\bullet}(M,M\smallsetminus N)\to H^{d+\bullet}(M)$ so that it fits in an exact sequence:

$$\cdots\to H^{k-1}(M\smallsetminus N)\to H^{k-d}(N)\xrightarrow{f_!}H^k(M)\to H^k(M\smallsetminus N)\to\cdots$$

The element $f_!(1)\in H^d(M)$ can be understood as assigning to a d-cycle on M its intersection number with N; we often refer to this as the *(cohomology) class* *defined by* N. Its image $f^*f_!(1) \in H^d(N)$ is the Euler class of the normal bundle of f.

If f is a proper submersion (so with with fiber dimension $-d \geq 0$), then the Gysin map $f_! : H^{\bullet -d}(N) \to H^\bullet(M)$ is 'integration along the fibers'. There is then a base change property: if $g : M' \to M$ is a smooth map of manifolds, then we have pull-back diagram

$$\begin{array}{ccc} N' & \xrightarrow{g'} & N \\ {\scriptstyle f'}\downarrow & & \downarrow{\scriptstyle f} \\ M' & \xrightarrow{g} & M \end{array}$$

with f' a proper submersion of manifolds and $g^*f_! = f'_!(g')^*$.

Remark 12.1. The Gysin map can be defined without requiring that f be proper, but then we need to restrict ourselves on N to cohomology with *proper f-support*, $H^\bullet_f(N)$: we only allow cochains such that the restriction of f to their support is proper. The aforementioned properties still hold and we then see that the general case is a combination of these two special spaces: if we factor f as the embedding $i_f : N \subset N \times M$ by its graph followed by the projection $\pi_M : N \times M \to M$: then $i_{f!}$ takes its values in $H^\bullet_{\pi_M}(M \times N)$ and we have $f_! = \pi_{M!} i_{f!}$.

If we are willing to take cohomology with $\mathbb{Q}$-coefficients, then the preceding is still true for rational homology manifolds. A *rational homology manifold of dimension m* is a Hausdorff space M that is locally the cone over a compact space and has the property that for every $p \in M$, $H_\bullet(M, M \smallsetminus \{p\}; \mathbb{Q})$ is isomorphic to $H_\bullet(\mathbb{R}^m, \mathbb{R}^m \smallsetminus \{0\}; \mathbb{Q})$ (in other words is of dimension one and concentrated in degree m). It is said to be *oriented* if we have specified a generator $\mu_p \in H_m(M, M \smallsetminus \{p\}; \mathbb{Q})$ which depends 'continuously' on p in the sense that if U is an open ball in M containing p, then the isomorphism $H_m(M, M \smallsetminus \{p\}; \mathbb{Q}) \cong H_m(M, M \smallsetminus U; \mathbb{Q}) \cong H_m(M, M \smallsetminus \{q\}; \mathbb{Q})$ maps μ_p to μ_q. In that case the orientation defines a duality isomorphism $\mu : H^k(M; \mathbb{Q}) \cong H^{lf}_{m-k}(M; \mathbb{Q})$.

This applies for example to the space underlying a complex-analytic orbifold. This also shows that if $E \to M$ is an oriented vector bundle of rank r in the orbifold sense, then its rational Euler class $e(E/M) \in H^r(M; \mathbb{Q})$ can be defined as $o^*o_!(1)$, where $o : M \to E$ is the zero section.

The moduli space of stable pointed curves as a category

Between the moduli spaces $\overline{\mathcal{M}}_{g,P}$ and products of those there are many interesting maps.

First, there is a rather trivial kind: a bijection $Q \cong P$ of finite sets gives rise to an evident identification $\overline{\mathcal{M}}_{g,Q} \cong \overline{\mathcal{M}}_{g,P}$. This implies that the symmetric group $\mathrm{Aut}(P)$ acts on $\overline{\mathcal{M}}_{g,P}$. More generally, every injection $P \to Q$ determines a surjective morphism of orbifolds $\overline{\mathcal{M}}_{g,Q} \to \overline{\mathcal{M}}_{g,P}$. Let us first see why this is so for the inclusion of P in a set that has one element more than P: $\tilde{P} = P \sqcup \{o\}$ (with part of the argument relegated to two exercises). If $(\tilde{C}, \tilde{P})$ is a stable pointed curve, then by forgetting o, the resulting $(\tilde{C}, P)$ need not be a stable, but nevertheless does give rise to a stable P-pointed genus g curve (we keep on assuming that $2g-2+|P|>0$, of course). The following two exercises make this precise.

Exercise 12.2. Show that $(\tilde{C}, P)$ is stable unless o lies on an irreducible component $\tilde{C}_o$ isomorphic to $\mathbb{P}^1$ which has exactly two points that are nodes or belong to P. Prove that in the last case, contraction of $\tilde{C}_o$ in $\tilde{C}$ yields a stable P-pointed curve (C, P).

Exercise 12.3. Next show a converse: let (C, P) be a stable pointed genus g curve. Prove that for every $q \in C$ there is naturally defined a stable $\tilde{P}$-pointed genus g curve $(\tilde{C}_q, \tilde{P})$ together with a surjective morphism $\tilde{C}_q \to C$ for which forgetting o yields the above contraction map. Show that the pair $(\tilde{C}_q, \tilde{P})$ is unique up to unique isomorphism and that the automorphism group of $(\tilde{C}_q, \tilde{P})$ can be identified with the group of automorphisms of (C, P) which fix q.

Exercise 12.3 can also be carried out 'with parameters': if $(\mathscr{C} \to B, x_P)$ is a family of stable P-pointed genus g curves, then $\mathscr{C}$ serves as a base for a naturally defined family of stable $\tilde{P}$-pointed genus g curves. This construction therefore defines a morphism of orbifolds $\overline{\mathscr{C}}_{g,P} \to \overline{\mathscr{M}}_{g,\tilde{P}}$. Exercise 12.2 provides an inverse and so we conclude:

Proposition 12.4. *We have a natural isomorphism $\overline{\mathscr{C}}_{g,P} \cong \overline{\mathscr{M}}_{g,\tilde{P}}$ of orbifolds. In particular, we have a morphism $f_{\tilde{P},P} : \overline{\mathscr{M}}_{g,\tilde{P}} \to \overline{\mathscr{M}}_{g,P}$ and every $p \in P$ defines a section x_p of it.*

Iteration then implies our assertion above that any inclusion $P \hookrightarrow Q$ of finite sets determines a surjective morphism $\overline{\mathscr{M}}_{g,Q} \to \overline{\mathscr{M}}_{g,P}$ (the order of contraction is indeed immaterial).

These moduli spaces can be connected for different genera as follows. First assume we are given a *stable pair* $(g-1, Q)$ (by which we simply mean that $2(g-1)-2+|Q|>0$) and a 2-element subset $D \subset Q$. Out of any stable pointed curve $(\tilde{C}, Q)$ of genus $g-1$ we construct a stable pointed curve $(\tilde{C}, Q)/D$ of genus g by simply identifying the two points of D so that this becomes a node. There is no difficulty in doing this in families and we thus get a finite morphism

$$i_D : \overline{\mathscr{M}}_{g-1,Q} \to \overline{\mathscr{M}}_{g,P},$$

where $P := Q - D$. The image is in fact an irreducible component of the Deligne-Mumford boundary of $\overline{\mathscr{M}}_{g,P}$ (that is often denoted by Δ_0).

Similarly, suppose we are given stable pairs (g', Q') and (g'', Q'') and elements $q \in Q'$ and $q'' \in Q''$. Put $g := g' + g''$, $D := \{q', q''\}$ and $P := Q' \sqcup Q'' \smallsetminus D$. If (C', Q') and (C'', Q'') are stable pointed curve of genus g' and g'' respectively, then joining these curves by identifying the members of D creates a P-pointed stable curve of genus g which at least one node. This defines a finite morphism

$$i_D : \overline{\mathscr{M}}_{g',Q'} \times \overline{\mathscr{M}}_{g'',Q''} \to \overline{\mathscr{M}}_{g,P}.$$

The image is an irreducible component of the Deligne-Mumford boundary (and the normal bundle of i_D may be identified with $\mathscr{L}_D$ as before). This irreducible component is sometimes denoted $\Delta_{(g',P')}$ $(=\Delta_{(g'',P'')})$.

Exercise 12.5. A special case of interest is when $g'' = 0$ and Q'' is a 3-element set $\{q'', q_0, o\}$. Then $\overline{\mathscr{M}}_{0,Q''}$ is a singleton, $g = g'$ and P is obtained from Q' by replacing q' by the 2-element set $\{q_0, o\}$, so that $\overline{\mathscr{M}}_{g,P}$ can be identified with $\overline{\mathscr{C}}_{g,Q'}$. Show that the resulting map $i_D : \overline{\mathscr{M}}_{g,Q'} \to \overline{\mathscr{M}}_{g,P}$ can then be understood as the section $x_{q'}$, at least if we rename q_0 as q'.

Vector bundles on moduli spaces of curves

If C is a projective nodal curve of genus g, then its *dualizing sheaf* ω_C is on the smooth part of C the sheaf of of differentials and a local section at a node is a differential on each branch with at most a simple pole at that node such that the two residues add up to 0. Notice that ω_C is in fact a line bundle on C. Its name is partly explained by the fact that it accomplishes Serre duality: if C is connected, then $H^1(C,\omega_C)$ is one-dimensional and for every locally free $\mathscr{O}_C$-module of finite rank $\mathscr{F}$, the obvious pairing $H^1(C,\mathscr{F}) \times H^0(C,\mathscr{H}om_{\mathscr{O}_C}(\mathscr{F},\omega_C)) \to H^1(C,\omega_C)$ is perfect. In particular, this perfectly pairs $H^1(C,\mathscr{O}_C)$ with $H^0(C,\omega_C)$.

Lemma 12.6. *Let $\pi : \hat{C} \to C$ be a partial normalization of C (with the branches at some nodes being separated), and denote by G the graph attached to π (the points in $\hat{C}$ lying over a node define oriented edges, connected components of $\hat{C}$ define vertices, normalized nodes define edges). Then we have a natural exact sequence*

$$0 \to H^0(\hat{C},\omega_{\hat{C}}) \to H^0(C,\omega_C) \to H^1(G;\mathbb{C}) \to 0,$$

so that for C connected, $\dim H^0(C,\omega_C) = g$.

Proof. We first do the case when π is the full normalization so that $\hat{C}$ is smooth. An element of $H^0(C,\omega_C)$ defines a meromorphic differential on $\hat{C}$. Its residues on every connected component sum up to zero and so do the two residues that over a node any given node. So if we think of the set $X(G)$ of oriented edges of G as a set of simplicial 1-chains, then these residue data define an element of $H^1(G;\mathbb{C}) \subset \mathbb{C}^{X(G)}$. On the other hand, on a closed connected Riemann surface we can arbitrarily prescribe the polar parts of a meromorphic differential as along as the sum of their residues is zero. This means that every element of $H^1(G;\mathbb{C})$ is so obtained. The genus g of C is indeed the sum of the genera of the connected components of $\hat{C}$ and the first Betti number of G.

The general case (of a partial blowup) is proven in the same way. □

This notion extends to families of such curves: if $f : \mathscr{C} \to B$ is a proper family of nodal curves, then we have defined the relatively dualizing sheaf $\omega_{\mathscr{C}/B}$: this is a line bundle which behaves well with respect to base change: if we pull-back the family $f : \mathscr{C} \to B$ along a morphism $\phi : B' \to B$, then the pull-back of $\omega_{\mathscr{C}/B}$ along $\phi^*\mathscr{C} \to \mathscr{C}$ is the relatively dualizing sheaf of the pull-back $\phi^*\mathscr{C} \to B'$. In particular (take B' a singleton), $\omega_{\mathscr{C}/B}$ restricts to the dualizing sheaf ω_{C_b} of every fiber. Moreover, $f_*\omega_{\mathscr{C}/B}$ is locally free. In case B is nonsingular of dimension d and $\mathscr{C}$ is nonsingular of dimension $d+1$, then we have a natural identification of $\omega_{\mathscr{C}/B}$ with the tensor product of line bundles $\mathscr{H}om_{\mathscr{O}_{\mathscr{C}}}(f^*\Omega^d_B, \Omega^{d+1}_{\mathscr{C}})$. For instance at a node in a one parameter family, we have local analytic coordinates such that f is there given by $(z_0, z_1) \mapsto z_0 z_1 = t$, and then a local generator of $\omega_{\mathscr{C}/B}$ is $dz_0 \wedge dz_1/f^*dt$.

We can do this universally: for the family $f : \overline{\mathscr{C}}_{g,P} \to \overline{\mathscr{M}}_{g,P}$ we have a line bundle $\omega_{g,P} := \omega_{\overline{\mathscr{C}}_{g,P}/\overline{\mathscr{M}}_{g,P}}$ on $\overline{\mathscr{C}}_{g,P}$ in the orbifold sense. This gives rise to the vector bundle of rank g on $\overline{\mathscr{M}}_{g,P}$, $\mathscr{E}_{g,P} := f_*\omega_{g,P}$ (called the *Hodge bundle*) and for every $p \in P$, a line bundle $\mathscr{L}_p := x_p^*\omega_{g,P}$, on $\overline{\mathscr{M}}_{g,P}$, both appropriately understood in an orbifold sense.

On $\overline{\mathscr{M}}_{g,\tilde{P}}$, we have similarly defined the line bundles $\tilde{\mathscr{L}}_{\tilde{p}}$, with $\tilde{p} \in \tilde{P} = P \sqcup \{o\}$. Since we have a natural identification $\overline{\mathscr{M}}_{g,\tilde{P}} \cong \overline{\mathscr{C}}_{g,P}$, the question arises how these bundles compare with $\omega_{g,P}$ and the $f^*\mathscr{L}_p$. This is somewhat subtle as it involves the natural map from $\overline{\mathscr{C}}_{g,\tilde{P}} \to f^*\overline{\mathscr{C}}_{g,P}$. This map is an isomorphism over a Zariski open subset, but unless we are in a very special situation there is some contraction. To see this, note that a fiber of $\overline{\mathscr{C}}_{g,P}/\overline{\mathscr{M}}_{g,P}$ is given by a single stable P-pointed curve $(C, x : P \hookrightarrow C)$ of genus g. Any $y \in C$ defines a stable $\tilde{P}$-pointed curve $(C^y, x^y : \tilde{P} \hookrightarrow C_y)$, together with a 'contraction map' $\pi^y : C^y \to C$ as follows: we have $C^y = C$ (with π^y being the identity) and x^y the obvious map unless y is a special point of C, that is, a node or some x_p, $p \in P$. In either case, the preimage of y is a copy of $\mathbb{P}^1$ with three special points on it, one of which the value of $x^y(o)$. In case y is a node, the other two points are nodes and if $y = x_p$, just one of the two points is a node and the other is $x^y(p)$. In either case, π^y is the contraction of the $\mathbb{P}^1$-copy. This makes it clear that we have $\pi^{y*}\omega_C = \omega_{C_y}$ unless $y = p$: then $\pi^{y*}\omega_C$ is the subsheaf of ω_{C^y} of sections that vanish on the contracted component. In particular it then vanishes on both $x^y(o)$ and $x^y(p)$. On the other hand the map $H^0(C, \omega_C) \to H^0(C^y, \omega_{C^y})$ induced by π^y is easily shown to be always an isomorphism.

We may express this universally by saying that on $\overline{\mathscr{M}}_{g,\tilde{P}}$ we have $\tilde{\mathscr{L}}_o \cong \omega_{g,P}(\sum_{p \in P} x_p)$, $\tilde{\mathscr{L}}_p \cong f^*\mathscr{L}_p(\sum_{p \in P} x_p)$ and that $\mathscr{E}_{g,\tilde{P}} \cong f^*\mathscr{E}_{g,P}$. The last property implies that $\mathscr{E}_{0,P}$ is the zero bundle, $\mathscr{E}_{1,P}$ is the pull-back of $\mathscr{E}_{1,1}$ and that for $g \geq 2$, $\mathscr{E}_{g,P}$ is the pull-back of the Hodge bundle $\mathscr{E}_g$ over $\overline{\mathscr{M}}_g$.

Weighted graphs

The system of maps defined above generate a categorical structure which contains what Getzler and Kapranov call a *modular operad*. This centers around the notion of a *stable weighted graph* which generalizes the graph that we introduced for a pants decomposition. It describes the complete isomorphism type of a simplex **a** of the curve complex of a pair (S,P) (with S allowed to be disconnected, but nonempty): this stable weighted graph is a quadruple $G=(X,\sim,\iota,g)$ that consists the following data: if $A\subset S^\circ$ is a representative closed 1-manifold, then X is the finite set that is disjoint union of P and the connected components of A *endowed with an orientation*, $\iota\in \mathrm{Aut}(X)$ is the automorphism that is the identity on P and reverses orientation of a connected component of A and $\sim$ is the equivalence relation on X whose set of equivalence classes can be identified with $\pi_0(S-P)$: for a connected component S' of $S-P$ take the union of $S'\cap P$ and the set of oriented connected components of A which can be regarded as a boundary component of S' with the induced orientation. Finally, the genus of a component defines a 'weight map' $g: X_0\to \mathbb{Z}_{\geq 0}$.

Any triple $(X,\sim,\iota)$, with X a nonempty finite set, $\sim$ an equivalence relation on X and $\iota\in\mathrm{Aut}(X)$ such that $\iota^2=1$, defines a graph with 'leaves' that has X as its set of oriented edges: the vertex set is the quotient $\pi_0: X\to X/\sim =: X_0$ (this assigns to an oriented edge its source), and the edge set (of unoriented edges) is the quotient $\pi_1: X\to X/(\iota)=: X_1$. So if a ι-orbit is contained in a $\sim$equivalence class defining the vertex $v\in X_1$, then when the orbit is regular, it defines a loop based at v and when it is not (this corresponds to an element of P in the situation above), it is a *leaf*: an outward pointing edge which has v as source vertex but has no target vertex. Note that this graph is connected if and only if any nonempty ι-invariant union of equivalence classes is all of X. The additional given of a map $g: X_0\to\mathbb{Z}_{\geq 0}$ defines a *weighted graph*, so this is a quadruple $G=(X,\sim,\iota,g)$. For obvious reasons we say that G is *stable* if for every $v\in X_1$, the sum of $g(v)$ and number of elements of $\pi_0^{-1}(v)$ is >2 and we define its genus g (assuming G is connected) as $b_1(G)+\sum_{v\in X_1} g(v_1)$. We say that a stable graph $G=(X,\sim,\iota,g)$ is P-pointed if we have specified a bijection of P with the fixed point set of ι. The simplest example is obtained for $X=P$ (assumed to be nonempty) and ι and $\sim$ trivial; we denote that weighted graph by $G_{g,P}$.

To a weighted stable graph G we associate the orbifold

$$\overline{\mathcal{M}}(G) := \prod_{v\in X_0} \overline{\mathcal{M}}_{g(v),\pi_0^{-1}(v)}.$$

So $\overline{\mathcal{M}}(G)$ does not involve ι: it only depends on $(X,\sim,g)$. If G is connected of genus g and P-pointed, then $\overline{\mathcal{M}}(G)$ parametrizes the stable curves that are homeomorphic to (S_A,P) relative to P. In particular, $\overline{\mathcal{M}}(G_{g,P})=\overline{\mathcal{M}}_{g,P}$.

The stable weighted graphs are the objects of a category $\mathfrak{G}^{st}$ whose morphisms are suggested by the operations we encountered: they will be such that a morphism

$r: G \to G'$ gives rise to a morphism $\mathcal{M}(r) : \overline{\mathcal{M}}(G) \to \overline{\mathcal{M}}(G')$. By definition, any morphism in $\mathfrak{G}^{st}$ is a composite of four types of these, being:

Cut and paste: Altering the definition of ι is regarded as an isomorphism. (This indeed determines an isomorphism $\overline{\mathcal{M}}(G) \cong \overline{\mathcal{M}}(G')$.)

Contraction: r collapses an internal edge, i.e., we remove from X a regular ι-orbit. If this orbit connects the vertices v_0 and v_1, then the new vertex v is assigned the weight $g(v_0)+g(v_1)$ or $g(v_0)+1$ according to whether $v_0 \neq v_1$ or not. The associated morphism $\overline{\mathcal{M}}(G) \to \overline{\mathcal{M}}(G')$ takes a two element subset D of X_1 and applies i_D to the associated factor(s) in $\prod_{v\in X_0} \overline{\mathcal{M}}_{g(v),\pi_1^{-1}(v)}$. (In our example this corresponds to erasing a connected component of A.)

Pruning: a morphism r is given by removing from X a fixed point p of ι. If that fixed point happens to lie on a trivalent vertex of genus zero, then make the result stable by passing to the quotient of X which identifies the corresponding triple in X to a singleton. It may be necessary to continue this process in order to arrive at a stable G' (so that in the end we collapse a trivalent tree in G that is connected with the rest in only one vertex; if we do not succeed in thus rendering G stable, then the morphism r is not defined). If the leaf is indexed by $p \in P$, then the associated morphism $\overline{\mathcal{M}}(G) \to \overline{\mathcal{M}}(G')$ is given on a factor by a projection of the type $f_{P,P-\{p\}}$.

Grafting a leaf: Given a ι-fixed point $p \in X$, then we define a graph $G^p = (x^p, \sim^p, \iota^p, g^p)$ as follows: $X^p = X \sqcup \{o, \tilde{p}, \infty\}$, $\sim^p$ is the equivalence relation which extends $\sim$ by putting $\{o, \tilde{p}, \infty\}$ in a new equivalence class, ι^p extends $\iota|X \smallsetminus \{p\}$ by stipulating that $\{o, \tilde{p}\}$ and $\{\infty, p\}$ are orbits of ι^p and g^p is the extension of g by zero. This defines a morphism $r: G \to G^p$. It is an isomorphism, for it is an inverse to 'filling the puncture operator' defined by forgetting o or $\tilde{p}$. The associated morphism $\overline{\mathcal{M}}(G) \to \overline{\mathcal{M}}(G^p)$ is indeed an isomorphism.

The morphisms of $\mathfrak{G}^{st}$ are composites of such maps. We have set things up in such a manner that $G \mapsto \overline{\mathcal{M}}(G)$ defines a covariant functor $\overline{\mathcal{M}}$ from $\mathfrak{G}^{st}$ to the category of projective orbifolds. For example, the section map x_p of 12.4 is the essentially the image of the composite $G_{g,P} \to G^p_{g,P} \to G_{g,\tilde{P}}$, where now $\tilde{P} = (P \smallsetminus \{p\}) \sqcup \{o, \tilde{p}\}$.

The notion of a tautological algebra

We begin with a definition.

Definition 12.7. A *tautological algebra functor* $\mathcal{T}^\bullet$ assigns to every moduli space $\overline{\mathcal{M}}_{g,P}$ a graded $\mathbb{Q}$-subalgebra $\mathcal{T}^\bullet(\overline{\mathcal{M}}_{g,P}) \subset H^\bullet(\overline{\mathcal{M}}_{g,P};\mathbb{Q})$ such that if for any stable weighted graph G we put

$$\mathcal{T}^\bullet(G) := \bigotimes_{v\in X_0} \mathcal{T}^\bullet(\overline{\mathcal{M}}_{g(v),\pi_0^{-1}(v)};\mathbb{Q}) \subset \bigotimes_{v\in X_0} H^\bullet(\overline{\mathcal{M}}_{g(v),\pi_0^{-1}(v)};\mathbb{Q}) \cong H^\bullet(\overline{\mathcal{M}}(G);\mathbb{Q}),$$

then for every morphism $r : G \to G'$ in $\mathfrak{G}^{\mathrm{st}}$, the map $\overline{\mathcal{M}}(r) : \overline{\mathcal{M}}(G) \to \overline{\mathcal{M}}(G')$ has the property that $\overline{\mathcal{M}}(r)^*$ takes $\mathcal{T}^\bullet(G')$ to $\mathcal{T}^\bullet(G)$ and $\overline{\mathcal{M}}(r)_!$ takes $\mathcal{T}^\bullet(G)$ to $\mathcal{T}^\bullet(G')$ (so it must contain $H^0(\mathcal{M}_{g,P};\mathbb{Q}) = \mathbb{Q}$ and be stable under the maps on cohomology of ordinary and of Gysin type induced by a morphism listed coming from $\mathfrak{G}^{\mathrm{st}}$).

An automorphism of P defines an automorphism of the graph $G_{g,P}$ and so $\mathcal{T}^\bullet(\overline{\mathcal{M}}_{g,P})$ will be then invariant under the permutation group of P.

We denote the image of $\mathcal{T}^\bullet(\overline{\mathcal{M}}_{g,P})$ in $H^\bullet(\mathcal{M}_{g,P};\mathbb{Q})$ by $\mathcal{T}^\bullet(\mathcal{M}_{g,P})$.

The obvious maximal choice for $\mathcal{T}^\bullet$ is to let it be all of $H^\bullet(\overline{\mathcal{M}}(G);\mathbb{Q})$. Another is $H^{even}(\overline{\mathcal{M}}(G);\mathbb{Q})$, but more interesting is the minimal one: an intersection of tautological functors is also one and so that there is a smallest such functor $\mathcal{R}$ with $\mathcal{R}(G) = \cap_{\mathcal{T}} \mathcal{T}^\bullet(G)$, where $\mathcal{T}^\bullet$ runs over all possible tautological algebra functors. We call $\mathcal{R}(G)$ the *tautological algebra* of $\overline{\mathcal{M}}(G)$. As $\mathcal{R}$ will have only even degrees, it will be commutative. We grade accordingly: $\mathcal{R}^k(G) \subset H^{2k}(\overline{\mathcal{M}}(G);\mathbb{Q})$.

The following lemma makes this more concrete. The orbifold line bundle $\mathcal{L}_p$ on $\overline{\mathcal{M}}_{g,P}$ has rational first Chern class that we shall denote by $\Psi_p(\overline{\mathcal{M}}_{g,P}) \in H^2(\overline{\mathcal{M}}_{g,P};\mathbb{Q})$.

Lemma-definition 12.8. *Consider the universal family $\overline{\mathcal{M}}_{g,\tilde{P}} \to \overline{\mathcal{M}}_{g,P}$. Then the orbifold line bundle $\mathcal{L}_p$ on $\overline{\mathcal{M}}_{g,P}$ has a rational first Chern class, denoted $\Psi_p(\overline{\mathcal{M}}_{g,P}) \in H^2(\overline{\mathcal{M}}_{g,P};\mathbb{Q})$. This Chern class equals $-x_p^* x_{p!}(1)$, hence lies in $\mathcal{R}^1(\overline{\mathcal{M}}_{g,P})$. We also have $c_1(\omega_{g,P}) = \Psi_o(\overline{\mathcal{M}}_{g,\tilde{P}}) - \sum_{p\in P} x_{p!}(1)$ and hence lies in $\mathcal{R}^1(\overline{\mathcal{M}}_{g,\tilde{P}})$.*

Proof. The first assertion is a general fact about orbifold line bundles; it also follows from the fact that if the base compact, a high tensor power of such a bundle will in fact be an ordinary line bundle. The second assertion is clear (use that x_p is induced by a $\mathfrak{G}^{\mathrm{st}}$-morphism) and the last assertion follows the line bundle isomorphism $\omega_{g,P} \cong \tilde{\mathcal{L}}_o(-\sum_{p\in P} x_p)$. □

We define for $r = 0,1,2,\dots$ the *rth kappa class* and the *rth λ class* in $H^{2r}(\overline{\mathcal{M}}_{g,P};\mathbb{Q})$ for $r = 0,1,2,\dots$ by

$$\kappa_r(\overline{\mathcal{M}}_{g,P}) := f_!(\Psi_o(\overline{\mathcal{M}}_{g,\tilde{P}})^{r+1}), \quad \lambda_r(\overline{\mathcal{M}}_{g,P}) := c_r(\mathcal{E}_{g,P}).$$

It is sometimes convenient to pass from the Chern classes of a vector bundle to its Chern characters. We recall that the rational Chern classes c_r and Chern characters ch_r of a vector bundle determine each other: if we write formally $1 + c_1 t + c_2 t^2 + \cdots + \cdots = (1+x_1)(1+x_2)\cdots$, then $\sum_{i\geq 0} \mathrm{ch}_i t^i = e^{x_1} + e^{x_2} + \cdots$. In other words, ch_r is the universal polynomial in the elementary symmetric functions $c_1, c_2, \dots, c_r$ of $x_1, x_2, \dots$ that yields $\sum_i x_i^r / r!$. So ch_r resp. c_r is a universal polynomial with rational coefficients in $c_1, \dots, c_r$ resp. $\mathrm{ch}_1, \dots, \mathrm{ch}_r$. The advantage of the Chern characters is that they behave additively with respect to short exact sequences of vector bundles.

We write $\mathrm{ch}_r(\overline{\mathscr{M}}_{g,P}) := \mathrm{ch}_r(\mathscr{E}_{g,P})$. Since the Hodge bundle is the pull-back of a Hodge bundle over $\overline{\mathscr{M}}_{1,1}$ (for $g = 1$) or $\overline{\mathscr{M}}_g$ (for $g \geq 2$), the same is true for the λ_r's and the ch_r's.

All these classes are tautological. It follows from our definition that the κ-classes lie in $\mathscr{R}^\bullet(\overline{\mathscr{M}}_{g,P})$. Its restriction to $\mathscr{M}_{g,P}$ is denoted by $\kappa_r(\mathscr{M}_{g,P}) \in H^{2r}(\mathscr{M}_{g,P};\mathbb{Q})$. When $P = \emptyset$, this is known as the rth *Miller-Morita-Mumford class*. Mumford [18] has shown that an application of the Grothendieck-Riemann-Roch formula implies that $\mathrm{ch}_r(\overline{\mathscr{M}}_{g,P})$ lies in $\mathscr{R}^r(\overline{\mathscr{M}}_{g,P})$ and is zero for r even. This implies of course that $\lambda_r(\overline{\mathscr{M}}_{g,P}) \in \mathscr{R}^r(\overline{\mathscr{M}}_{g,P})$.

Definition 12.9. Let χ assign to every stable pair (g,P) a class $\chi(\overline{\mathscr{M}}_{g,P}) \in \mathscr{R}^\bullet(\overline{\mathscr{M}}_{g,P})$. Then for a stable weighted graph G, put

$$\chi(G) := \sum_{v \in X_o(G)} \pi_v^* \chi(\overline{\mathscr{M}}_{g(v),\pi_0^{-1}v}),$$

where $\pi_v : \overline{\mathscr{M}}(G) \to \overline{\mathscr{M}}_{g(v),\pi_0^{-1}v}$ is the projection. We say that χ is *primitive* if for every morphism $r : G \to G'$ in $\mathfrak{G}^{\mathrm{st}}$ that comes from cut and paste, contraction or grafting a leaf, $H^\bullet(\overline{\mathscr{M}}(r))$ takes $\chi(G')$ to $\chi(G)$.

Proposition 12.10. *The assignments defined κ_r, Ψ_o and ch_r are primitive.*

Proof. We need to go through the definitions. We do this for κ_r only as the case for Ψ_o and ch_r is similar (the last case uses the additivity property of the Chern character). For the cut and paste case, we must show that the pasting maps $i_D : \overline{\mathscr{M}}_{g-1,Q} \to \overline{\mathscr{M}}_{g,P}$ and $i_D : \overline{\mathscr{M}}_{g',Q'} \times \overline{\mathscr{M}}_{g'',Q''} \to \overline{\mathscr{M}}_{g,P}$ have the property that they take $\kappa_r(\overline{\mathscr{M}}_{g,P})$ to $\kappa_r(\overline{\mathscr{M}}_{g-1,Q})$ resp. $\kappa_r(\overline{\mathscr{M}}_{g',Q'}) \otimes 1 + 1 \otimes \kappa_r(\overline{\mathscr{M}}_{g'',Q''})$.

The first pasting map involves the commutative diagram of universal families

$$\begin{array}{ccc}
\overline{\mathscr{M}}_{g-1,\tilde{Q}} & \xrightarrow{\tilde{i}_D} & \overline{\mathscr{M}}_{g,\tilde{P}} \\
\downarrow{\scriptstyle \tilde{f}} & & \downarrow{\scriptstyle f} \\
\overline{\mathscr{M}}_{g-1,Q} & \xrightarrow{i_D} & \overline{\mathscr{M}}_{g,P}
\end{array}$$

where $\tilde{P} = P \sqcup \{o\}$ and $\tilde{Q} = Q \sqcup \{o\}$. This is in fact a pull-back diagram. Recall that $\mathscr{L}_o|\overline{\mathscr{M}}_{g,\tilde{P}} \cong \omega_{g,P}(\sum_{p\in P} x_p)$ and that likewise $\mathscr{L}_o|\overline{\mathscr{M}}_{g-1,\tilde{Q}} \cong \omega_{g,Q}(\sum_{q\in Q} x_q)$. The definition of the relative dualizing sheaf shows that then $\omega_{g,Q}(\sum_{q\in Q} x_q)$ is the pull-back of $\omega_{g,P}(\sum_{p\in P} x_p)$ and so it follows that $\tilde{i}_D^*$ takes $\Psi_o(\overline{\mathscr{M}}_{g,\tilde{P}})$ to $\Psi_o(\overline{\mathscr{M}}_{g-1,\tilde{Q}})$. By the base change property, i_D^* then maps $\kappa_r(\overline{\mathscr{M}}_{g,P}) = f_!(\Psi_o(\overline{\mathscr{M}}_{g,\tilde{P}})^{r+1})$ to $f_!(\Psi_o(\overline{\mathscr{M}}_{g-1,\tilde{P}})^{r+1}) = \kappa_r(\overline{\mathscr{M}}_{g-1,Q})$.

The discussion for the second pasting map somewhat more complicated. We have the universal families $\overline{\mathscr{M}}_{g',\tilde{Q}'} \to \overline{\mathscr{M}}_{g',Q'}$ and $\overline{\mathscr{M}}_{g'',\tilde{Q}''} \to \overline{\mathscr{M}}_{g'',Q''}$. Each of these can be pulled back over the product $\overline{\mathscr{M}}_{g',Q'} \times \overline{\mathscr{M}}_{g'',Q''}$ and then we can form their disjoint union

$$\tilde{f} := (f' \times 1) \sqcup (1 \times f'') : \pi'^* \overline{\mathcal{M}}_{g',\tilde{Q}'} \sqcup \pi''^* \overline{\mathcal{M}}_{g'',\tilde{Q}''} \to \overline{\mathcal{M}}_{g',Q'} \times \overline{\mathcal{M}}_{g'',Q''}.$$

A quotient of the total space is obtained by taking the union along $\overline{\mathcal{M}}_{g',\tilde{Q}'} \times \overline{\mathcal{M}}_{g'',\tilde{Q}''}$ (embedded in each summand by the the section defined by o) and this is of course the pull-back along i_D of the universal family over $\overline{\mathcal{M}}_{g,P}$. In particular, we have a commutative diagram

$$\begin{array}{ccc} \pi'^* \overline{\mathcal{M}}_{g',\tilde{Q}'} \sqcup \pi''^* \overline{\mathcal{M}}_{g'',\tilde{Q}''} & \xrightarrow{\tilde{i}_D} & \overline{\mathcal{M}}_{g,\tilde{P}} \\ \tilde{f} \downarrow & & \downarrow f \\ \overline{\mathcal{M}}_{g',Q'} \times \overline{\mathcal{M}}_{g'',Q''} & \xrightarrow{i_D} & \overline{\mathcal{M}}_{g,P} \end{array}$$

which is almost Cartesian. The pull-back of $\Psi_o(\overline{\mathcal{M}}_{g,\tilde{P}})^{r+1}$ is on one component equal to $\Psi_o(\overline{\mathcal{M}}_{g',\tilde{Q}'})^{r+1} \otimes 1$ and $1 \otimes \Psi_o(\overline{\mathcal{M}}_{g'',\tilde{Q}''})^{r+1}$ on the other. This implies that

$$\tilde{f}_! \tilde{i}_D^* \Psi_o(\overline{\mathcal{M}}_{g,\tilde{P}})^{r+1} = \kappa_r(\overline{\mathcal{M}}_{g',Q'}) \otimes 1 + 1 \otimes \kappa_r(\overline{\mathcal{M}}_{g'',Q''}).$$

Hence it remains to show that the left hand side can be identified with $i_D^* f_! \Psi_o(\overline{\mathcal{M}}_{g,\tilde{P}})^{r+1}$, in other words, that we can pretend that base change applies here. This can be justified by the fact $\mathcal{L}_o$ is canonically trivial along the common section of the two components so that its Chern class has its support away from that locus and invoking Remark 12.1.

The proof that a grafting isomorphism has the desired property is easily reduced to the cut and paste case by observing that we then have the singleton $\overline{\mathcal{M}}_{0,3}$ as a factor. The case of a contraction is also straightforward. □

Proposition 12.11. *The tautological algebra $\mathcal{R}^\bullet(\overline{\mathcal{M}}_{g,P})$ is as a $\mathbb{Q}$-algebra generated by the classes $\{\kappa_r(\overline{\mathcal{M}}_{g,P})\}_{r \geq 1}$, $\{\Psi_p(\overline{\mathcal{M}}_{g,P})\}_{p \in P}$, and the classes of the boundary strata closures (i.e., the images of the element of 1 under their Gysin maps). In particular, $\mathcal{R}^\bullet(\mathcal{M}_{g,P})$ resp. $\mathcal{R}^\bullet(\mathcal{M}_g)$ is as a $\mathbb{Q}$-algebra generated by the kappa classes and the psi classes resp. the kappa classes.*

Proof. We must verify that the $\mathbb{Q}$-subalgebra $\mathcal{R}_o^\bullet(\overline{\mathcal{M}}_{g,P}) \subset H^\bullet(\overline{\mathcal{M}}_{g,P};\mathbb{Q})$ generated by the classes mentioned defines a tautological algebra functor. That is relatively straightforward. □

The tautological algebra in genus zero, $\mathcal{R}^\bullet(\overline{\mathcal{M}}_{0,P})$ has been determined by S. Keel [9] and is quite nice. First notice that the boundary divisors of $\overline{\mathcal{M}}_{0,P}$ are indexed by the splittings of P into two sets with least two elements (so this assumes $|P| \geq 4$): for every such splitting $S = \{S'|S''\}$ we have an embedding $i_S : \overline{\mathcal{M}}_{0,\tilde{S}'} \times \overline{\mathcal{M}}_{0,\tilde{S}''} \hookrightarrow \overline{\mathcal{M}}_{0,P}$ whose image is a boundary divisor. The following lemma gives in $\mathcal{R}^\bullet(\overline{\mathcal{M}}_{0,P})$ multiplicative relations in degree 2 and linear relations in degree 1.

Lemma 12.12. *Assume that $|P| \geq 4$ so that $\overline{\mathcal{M}}_{0,P}$ is defined.*

(*mult*) *Suppose S and T are splittings of P into two subsets which possess at least two elements. If each member of S has nonempty intersection with each member of T, then $i_{S!}(1)i_{T!}(1) = 0$.*

(*add*) *Let $p \in P$, $Q \subset P - \{p\}$ a 2-element subset and denote by $\mathcal{S}(p|Q)$ the collection of splittings of P into two sets with at least two elements which separate p from Q. Then $\Psi_p = \sum_{S \in \mathcal{S}(p|Q)} i_{S!}(1)$.*

Proof. The first identity follows from the geometry: if the codimension one strata parametrized by i_S and i_T meet, then they do so along a codimension two stratum. But a codimension two stratum defines a partition of P into three parts which refines both S and T. This implies that S and T have a common member.

The proof of the second identity is somewhat more subtle. We define a section s of $\mathcal{L}_p$ with the help of Q as follows. Denote the two elements of Q by q_0 and q_∞. If $c \in \overline{\mathcal{M}}_{0,P}$, then represent it by a stable P-pointed curve (C,P) of genus zero. You may verify that there is a unique morphism $f_c : C \to \mathbb{P}^1$ with $f_c(p) = 1$, $f_c(q_0) = 0$, $f_c(q_\infty) = \infty$. We may then view $s(c) := f_c^*(\frac{dz}{z})|T_p^*C$ as an element of the fiber of $\mathcal{L}_p$ over c. It turns out that the naturality of this construction leads to the conclusion that s indeed defines a section of $\mathcal{L}_p$. So its zero divisor defines the class Ψ_p. But $s(c) = 0$ if and only if p lies on a component of C that is contracted by f_c and this happens precisely when c lies on the image of an i_S which separates p from q_0, q_∞. The multiplicity of vanishing is 1 and so the identity follows. □

Remark 12.13. Keel [9] has shown that $\mathcal{R}^\bullet(\overline{\mathcal{M}}_{0,P}) = H^\bullet(\overline{\mathcal{M}}_{0,P};\mathbb{Q})$ and that as an algebra, it is generated by the classes of the boundary divisors $i_{S!}(1)$ and the Ψ_p's subject to the relations above (in particular, we may eliminate the psi classes). It is not true in general that $\mathcal{R}^\bullet(\overline{\mathcal{M}}_{g,P})$ is all of $H^\bullet(\overline{\mathcal{M}}_{g,P};\mathbb{Q})$. In fact, $\mathcal{R}^k(\overline{\mathcal{M}}_{g,P})$ is algebraic: it is of Hodge bidegree (k,k). But $H^{11,0}(\overline{\mathcal{M}}_{1,11};\mathbb{C})$ is known to be nonzero, and hence not contained in $\mathcal{R}^k(\overline{\mathcal{M}}_{g,P})$. Pikaart [19] showed that this property propagates to larger genera. In [4] Hain and Looijenga had asked whether $\mathcal{R}^\bullet(\overline{\mathcal{M}}_{g,P})$ makes up a nondegenerate subspace of $H^\bullet(\overline{\mathcal{M}}_{g,P};\mathbb{Q})$ with respect to the intersection pairing, but this question was recently answered negatively by Petersen and Tommasi [7].

A theorem of Kontsevich

Let (g,P) be a stable pair with $P \neq \emptyset$. Consider for map $k : P \to \{0,1,2,\dots\}$, the intersection numbers

$$\langle \prod_{p \in P} \psi_p^{k_p}(\overline{\mathcal{M}}_{g,P}), [\overline{\mathcal{M}}_{g,P}] \rangle \in \mathbb{Q}.$$

Clearly this number is zero if $\sum_{p \in P} k_p \neq 3g - 3 + |P|$ (we can of course write this as $\sum_{p \in P}(k_p - 1) \neq 3g - 3$). It is also invariant under $\mathrm{Aut}(P)$. We let these numbers be the coefficients of a polynomial in the reciprocals of variables λ_p, $p \in P$, by putting:

$$F_{g,P} := \sum_{\sum_{p\in P}(k_p-1)=3g-3} \langle \prod_{p\in P} \psi_p^{k_p}(\overline{\mathcal{M}}_{g,P}) \frac{(2k_p-1)!!}{\lambda_p^{2k_p+1}}, [\overline{\mathcal{M}}_{g,P}]\rangle \in \mathbb{Q}[\lambda_p^{-1} : p \in P],$$

where $(2k-1)!! = 1.3.5.\cdots.(2k-1)$. A theorem of Kontsevich [10] (which leads to a confirmation of a conjecture of Witten [21]), asserts that this is equal to the following rational expression:

$$F_{g,P} = \sum_{(G,\phi)} \frac{2^{-|X_0(G)|}}{|\mathrm{Aut}(G,\phi)|} \prod_{e\in X_1(G)} \frac{2}{\lambda(e)},$$

where the sum is over the set of isomorphism classes of P-marked ribbon graphs (G,ϕ) of genus g for which every vertex has valency 3. We recall that a P-marking of the ribbon graph G in this case means that a bijection ϕ between P and the set $X_\infty(G)$ of boundary cycles has been specified. The identity displayed above is then to be understood as follows: $\lambda(e)$ is a formal variable assigned to each edge $e \in X_1(G)$ and λ_p is the sum of the $\lambda(e)$ that make up the boundary cycle $\phi(p)$. It is then implicitly asserted that for a fixed G, the corresponding subsum of the right hand side over all its P-markings only depends on the λ_p's. In view of the discussion in Section 6 (especially the remark following Definition 6.10), the right hand sum is over all the maximal simplices of the orbifold triangulation if $\mathbb{M}_{g,P}$. Indeed, the identity is essentially proved, by writing the intersection number on the left as an integral over $\mathbb{M}_{g,P}$ by representing each ψ_p as a simplicial 2-form on $\mathbb{M}_{g,P}$; then on the right hand side we find the contribution of the integral over each maximal simplex.

The expressions $F_{g,P}$ can in their turn be organized (for various g and $|P|$) into a single power series in a (countably) infinite number of variables. The above identity then takes the form of a very explicit infinite system of differential equations (a KdV hierarchy) whose associated recursive relations enable a quick computation. Other proofs have been found since. For example, one based on symplectic techniques is due to Mirzakhani [16] and a more algebro-geometric proof has been found by Kazarian–Lando [8].

Faber's conjectures

Rather than to recall these conjectures in their original form, let us begin stating what is essentially the proven part. We first observe that Proposition 12.10 implies that the lambda classes behave under the pasting maps as follows: for $i_D : \overline{\mathcal{M}}_{g-1,Q} \to \overline{\mathcal{M}}_{g,P}$ we have that $i_D^*\lambda_r(\overline{\mathcal{M}}_{g,P}) = \lambda_r(\overline{\mathcal{M}}_{g-1,Q})$ and for $i_D : \overline{\mathcal{M}}_{g',Q'} \times \overline{\mathcal{M}}_{g'',Q''} \to \overline{\mathcal{M}}_{g,P}$ we have $i_D^*\lambda_r(\overline{\mathcal{M}}_{g,P}) = \sum_{k+l=r} \lambda_k(\overline{\mathcal{M}}_{g',Q'}) \otimes \lambda_l(\overline{\mathcal{M}}_{g'',Q''})$. This implies that $\lambda_g(\overline{\mathcal{M}}_g)\lambda_{g-1}(\overline{\mathcal{M}}_g) \in H^{4g-2}(\overline{\mathcal{M}}_g;\mathbb{Q})$ has the property that its restriction every boundary divisor of $\overline{\mathcal{M}}_g$ (which is the finite image of $\overline{\mathcal{M}}_{g-1,2}$ or

$\overline{\mathcal{M}}_{k,1} \times \overline{\mathcal{M}}_{g-k,1}$ for some $k = 1,2,\ldots,\lfloor g/2 \rfloor$) is zero. This means that it lifts to a class in $H_c^{4g-2}(\mathcal{M}_g;\mathbb{Q})$. Poincaré duality identifies the latter with the dual of $H^{2g-4}(\mathcal{M}_g;\mathbb{Q})$.

(i) (Morita [17]) The $\mathbb{Q}$-algebra $\mathcal{R}^\bullet(\mathcal{M}_g)$ is generated by the κ_r, with $r \leq \lfloor \frac{1}{3}g \rfloor$.
(ii) (Looijenga [12]) The $\mathbb{Q}$-algebra $\mathcal{R}^\bullet(\mathcal{M}_g)$ is zero in degree $> g-2$ and $\mathcal{R}^{g-2}(\mathcal{M}_g)$ is spanned by a class of the *hyperelliptic locus* $\mathcal{H}_g$.
(iii) (Faber [2]) Multiplication by $\lambda_g\lambda_{g-1}$ takes a nonzero value on $\kappa_{g-2}(\mathcal{M}_g)$ and defines an isomorphism $\mathcal{R}^{g-2}(\mathcal{M}_g) \cong \mathcal{R}^{3g-3}(\overline{\mathcal{M}}_g) \cong \mathbb{Q}$.

Here the *hyperelliptic locus* $\mathcal{H}_g$ is the locus in $\mathcal{M}_g$ parameterizing the isomorphism classes of hyperelliptic curves of genus g. (Since a hyperelliptic curve of genus g is given up to isomorphism as a double cover of $\mathbb{P}^1$ ramified in $2g+2$ points, we have an identification of $\mathcal{H}_g$ with the $2g-1$-dimensional variety $\mathcal{M}_{0,2g+2}$, although not as orbifolds, because of the hyperelliptic involution. The hyperelliptic locus is closed and of codimension $g-2$ in $\mathcal{M}_g$.)

The result of Faber is based on the theorem of Kontsevich. Still open is:

Conjecture 12.14 (Faber [3]). *The $\mathbb{Q}$-algebra $\mathcal{R}^\bullet(\mathcal{M}_g)$ is Gorenstein with socle $\mathcal{R}^{g-2}(\mathcal{M}_g)$. Equivalently, the pairing*

$$(a,b) \in \mathcal{R}^k(\mathcal{M}_g) \times \mathcal{R}^{g-2-k}(\mathcal{M}_g) \mapsto \lambda_g\lambda_{g-1}ab \in \mathcal{R}^{3g-3}(\overline{\mathcal{M}}_g) \cong \mathbb{Q}$$

is perfect.

There are a number of variants, one of which involves $\mathcal{R}^\bullet(\mathcal{M}_g^c)$ (supposedly to be Gorenstein with socle in degree $2g-3$, the role of $\lambda_g\lambda_{g-1}$ taken by λ_g). However, a number of recent results (for example [7]) suggest that the above conjecture may fail for large g.

Primitivity of the kappa classes

The maps appearing in the Harer stability theorem and in the above discussion can be made rather concrete in terms of the Harvey bordification. Recall that $\mathcal{M}_{g,P}^+$ is a compact manifold with corners whose interior is $\mathcal{M}_{g,P}^+$. So $\mathcal{M}_{g,P} \subset \mathcal{M}_{g,P}$ induces an isomorphism on cohomology. For $g' > 0$ we have the boundary stratum of $\overline{\mathcal{M}}_{g+k,P}$ parameterized by $\mathcal{M}_{g,\tilde{P}} \times \mathcal{M}_{g',\{o\}} \to \overline{\mathcal{M}}_{g+g',P}$. The preimage of this parameterization over $\mathcal{M}_{g+g',P}^+$ is given by a map

$$\mathcal{M}_{g,P,\vec{o}} \times \mathcal{M}_{g',\vec{o}} \to \mathcal{M}_{g+g',P}^+$$

(which actually factors through the orbit space of $\mathcal{M}_{g,P,\vec{o}} \times \mathcal{M}_{g',\vec{o}}$ for the $\mathbb{C}^\times$-action defined by $\lambda \in \mathbb{C}^\times : (w,w') \mapsto (\lambda w, \lambda^{-1}w')$). Then the $\mathbb{Q}$-stability theorem amounts to the assertion that the projection map $\mathcal{M}_{g,P,\vec{o}} \to \mathcal{M}_{g,P}$ and the maps

$\mathcal{M}_{g,P,\vec{o}} \times \{w'\} \to \mathcal{M}^+_{g+g',P}$, with $w' \in \mathcal{M}_{g',\vec{o}}$, induce isomorphisms on rational cohomology in degree $< \frac{2}{3}g$. With the help of the primitivity property Proposition 12.10 we observe that each of these maps respects the kappa class κ_r and so κ_r defines in fact a stable cohomology class $H^{2r}(\mathrm{Mod}_\infty)$. For $P = \emptyset$, the map $\mathcal{M}_{g,\vec{o}} \times \mathcal{M}_{g',\vec{o}} \to \mathcal{M}^+_{g+g'}$ yields in the stable range the coproduct for the Hopf algebra structure and again Proposition 12.10 shows that $\Delta(\kappa_r) = \kappa_r \otimes 1 + 1 \otimes \kappa_r$. We conclude that κ_r gives rise to a primitive element of $H^{2r}(\mathrm{Mod}_\infty, \mathbb{Q})$: for $g > \frac{3}{2}r$, the class $\kappa_r \in H^{2r}(\mathcal{M}_g; \mathbb{Q}) \cong H^{2r}(\mathrm{Mod}_g; \mathbb{Q}) \cong H^{2r}(\mathrm{Mod}_\infty; \mathbb{Q})$ is independent of g and is primitive.

Proposition 12.15 (E.Y. Miller [14]). *Every κ_r is nonzero and so the natural map of graded algebras $\mathbb{Q}[K_1, K_2, \dots] \to H^\bullet(\mathrm{Mod}_\infty; \mathbb{Q})$ which sends K_r to κ_r (so that $\deg K_r = 2r$) is an embedding.*

Miller's proof is a mixture of topology and algebraic geometry and extends the technique employed first by Atiyah to prove the non-vanishing of κ_1. We could also refer to Faber's result cited above instead: since $\kappa_g(\mathcal{M}_{g+2})$ is nonzero, it follows that κ_g is nonzero, but as mentioned this proof relies on the not-so-trivial theorem of Kontsevich.

Mumford made in [18] the conjecture that this is in fact an isomorphism. This has been proved by Madsen and Weiss [13], (see for instance Hatcher [6] for an exposition) who in fact obtained a considerably stronger statement in terms of homotopy theory. From the preceding it is not hard to deduce that for a finite set P, the psi classes stabilize to $(\psi_p \in H^2(\mathrm{Mod}_{\infty,P}))_{p \in P}$ (ψ_p is in fact the Euler class of the infinite central extension of $\mathrm{Mod}_{\infty,P}$ defined by connected group of the compactly supported diffeomorphisms that fix P pointwise and fix a nonzero tangent vector at p) and that $H^\bullet(\mathrm{Mod}_{\infty,P}; \mathbb{Q})$ is the free polynomial algebra generated by the kappa classes and the psi classes.

References

[1] E. Arbarello, M. Cornalba, Ph.A. Griffiths, *Geometry of algebraic curves, Volume II,* Grundl. d. Math. Wiss. **268**, Springer, Heidelberg (2011).

[2] C. Faber, *A non-vanishing result for the tautological ring of $\mathcal{M}_g$,* available from `arXiv:math/9711219 [math.AG]` (1997).

[3] C. Faber, *A conjectural description of the tautological ring of the moduli space of curves,* In: Moduli of curves and abelian varieties, 109–129, Aspects Math., E**33**, Vieweg, Braunschweig (1999).

[4] R. Hain, E. Looijenga, *Mapping class groups and moduli spaces of curves* in: Algebraic geometry–Santa Cruz 1995, 97–142, Proc. Sympos. Pure Math. **62** Part 2, Amer. Math. Soc., Providence, RI (1997).

[5] J. Harer, *The virtual cohomological dimension of the mapping class group of an orientable surface,* Invent. Math. **84** 157–176 (1986).

[6] A. Hatcher, *A short exposition of the Madsen-Weiss theorem,* A short exposition of the Madsen-Weiss theorem, available at <arXiv:1103.5223> (2011).
[7] D. Petersen, O. Tommasi, *The Gorenstein conjecture fails for the tautological ring of* $\overline{\mathcal{M}}_{2,n}$, Invent. Math. **196** 139–161 (2014).
[8] M.E. Kazarian, S. Lando, *An algebro-geometric proof of Witten's conjecture,* J. Amer. Math. Soc. **20**, 1079–1089 (2007).
[9] S. Keel, *Intersection theory of moduli space of stable n-pointed curves of genus zero,* Trans. Amer. Math. Soc. **330** 545–574 (1992).
[10] M. Kontsevich, *Intersection theory on the moduli space of curves and the matrix Airy function,* Comm. Math. Phys. **147**, 1–23 (1992).
[11] E. Looijenga, *A minicourse on moduli of curves,* School on Algebraic Geometry (Trieste, 1999), 267–291, ICTP Lect. Notes **1**, Abdus Salam Int. Cent. Theoret. Phys., Trieste (2000).
[12] E. Looijenga, *On the tautological ring of* $\mathcal{M}_g$, Invent. Math. **121** 411–419, (1995).
[13] I. Madsen, M. Weiss, *The stable moduli space of Riemann surfaces: Mumford's conjecture,* Ann. of Math. texbf165 843–941 (2007).
[14] E. Miller, *The homology of the mapping class group,* J. Differential Geom. **24**, 1–14 (1986).
[15] J.W. Milnor, J.C. Moore, *On the structure of Hopf algebras,* Ann. of Math. **81** 211–264 (1965).
[16] M. Mirzakhani, *Weil-Petersson volumes and intersection theory on the moduli space of curves,* J. Amer. Math. Soc. **20**, 1–23(2007).
[17] S. Morita, *Generators for the tautological algebra of the moduli space of curves,* Topology **42**, 787–819 (2003).
[18] D. Mumford, *Towards an enumerative geometry of the moduli space of curves,* in: Arithmetic and geometry II, 271–328, Progr. Math. **36**, Birkhäuser, Boston, MA (1983).
[19] M. Pikaart, *An orbifold partition of* $\overline{\mathcal{M}}_g^n$, in: The moduli space of curves (Texel Island, 1994), 467–482, Progr. Math. **129**, Birkhäuser Boston, MA (1995).
[20] N. Wahl, *Homological stability for mapping class groups of surfaces,* in: Handbook of moduli, Vol. III, 547–583, Adv. Lect. Math. **26**, Int. Press, Somerville, MA (2013).
[21] E. Witten, *Two-dimensional gravity and intersection theory on moduli space,* in: Surveys in differential geometry (Cambridge, MA, 1990), 243–310, Lehigh Univ., Bethlehem, PA (1991).

Part II

Introduction to Tropical Curves, Tropical Moduli and Teichmüller Spaces

by Lizhen Ji

In this second part of the book, we give an introduction to the rapidly developing subject of tropical geometry by focusing on tropical curves and their moduli spaces and Teichmüller spaces, and by explaining or rather summarizing some of their applications in many different subjects. We will also discuss some new results. One is concerned with putting a natural tropical structure on the tropical Teichmüller space $\mathscr{T}_n^{tr}$ and the tropical moduli space $\mathscr{M}_n^{tr}$ of compact smooth tropical curves of genus n, and another with constructing complete invariant metrics on the outer space X_n of compact normalized metric graphs of genus n, which can be identified with a special subset of the tropical Teichmüller space $\mathscr{T}_n^{tr}$.

We want to emphasize close connections between moduli space $\mathscr{M}_n^{tr}$ of compact smooth tropical curves of genus n and the outer automorphism group $\mathrm{Out}(F_n)$ of the free group F_n. More specifically, we want to convey the following basic points:

1. Tropical geometry can be used to study the outer automorphism group $\mathrm{Out}(F_n)$ of free group F_n and the outer space X_n of marked compact normalized metric graphs of genus n, which are basic objects in geometric group theory.
2. Conversely, the outer space X_n, or the tropical Teichmüller space $\mathscr{T}_n^{tr}$, which is a slight extension of X_n obtained without normalizing the metrics on graphs, can be used to define and understand better structures of the moduli space $\mathscr{M}_n^{tr}$ of tropical curves.
3. There is a lot of fruitful similarities between the moduli space $\mathscr{M}_g$ of compact Riemann surfaces Σ_g of genus g, the Teichmüller space $\mathscr{T}_g$, the mapping class group Mod_g, and their tropical counterparts $\mathscr{M}_n^{tr}$, $\mathscr{T}_n^{tr}$, and $\mathrm{Out}(F_n)$.
4. Besides a fruitful analogy, there is also direct interaction between the moduli space $\mathscr{M}_n^{tr}$ of tropical curves and arithmetic locally symmetric spaces such as $\mathrm{GL}(n,\mathbb{Z})\backslash\mathrm{GL}(n,\mathbb{R})/\mathrm{O}(n)$.

Though the tropical geometry is the algebraic geometry over the tropical semifield $\mathbb{T}$, tropical varieties have some rather different features from usual algebraic varieties defined over fields such as $\mathbb{C}$ and $\mathbb{R}$. On the other hand, they share some similarities, and tropical varieties can often be used as approximations to algebraic varieties over $\mathbb{C}$ and other fields.

We will start from basics of the tropical semifield $\mathbb{T}$ and illustrate definitions of plane tropical curves and their properties through many examples. We will also discuss briefly various motivations of tropical varieties and applications, and emphasize some applications to geometric group theory. For example, in connection with the Teichmüller space $\mathscr{T}_g$ of compact Riemann surfaces Σ_g of genus g and the action of mapping class group Mod_g on it, one application is concerned with understanding the piecewise linear structures at the infinity of the Thurston compactification of the Teichmüller space $\mathscr{T}_g$.

In studying the problem on constructing complete geodesic metrics on the outer space X_n, we will also describe some results on the moduli space $\mathscr{A}_n^{tr}$ of principally polarized tropical abelian varieties and the tropical Jacobian map $\mathscr{M}_n^{tr} \to \mathscr{A}_n^{tr}$. Compactification of the tropical moduli space $\mathscr{M}_n^{tr}$ will also be considered.

1 Introduction

1.1 A general overview

Recently, there has been an explosion of a new subject called *tropical geometry*. Briefly, the tropical geometry is the geometry about piecewise-linear spaces with integral structures, or polyhedral complexes with rational slopes and integral structures, which satisfy some balanced conditions.

More generally, tropical varieties also include smooth manifolds which locally are linear spaces with some integral structures. This includes tropical abelian varieties.

The adjective *tropical* comes from the fact that the tropical semiring (or semifield) $\mathbb{T}$ with plus and max operations was initially studied by a Brailian computer scientist Imre Simon and was named by the French mathematicians after the tropical region in honor of Simon, and tropical geometry is the algebraic geometry over the tropical semifield $\mathbb{T}$.

Tropical varieties can be used as a convenient and powerful tool (or an approximation) to understand algebraic varieties over $\mathbb{C}$ and other fields. One reason is that tropical varieties can be considered as shadows or highly singular degenerations of algebraic varieties over fields, and hence reflect large-scale properties of the varieties. The piecewise-linear and combinatorial structures of tropical varieties simplify problems but still keep track of enough information about the original algebraic varieties for some applications.

Basically, tropical varieties can be used to understand the usual algebraic varieties in two ways:

1. Describe structures at infinity (or asymptotic directions) of affine varieties and other noncompact varieties.
2. Arise as shadows (degenerations, or approximation) of the usual algebraic varieties via the notion of amoebas.

For example, the case (1) occurs in the logarithmic limits of an algebraic variety in $(\mathbb{C}^\times)^n$ [10] (see §5.3 below), and the case (1) can also be used to explain the piecewise linear structure on the Thurston boundary of the Teichmüller space $\mathcal{T}_g$ (§7.5). The case (2) is best seen in the theory of amoebas, though there is an important difference between amoebas over the complex numbers and non-archimedean fields. Similarly, the case (2) is also related to patchworking of real algebraic varieties in order to determine and classify relative position of connected components of real algebraic plane curves [130] (see also [27]).

On the other hand, for other applications, putting piecewise linear structures into the framework of algebraic geometry over the tropical semifield $\mathbb{T}$ gives one a new perspective. This is particularly true for our discussion below to interpret *metric graphs* as *smooth tropical curves.*

The fact that these piecewise-linear spaces with integral structures (for example tropical varieties) are useful to understand nonlinear algebraic varieties and other nonlinear structures is not surprising. For example, it has been known since Newton that Newton polygons provide an effective approach to understand roots of polynomials and singularities of plane curves (see §4.2 below), and simplicial polyhedrals and fans in $\mathbb{R}^n$ are the essential data for toric varieties. As we shall see later, tropical curves are closely related to Newton polygons, and some tropical varieties are also related to toric varieties and fans. In some sense, tropical varieties can be considered as an enrichment of Newton polygons and fans.

In spite of these reasonable explanations for the usefulness of tropical geometry, tropical varieties first appeared in other contexts, for example, in theories of groups and ideals [10].

Though there exist definitions of tropical varieties of higher dimension [106] and [95], tropical curves are the best understood class of tropical varieties in tropical geometry. The recent successful applications of tropical curves to enumerative algebraic geometry, in particular, the Gromov-Witten theory [58] and [105], brought tropical geometry to the attention of the broader mathematics community.

On the other hand, a version of tropical curves and the moduli space of marked tropical curves (or the tropical Teichmüller space) was introduced in 1986 by Culler and Vogtmann in 1986 [42]. They were motivated by applications in geometric group theory, i.e., in order to construct a contractible space on which the group $\mathrm{Out}(F_n)$, the outer automorphism group of the free group F_n, acts properly. The desired space is the outer space X_n of *marked, normalized* metric graphs of genus n, and $\mathrm{Out}(F_n)$ acts on it by changing markings.

Briefly, metric graphs can be canonically identified with smooth tropical curves, and hence the outer space X_n is the tropical analogue of Teichmüller space $\mathcal{T}_g$. The outer automorphism group $\mathrm{Out}(F_n)$ is the analogue of the mapping class group Mod_g of a compact oriented surface S_g of genus g, since the latter can be identified with the outer automorphism group of the surface group $\pi_1(S_g)$.

Motivated by applications of the action of Mod_g on $\mathcal{T}_g$, the outer space X_n was used to determine the virtual cohomology dimension of $\mathrm{Out}(F_n)$ in [42], to prove the virtual duality property of $\mathrm{Out}(F_n)$ in [14], and to understand other properties of $\mathrm{Out}(F_n)$.

Once metric graphs are identified with smooth tropical curves, the quotient $\mathrm{Out}(F_n)\backslash X_n$ is embedded into the moduli space $\mathcal{M}_n^{tr}$ of compact smooth tropical curves of genus n.

The connection between metric graphs and the group $\mathrm{Out}(F_n)$ on the one hand, and tropical curves and their moduli spaces on the other hand, has been beneficial to both sides.

Recall that Teichmüller space $\mathcal{T}_g$ of compact Riemann surfaces was introduced to study the moduli space $\mathcal{M}_g$ of compact Riemann surfaces of genus g (or algebraic curves over $\mathbb{C}$). See the first part of this book, and also [2] and [71] for the history and more references. In this second part of the book, we will show that the tropical Teichmüller space $\mathcal{T}_n^{tr}$ also provides an important approach to understand the moduli space $\mathcal{M}_n^{tr}$ of tropical curves. For example, we will show that $\mathcal{T}_n^{tr}$ admits a natural local structure of tropical variety and hence $\mathcal{M}_n^{tr}$ admits a structure of tropical orbifold. See Theorem 8.27 in §8 below for detail.

It is known that similarity between Teichmüller spaces and symmetric spaces of noncompact has been very fruitful (see the summary and references in [69]). It seems reasonable to believe that similarity between the tropical Teichmüller space $\mathcal{T}_n^{tr}$ and symmetric spaces is also useful for the study of moduli space $\mathcal{M}_n^{tr}$ of tropical curves. In this paper, we will explain some results to support these implications.

For example, one particular motivation for us to study tropical geometry is to *construct complete geodesic metrics on the outer space X_n which are invariant under* $\mathrm{Out}(F_n)$, and it will be explained in Problems 1.1 and 1.3 below. See also §12.1 below. In this result, a direct connection between tropical moduli space $\mathcal{M}_n^{tr}$ and the locally symmetric space $\mathrm{SL}(n,\mathbb{Z})\backslash\mathrm{SL}(n,\mathbb{R})/\mathrm{SO}(n)$ is needed, which is motivated by the period map for Riemann surfaces.

Another example is to use some results on partial compactifications of symmetric spaces and compactifications of locally symmetric spaces to understand compactifications of the moduli space $\mathcal{M}_n^{tr}$ and partial compactifications of the tropical Teichmüller space $\mathcal{T}_n^{tr}$.

There are many analogues between algebraic varieties over the field $\mathbb{C}$ and tropical algebraic varieties. It is known that the Jacobian variety of a projective algebraic curve over $\mathbb{C}$ defines the important *period or Jacobian map*

$$\Pi : \mathcal{M}_g \to \mathcal{A}_g,$$

where $\mathcal{M}_g$ is the moduli space of compact Riemann surfaces of genus g, and $\mathcal{A}_g$ is the moduli space of principally polarized abelian varieties of dimension g. The map Π has played a crucial role in the study of both moduli spaces $\mathcal{M}_g$ and $\mathcal{A}_g$.

As we will see, there is an analogous *tropical period or Jacoban map*

$$\Pi^{tr} : \mathcal{M}_n^{tr} \to \mathcal{A}_n^{tr},$$

where the space $\mathcal{A}_n^{tr}$ is the moduli space of principally polarized tropical abelian varieties of dimension n. This tropical period map Π^{tr} will be used crucially in solving the above problem on constructing complete metrics on the outer space (see §2.4).

We can simply define a principally polarized tropical abelian variety as a flat Riemannian torus, and write down the tropical period of a compact smooth tropical curve explicitly after they are viewed as metric graphs, as it is done in some papers. But it is a mystery why they are defined in such a way, or why they are called periods of tropical curves. We will *use tropical geometry to explain this.* This explanation will help to clarify a natural local tropical structure on the tropical Teichmüller space $\mathcal{T}_n^{tr}$ in §8.5 and the induced tropical orbifold structure on the moduli space $\mathcal{M}_n^{tr}$ in §8.4.

As it is known, the moduli space $\mathcal{M}_g$ of compact Riemann surfaces of genus $g \geq 1$ is noncompact and admits several compactifications, for example, the

Deligne-Mumford compactification $\overline{\mathcal{M}_g}^{DM}$, which is important for many applications.

Similarly, the moduli space $\mathcal{M}_n^{tr}$ of smooth tropical curves of genus $n \geq 1$ is noncompact, and we consider several compactifications of $\mathcal{M}_n^{tr}$.

We note that in geometric group theory, slightly before [42], some special tropical varieties (but not tropical curves) were introduced in [15] by following [10], and were also applied to study some invariants of infinite groups.

One purpose of this second part of the book is to explain the above results by starting from basics. One reason for doing this is that tropical curves and the tropical geometry are developing rapidly with many applications in different subjects.

Tropical geometry is a beautiful subject, and it is beneficial for everyone to learn something about tropical varieties and their connections with and applications to other subjects. But in many papers and some books, tropical curves are simply defined to be metric graphs without too much or any explanation. At other places, only plane tropical curves are discussed. Though they are important and of special interest, as we will see, plane tropical curves are not sufficient *even* for the applications to the outer space X_n and the group $\mathrm{Out}(F_n)$ mentioned above.

Relations between these different definitions of tropical curves and varieties may not be so clear or easy to the beginners or nonexperts. For example, it is not obvious how the tropical semifield $\mathbb{T}$ enters into the identification of tropical curves with metric graphs, and how plane tropical curves can be identified with metric graphs. Even after we have defined plane tropical curves, definitions of abstract tropical curves are not automatic. We will try to explain all these connections.

In summary, we want to convey the following basic points in this second part of the book:

1. Tropical curves and tropical varieties natural occur in many different contexts, since both linear and integral structures are both intrinsically interesting objects in themselves and effective approximations to nonlinear varieties and spaces.
2. Tropical geometry can be used to study basic objects in geometric group theory such as the outer automorphism group of free group $\mathrm{Out}(F_n)$ and the outer space X_n of marked compact normalized metric graphs of genus n.
3. Conversely, the outer space X_n, or the tropical Teichmüller space $\mathcal{T}_n^{tr}$, which is a slight extension of X_n obtained without normalizing the metrics on graphs, can be used to understand better the moduli space $\mathcal{M}_n^{tr}$ of compact smooth tropical algebraic curves of genus n, for example, as a tropical orbifold.
4. There is a lot of fruitful similarities between the moduli space $\mathcal{M}_g$ of compact Riemann surfaces Σ_g of genus g, the Teichmüller space $\mathcal{T}_g$, the mapping class group Mod_g, and their tropical counterparts $\mathcal{M}_n^{tr}$, $\mathcal{T}_n^{tr}$, and $\mathrm{Out}(F_n)$.
5. Besides a fruitful analogy, there is also direct interaction between arithmetic locally symmetric spaces and the moduli space $\mathcal{M}_n^{tr}$ of tropical curves.

There are several recent books on tropical geometry [96] [106]. There are also many surveys and expository writings on tropical geometry, for example [27], [28] [30], [124], [99], [100], [103], [104], [105], [51], [94], [19], [77], and [58]. In some sense, our discussion in this book is tangential to the main points of these books and papers.

1.2 A specific motivation from geometric group theory

In this section, we explain the problem in geometric group theory which motivated our study in tropical geometry.

Let F_n be the free group on n generators, $n \geq 2$, and $\mathrm{Out}(F_n) = \mathrm{Aut}(F_n)/\mathrm{Inn}(F_n)$ its outer automorphism group. The group $\mathrm{Out}(F_n)$ is one of the most basic groups in combinatorial group theory and it has been extensively studied. One reason is that it is related to a basic question: given a finitely generated group G, how to get all possible generating sets (or different generating sets with various properties) of G [96, p. 120]? One key ingredient in understanding properties of $\mathrm{Out}(F_n)$ is to use its action on the outer space X_n of marked metric graphs and properties of X_n. In some sense, group action is essentially the only way to understand a group. See §2 below for definition of the outer space X_n and its applications to $\mathrm{Out}(F_n)$.

According to the celebrated Erlangen program of Klein (see [73] for some discussions and references), an essential part of the geometry of a space is concerned with invariants of the group of automorphisms (or symmetries) of geometric structures of the space (for example, isometries of a metric space).

Similarly, an essential part of the *geometry of a group* is to *find and understand spaces on which the group acts and preserves suitable additional structures of the space.* It is often natural to require that the space is a metric space and the action is by isometries.

We note that for any connected topological manifold, its group of homeomorphisms always acts transitively on it. Similarly, the group of diffeomorphisms of any smooth manifold also acts transitively on the manifold. Therefore, general homeomorphisms and diffeomorphisms are not really symmetries. On the other hand, if the group of isometries of a metric space acts transitively on it, the metric space is rather special. Therefore, isometries are genuine symmetries in some sense.

In many known examples, the space is also simply connected. In classical geometry, metric spaces are often complete Riemannian manifolds such as the Euclidean space, the sphere or the hyperbolic space. But it is also important to consider metric spaces which are not manifolds. Examples include Tits buildings for real semisimple Lie groups and Bruhat-Tits buildings for p-adic semisimple Lie groups (see [70] for some detail and references). These are simplicial complexes with a natural *complete* Tits metric so that the groups act isometrically and simplicially on them. Of course, the rich combinatorial structure also makes their geometry interesting.

The outer space X_n is a finite-dimensional infinite simplicial complex and hence admits a natural simplicial metric d_0 (Proposition 2.10), where each n-simplex is realized as the standard simplex in $\mathbb{R}^{n+1}$ with vertices $(1,0,\ldots,0),\ldots,$ $(0,\ldots,0,1)$ and is given the metric induced from the Euclidean metric of $\mathbb{R}^{n+1}$. Clearly, d_0 is invariant under $\mathrm{Out}(F_n)$. Since simplices of X_n have missing simplicial faces, d_0 is not a complete metric on X_n.

Since the completeness of metrics is a basic condition which is important for many applications, one natural problem is the following (see [23, Question 2] for more discussion).

Problem 1.1. *Construct complete geodesic metrics on the outer space X_n which are invariant under* $\mathrm{Out}(F_n)$.

As explained above, solutions to this problem provide geometries of $\mathrm{Out}(F_n)$ in a certain sense. Another motivation for this problem comes from the analogy with other important groups in geometric group theory: arithmetic subgroups Γ of semisimple Lie groups G and mapping class groups Mod_g of compact surfaces of genus g. In some sense, more work has been done on these latter two families of groups, and a lot of work on $\mathrm{Out}(F_n)$ is motivated by them.

For our purpose, we note that an arithmetic group Γ acts properly on the symmetric space G/K, and the mapping class group Mod_g acts on the Teichmüller space $\mathcal{T}_g$ of compact Riemann surfaces of genus g. The outer space X_n was introduced as an analogue for $\mathrm{Out}(F_n)$. These actions have played a crucial and similar role in understanding properties of the corresponding classes of groups.

Though these three classes of groups and spaces share many common properties, there is one important difference between them, which might have not been emphasized enough: the symmetric spaces X and the Teichmüller space $\mathcal{T}_g$ are smooth manifolds, but the outer space X_n is *not a manifold*. Therefore, differential geometric analysis can be carried out on the first two spaces but probably not so easily on the third one.

The symmetric space $X = G/K$ admits a complete G-invariant Riemannian metric which is unique up to scaling on irreducible factors of X, and arithmetic subgroups Γ act isometrically and properly on X. Consequently, the locally symmetric spaces $\Gamma\backslash X$ also admit a complete metric. It is known that X is a contractible, nonpositively curved Riemannian manifold, and that it is an important example of Hadamard manifolds. Furthermore, the locally symmetric spaces $\Gamma\backslash X$ have finite volume.

In some sense, the invariant metric is the only interesting metric on X due to its large symmetry group. As it is known, this invariant metric makes structures of the symmetric space X and locally symmetric spaces $\Gamma\backslash X$ much richer. For example, even the formulation of many questions about them, such as the Mostow strong rigidity and other metric rigidity results, depends on this metric.

The Teichmüller space $\mathcal{T}_g$ is also contractible, and admits several complete Riemannian and Finsler metrics such as the Bergman metric (when $\mathcal{T}_g$ is realized as a bounded domain in $\mathbb{C}^{3g-3}$), Kobayashi metric and Teichmüller metric,

Kähler-Einstein metric, and Mod_g acts isometrically and properly on them. Metrics are also crucial for studying $\mathcal{T}_g$. For example, after introducing the space $\mathcal{T}_g$, Teichmüller defined a natural topology on $\mathcal{T}_g$ using the Teichmüller metric.

The Teichmüller space $\mathcal{T}_g$ also admits the incomplete negatively curved Weil-Petersson metric. For all these natural metrics on $\mathcal{T}_g$, the quotient $\mathrm{Mod}_g\backslash\mathcal{T}_g$ has finite volume, which follows from the quasi-isometric asymptotics of these metrics in [92]. An interesting feature is that the known complete metrics on $\mathcal{T}_g$ are not nonpositively curved, but the negatively curved Weil-Petersson metric is incomplete. A natural and folklore conjecture is the following

Conjecture 1.2. *The Teichmüller space $\mathcal{T}_g$ does not admit any complete nonpositively curved metric which is invariant under* Mod_g.

See [26] for more detail about this conjecture. These results on metrics on the symmetric space X and the Teichmüller space make comparison between symmetric spaces and Teichmüller spaces more fruitful and interesting.

Though the outer space X_n is not a manifold, it is a locally finite simplicial complex and hence is a canonically stratified space with a smooth structure in the sense of [120]. Therefore, the following problem also seems to be natural in view of the above analogy.

Problem 1.3. *Construct piecewise smooth Riemannian metrics on the outer space X_n which are invariant under* $\mathrm{Out}(F_n)$ *and whose induced length metrics are complete geodesic metrics. Furthermore, the quotient* $\mathrm{Out}(F_n)\backslash X_n$ *has finite volume.*

With respect to the smooth structure on X_n as a canonically stratified space, it is natural to require that the Riemannian metric on X_n be smooth in the sense of [120, Chapter 2]. If the second condition of finite volume is satisfied, it might give further justification to the *philosophy that* $\mathrm{Out}(F_n)$ *is an analogue of lattices of Lie groups*, which has played a guiding role in studying and understanding the group $\mathrm{Out}(F_n)$ (see [133], [134], and [12]).

We will explain how to use tropical curves and their tropical Jacobians to construct several invariant complete geodesic metrics and complete piecewise-smooth Riemannian metrics on X_n and hence solve Problems 1.1 and 1.3 in some sense.

The basic idea is to identify metric graphs with tropical curves and hence relate the outer space X_n to the tropical Teichmüller space $\mathcal{T}_n^{tr}$. It turns out that the tropical analogy of the Siegel upper half space is the symmetric space $\mathrm{GL}(n,\mathbb{R})/\mathrm{O}(n)$ of positive definite matrices. Then the tropical Jacobian map allows us to pull back the invariant metric of the symmetric space $\mathrm{GL}(n,\mathbb{R})/\mathrm{O}(n)$ to a pseudo-metric on the outer space X_n, which together with the simplicial metric d_0 on X_n will produce desired complete metrics. For precise statements, see Theorems 12.5, 12.11, 12.14, and Propositions 12.16, 12.18 below.

Remark 1.4. It should be pointed out that the tropical Jacobian map, or rather the period map of metric graphs, can be defined directly for compact metric graphs.

But without the interpretation through tropical curves and their Jacobians, it is a mystery why it is defined in this way. There are infinitely many different ways to define a map from the outer space X_n to the symmetric space $\mathrm{GL}(n,\mathbb{R})/\mathrm{O}(n)$ (see Remark 2.25 below). In any case, the connection between metric graphs and tropical curves is interesting and our discussion here also clarifies some questions in tropical geometry, for example, the existence of a natural tropical structure on the tropical moduli space $\mathscr{M}_n^{tr}$ and the tropical Teichmüller space $\mathscr{T}_n^{tr}$.

Remark 1.5. It is known that X_n admits a non-symmetric geodesic metric, called the Lipschitz (or Thurston) metric. Briefly, the non-symmetric Lipschitz metric is not complete, and its associated (symmetric) is not geodesic. For its definition, some properties and applications, see [48], [6], [5], and [13].

The analogy with Teichmüller space suggests the following

Conjecture 1.6. *The outer space X_n does not admit any complete* CAT(0)*-metric which is invariant under* $\mathrm{Out}(F_n)$.

Besides the analogy, there is some supporting evidence. For example, it was shown in [22] that the standard spine of X_n does not admit any invariant piecewise-Euclidean or piecewise-hyperbolic metric of nonpositive curvature. It was shown in [55] and [24] that $\mathrm{Out}(F_n)$ is not a CAT(0)-group, i.e., it cannot act isometrically and cocompactly on a CAT(0)-space. Since the quotient $\mathrm{Out}(F_n)\backslash X_n$ is not compact, the above result does not exclude the possibility that X_n might admit a CAT(0)-metric which is invariant under $\mathrm{Out}(F_n)$.

1.3 Interaction between moduli spaces of tropical curves and outer space of metric graphs

In the previous subsection, we discussed an application of tropical algebraic curves to the group $\mathrm{Out}(F_n)$ and the outer space X_n. The second purpose of the second part of this book is to give an application in the opposite direction. Instead of normalizing the total length of each metric graph to be equal to 1, we consider all marked compact metric graphs of genus n and obtain an enlarged outer space, which turns out to be the tropical Teichmüller space $\mathscr{T}_n^{tr}$ of compact smooth tropical curves of genus n. Then the moduli space $\mathscr{M}_n^{tr}$ of smooth compact tropical curves of genus n is equal to the quotient space $\mathrm{Out}(F_n)\backslash\mathscr{T}_n^{tr}$.

It turns that the markings on tropical curves allow one to obtain well-defined edge length functions on $\mathscr{T}_n^{tr}$, which can be used to define a natural local tropic variety structure on $\mathscr{T}_n^{tr}$. Using the identification $\mathscr{M}_n^{tr} = \mathrm{Out}(F_n)\backslash\mathscr{T}_n^{tr}$, we can give $\mathscr{M}_n^{tr}$ the structure of a tropical orbifold, and show that some finite covers of $\mathscr{M}_n^{tr}$ are tropical varieties. See §8 below for more detail.

This is also similar to the relation between the moduli space $\mathscr{M}_g$ of Riemann surfaces and the Teichmüller space $\mathscr{T}_g$. The complex structure on $\mathscr{M}_g$ was first proved by constructing a complex structure on $\mathscr{T}_g$ and then realizing $\mathscr{M}_g$ as a quotient space $\mathrm{Mod}_g\backslash\mathscr{T}_g$ of $\mathscr{T}_g$. See [71] for detail and references.

Acknowledgments. The idea of using the period map of metric graphs in Equation (2.7) to pull back and define a metric on the outer space was suggested by Enrico Leuzinger, partly motivated by the work [72], and I would like to thank him for sharing this idea. Though the period map fails to be injective (unlike the case of Riemann surfaces), the simple idea of combining the period map with the canonical simplicial metric d_0 on the outer space X_n is crucial to overcome this problem. This work is partially Supported by NSF grant DMS-1104696 and Simons Fellowship from the Simons Foundation (grant #305526).

2 Definitions and basic facts on $\mathrm{Out}(F_n)$ and the outer space X_n

In this section, we recall some basic definitions and results on the outer automorphism group $\mathrm{Out}(F_n)$ and the outer space X_n, and on their relations with arithmetic groups and mapping class groups.

2.1 Combinatorial group theory and motivations from mapping class groups

The outer automorphism group $\mathrm{Out}(F_n)$ and arithmetic subgroups of semisimple Lie groups are basic examples in combinatorial group theory, and they share many similar properties.

Probably the most basic arithmetic subgroup of a semisimple Lie group is

$$\mathrm{SL}(n,\mathbb{Z}) \subset \mathrm{SL}(n,\mathbb{R}), \quad n \geq 2.$$

A closely related group is the arithmetic subgroup $\mathrm{GL}(n,\mathbb{Z})$ of a reductive Lie group $\mathrm{GL}(n,\mathbb{R})$. Note that $\mathrm{GL}(n,\mathbb{Z})$ contains $\mathrm{SL}(n,\mathbb{Z})$ as a subgroup of index 2, and hence $\mathrm{GL}(n,\mathbb{Z})$ and $\mathrm{SL}(n,\mathbb{Z})$ are essentially the same.

The group $\mathrm{Out}(F_n)$ is related to $\mathrm{GL}(n,\mathbb{Z})$ as follows. Since the abelianization of F_n is $\mathbb{Z}^n = F_n/[F_n,F_n]$, the free abelian group, and

$$\mathrm{Out}(\mathbb{Z}^n) = \mathrm{Aut}(\mathbb{Z}^n) = \mathrm{GL}(n,\mathbb{Z}),$$

there is a homomorphism

$$A : \mathrm{Out}(F_n) \to \mathrm{GL}(n,\mathbb{Z}).$$

It is known that when $n = 2$,

$$A : \mathrm{Out}(F_2) \cong \mathrm{GL}(2,\mathbb{Z})$$

is an isomorphism, and for every $n \geq 3$, the homomorphism $A : \mathrm{Out}(F_n) \to \mathrm{GL}(n,\mathbb{Z})$ is *surjective with an infinite kernel.* The surjectivity follows by noting that a set of generators of $\mathrm{GL}(n,\mathbb{Z})$ is contained in the image. See [133, p. 2] for references. This shows that the two families $\mathrm{GL}(n,\mathbb{Z})$ and $\mathrm{Out}(F_n)$ have the same starting group.

Let S_g be a compact oriented surface of genus g, $\mathrm{Diff}^+(S_g)$ its group of orientation preserving diffeomorphisms, and $\mathrm{Diff}^0(S_g)$ the identify component of $\mathrm{Diff}^+(S_g)$. Then the quotient group

$$\mathrm{Mod}_g = \mathrm{Diff}^+(S_g)/\mathrm{Diff}^0(S_g)$$

is the *mapping class group* of S_g.

By the Dehn-Nielson theorem (see [68] [46]),

$$\mathrm{Mod}_g = \mathrm{Out}(\pi_1(S_g)).$$

This is suggestive of the relation between Mod_g and $\mathrm{Out}(F_n)$, though of course $\pi_1(S_g)$ is not isomorphic to F_n.

Remark 2.1. The standard representation of $\pi_1(S_g)$ has $2g$ generators and 1 relation, but this does not prove that $\pi_1(S_g)$ is not a free group. Since S_g is a closed aspherical manifold, the Poincaré duality shows that the cohomology dimension of $\pi_1(S_g)$ is equal to 2, while the cohomological dimension of F_n is equal to 1. Therefore, they are not isomorphic.

Since the abelianization of $\pi_1(S_g)$ is $H_1(S_g,\mathbb{Z})$, which is isomorphic to $\mathbb{Z}^{2g}$, we also get a homomorphism $\mathrm{Mod}_g \to \mathrm{GL}(2g,\mathbb{Z})$. Since elements of Mod_g preserve the intersection form on $H_1(S_g,\mathbb{Z})$, which is a symplectic form, the image of Mod_g lies in $\mathrm{Sp}(2g,\mathbb{Z})$. Therefore, we have a homomorphism

$$j : \mathrm{Mod}_g \to \mathrm{Sp}(2g,\mathbb{Z}).$$

It is known that j is surjective with an infinite kernel when $g \geq 2$, and j is an isomorphism when $g = 1$, i.e.,

$$\mathrm{Mod}_1 \cong \mathrm{SL}(2,\mathbb{Z}).$$

This shows that the two families Mod_g and $\mathrm{SL}(n,\mathbb{Z})$ have the same origin. Consequently, the three families of groups

$$\mathrm{Out}(F_n),\ \ \mathrm{SL}(n,\mathbb{Z}),\ \ \mathrm{Mod}_g, n \geq 2, g \geq 1,$$

have essentially the same starting group $\mathrm{SL}(2,\mathbb{Z})$, and hence should share some properties.

Remark 2.2. If $S_{g,n}$ is a compact oriented surface of genus g with n punctures, then $\pi_1(S_{g,n})$ is a free group F_N, where N depends on g and n, for example $N = 2g+n-1$ when $n \geq 2$, and the mapping class group $\mathrm{Mod}_{g,n}$ of $S_{g,n}$ can be canonically realized as a subgroup of $\mathrm{Out}(F_N)$. Specifically, loops around each puncture of $S_{g,n}$ correspond to a conjugacy class of nontrivial elements of $\pi_1(S_{g,n})$. Then the mapping class group $\mathrm{Mod}_{g,n}$ of $S_{g,n}$ consists of elements of $\mathrm{Out}(F_N)$ which stabilize the collection of n conjugacy classes corresponding to the punctures of $S_{g,n}$. (We note that elements of $\mathrm{Diff}^+(S_{g,n})$ can permute the ends of $S_{g,n}$.) See [133, p. 2] for references. This inclusion $\mathrm{Mod}_{g,n} \subset \mathrm{Out}(F_N)$ can be used to show that $\mathrm{Mod}_{g,n}$ is of type VFL (virtually finite free). See [42, p. 114].

The above discussions establish close relations between $\mathrm{Out}(F_n)$, the mapping class group Mod_g, and arithmetic groups such as $\mathrm{GL}(n,\mathbb{Z})$ and $\mathrm{Sp}(2g,\mathbb{Z})$.

In summary, the most basic example of arithmetic subgroups,

$$\mathrm{SL}(2,\mathbb{Z}) \cong \mathrm{Sp}(2,\mathbb{Z}),$$

admits three important types of generalization:

1. Arithmetic subgroups Γ of noncompact semisimple Lie groups G such as $\mathrm{SL}(n,\mathbb{Z}) \subset \mathrm{SL}(n,\mathbb{R})$, $n \geq 2$, and $\mathrm{Sp}(2g,\mathbb{Z}) \subset \mathrm{Sp}(2g,\mathbb{R})$, $g \geq 1$.
2. Mapping class groups Mod_g of compact surfaces and $\mathrm{Mod}_{g,n}$ of noncompact surfaces, $g,n \geq 1$.
3. Outer automorphism group $\mathrm{Out}(F_n)$ of the free group F_n, $n \geq 2$.

The common root of these groups suggests that they should share many properties. Indeed, they do share group theoretical properties and cohomological properties, many of which are proved using actions on symmetric spaces, Teichmüller spaces and outer spaces as mentioned before. For some statements and references, see [69], [68], [65], [134], and [12].

2.2 Geometric group theory, metric graphs and outer space X_n

One important approach to study groups is to construct and understand their actions on suitable spaces. This is reasonable in view of how groups were first defined as permutation groups. The theory of transformation groups promoted by Klein and Lie is really about group actions. Groups derive their power from and unwrap their mystery by group actions.

Given any discrete group Γ, one canonical nontrivial action of Γ is the action on itself Γ or $L^2(\Gamma)$. For various topological applications, we often want actions on finite dimensional topological spaces. If Γ is an infinite group, it is often difficult to find nontrivial actions of Γ on spaces with some finiteness properties.

We recall that for any discrete group Γ, a model $\underline{E}\Gamma$ for the the *universal space for proper actions* of Γ is a Γ-space such that

1. Γ acts properly on $\underline{E}\Gamma$,
2. for every finite subgroup $F \subset \Gamma$, the set of fixed point $\underline{E}\Gamma^F$ is nonempty and contractible. Since the trivial subgroup is finite, $\underline{E}\Gamma$ is contractible.

It is known that $\underline{E}\Gamma$ exists and is unique up to Γ-homotopy equivalence. A general construction essentially due to Milnor gives infinite dimensional models of $\underline{E}\Gamma$. But for many applications, it is important to get *finite dimensional models* and even require that the quotient $\Gamma\backslash\underline{E}\Gamma$ is *compact*.

As mentioned before, an arithmetic subgroup Γ of a semisimple Lie group G acts *isometrically* and *properly* on the symmetric space $X = G/K$. The following result shows the importance of the invariant metric on the symmetric space X.

Proposition 2.3. *For any arithmetic subgroup $\Gamma \subset G$, or more generally any discrete subgroup $\Gamma \subset G$, the symmetric space X is a finite dimensional model of the universal space $\underline{E}\Gamma$ for proper actions of Γ.*

Proof. Since X is a contractible nonpositively curved complete Riemannian manifold, the Cartan fixed point theorem implies that for every finite subgroup $F \subset \Gamma$, the set of fixed points X^F is nonempty. Briefly, for any point $x \in X$, the orbit $F \cdot x$ is a finite subset. Since X is nonpositively curved, simply connected and complete, the finite subset $F \cdot x$ has a unique center of gravity. Since F leaves $F \cdot x$ invariant, it must fix the unique center. Since X^F is a totally geodesic submanifold of X, it is also contractible. □

An important space for the group Mod_g to act on is the Teichmüller space $\mathcal{T}_g$. In some sense, Mod_g was introduced for this action, and the Teichmüller space $\mathcal{T}_g$ was introduced by Teichmüller to study the moduli space $\mathcal{M}_g$ of compact Riemann surfaces of genus g. (Note that the earlier variant, the so-called Fricke space was introduced by Poincaré and Klein to prove the uniformization theorem of Riemann surfaces. See [71].)

More precisely, $\mathcal{T}_g$ is the space of marked complex structure of S_g, i.e., the set of equivalence classes of marked compact Riemann surfaces of genus g. Recall that a *marked Riemann surface* is a Riemann surface Σ_g together with a homotopy equivalence class of diffeomorphism $\Sigma_g \to S_g$, where S_g is a fixed compact oriented surface of genus g. Two marked Riemann surfaces are called *equivalent* if there is a biholomorphic map between them which commutes with the markings.

Since the action of Mod_g on $\mathcal{T}_g$ comes from changing the markings of surfaces, the quotient $\mathrm{Mod}_g\backslash\mathcal{T}_g$ is $\mathcal{M}_g$. See [71] for a history and some references.

It is an important result of Teichmüller that the Teichmüller space $\mathcal{T}_g$ is a contractible manifold and that the action of Mod_g on $\mathcal{T}_g$ is proper and isometric with respect to the Teichmüller metric (and also many other metrics). It is also known but more difficult to prove

Proposition 2.4. *The Teichmüller space $\mathcal{T}_g$ is a finite dimensional model of $\underline{E}\mathrm{Mod}_g$.*

Proof. The positive solution of the Nielsen realization problem in [78] says that for every finite subgroup $F \subset \mathrm{Mod}_g$, the fixed point set $\mathcal{T}_g^F$ is nonempty. Since $\mathcal{T}_g^F$ is the Teichmüller space of Riemann surfaces with automorphism group containing F, it is also contractible. We can also use earthquake paths and the geodesic convexity of the Weil-Petersson to prove the contractibility of $\mathscr{T}_g^F$ as in the previous proposition. □

Besides these two results, there are many others showing similarities between arithmetic groups and mapping class groups. See [69] for brief summaries and references.

Motivated by the definition of Teichmüller space $\mathscr{T}_g$ and the action of Mod_g on $\mathscr{T}_g$, Culler and Vogtmann [42] introduced the *outer space* X_n of marked metric graphs of genus g.

Specifically, a *metric graph* (G,ℓ) is a graph G together with a positive length $\ell(e)$ assigned to each edge e. Note that the fundamental group of a finite graph is a finitely generated free group. We only consider connected graphs G with *genus* equal to n, i.e., $\pi_1(G) \cong F_n$, which is equivalent to the condition that $H_1(G,\mathbb{Z}) \cong \mathbb{Z}^n$. We also assume that *$G$ has no vertices of valence 1 or 2*. (Note that erasing vertices of valence 1 or 2 does not change the homotopy type of the underlying topological space of the graph, and erasing vertices of valence 2 does not change the homeomorphism type.) If we do not allow separating edges in metric graphs, which are also called bridges, then we obtain the *reduced outer space* of metric graphs. Later when we identify tropical curves with metric graphs, the resulting metric graphs can have bridges. Therefore, in this book, we will use the *nonreduced outer space*, and call it the outer space.

Let R_n be the wedge of n circles, also called *a rose with n petals.* This is a graph of genus n with the smallest number of edges. Then *a marking* on a metric graph (G,ℓ) is an homotopy equivalence $h : G \to R_n$, which amounts to a choice of a set of ordered generators of $\pi_1(G)$. *A marked metric graph* is a metric graph (G,ℓ) together with a marking h, denoted by (G,ℓ,h).

Two marked metric graphs $(G,\ell,h),(G',\ell',h')$ are called *equivalent* if there exists an isometry between (G,ℓ) and (G',ℓ') that commutes with the markings h,h' up to homotopy, i.e., having the same set of ordered generators of their fundamental groups.

Usually in connection with the study on $\mathrm{Out}(F_n)$, we *normalize* (or *scale*) the metric ℓ of each metric graph (G,ℓ) so that the total sum of lengths of all edges is equal to 1. On the other hand, when we use metric graphs to study tropical curves, we will also use metric graphs whose edge lengths are not normalized. This is more natural in view of tropical geometry.

Definition 2.5. With the above definitions, the outer space X_n is the set of equivalence classes of marked metric graphs of total edge length 1,

$$X_n = \{(G,\ell,h)\}/\sim. \tag{2.1}$$

Since elements of $\mathrm{Out}(F_n)$ can be realized by homotopy equivalences of the rose R_n, $\mathrm{Out}(F_n)$ acts on X_n by changing the markings of metric graphs. Specifically, we have the following observation of Nielsen.

Lemma 2.6. *Every automorphism $\varphi \in \mathrm{Aut}(F_n)$ can be realized by a homotopy equivalence $f_\varphi : R_n \to R_n$, i.e., $f_{\varphi}* : \pi_1(R_n) \to \pi_1(R_n)$ is equal to φ, such that when φ is an inner automorphism, f_φ is homotopy equivalent to the identity map.*

Proof. Let the n petals of R_n be denoted by $a_1,\dots,a_n$, and the corresponding elements of $F_n = \pi_1(F_n)$ by $x_1,\dots,x_n$, which clearly form a set of free generators of F_n. Given any automorphism φ of F_n, for each x_i, write $\varphi(x_i)$ as a reduced word in $x_1,\dots,x_n$. Define $f_\varphi : R_n \to R_n$ to map a_i to the closed edge path from the base point representing the reduced word $\varphi(x_i)$. Since the inverse φ^{-1} is also an automorphism of F_n, we obtain a map $f_{\varphi^{-1}} : R_n \to R_n$. Note that for any petal a_i of R_n, $f_{\varphi^{-1}}(f_\varphi(a_i))$ is an edge path of a word in $x_1,\dots,x_n$ which is freely reduced to x_i. Therefore, $f_{\varphi^{-1}}(f_\varphi(a_i))$ is homotopic to a_i. Since these petals a_i of R_n are disjoint except for the end point, these homotopy equivalences for a_i are compatible, and $f_{\varphi^{-1}} \circ f_\varphi$ is homotopic to the identity map. This implies that f_φ is an homotopy equivalence.

Suppose that φ is an inner automorphism, i.e., the conjugation by an element g of F_n. Let Ω be the edge path representing the reduced word for g, and Ω^{-1} be that for the inverse g^{-1}. Then for any a_i, $f_\varphi(a_i)$ is the concatenation of Ω^{-1}, a_i, and Ω. By retracting along the edge path Ω, we can see that $f_\varphi(a_i)$ is homotopic to a_i. Therefore, f_φ is homotopy equivalent to the identity map. □

By the above lemma, each element of $[\varphi] \in \mathrm{Out}(F_n)$ is represented by a homotopy equivalence $f_\varphi : R_n \to R_n$, which is unique up to homotopy equivalence in the space of homotopy equivalences between R_n and R_n. For any marked metric graph (G,ℓ,h) and any element $[\varphi] \in \mathrm{Out}(F_n)$, define

$$[\varphi] \circ (G,\ell,h) = (G,\ell,f_\varphi \circ h).$$

By the above comment on the uniqueness of f_φ, the new marking $f_\varphi \circ h$ on the metric graph (G,ℓ) is well-defined. This gives the *action of* $\mathrm{Out}(F_n)$ *on* X_n:

$$\mathrm{Out}(F_n) \times X_n \to X_n. \tag{2.2}$$

Proposition 2.7. *The outer space X_n is a locally finite simplicial complex of dimension $3n-4$, and its simplices are partially open, i.e., every simplex of X_n has some simplicial faces missing.*

Briefly, for each marked graph (G,h), let $e_1,\dots,e_k$ be its edges. Then different values of the edge lengths $\ell(e_1),\dots,\ell(e_k)$ under the conditions $\ell(e_1),\dots,\ell(e_k) > 0$, $\sum_{i=1}^k \ell(e_i) = 1$, fill out an *open* $(k-1)$-simplex $\Sigma_{(G,h)}$. Its simplicial faces correspond to vanishing of some edge lengths $\ell(e_i)$. For any subset $\{e_{i_1},\dots,e_{i_r}\}$ of edges, the simplicial face of $\Sigma_{(G,h)}$ defined by $\ell(e_{i_1}) = \dots = \ell(e_{i_r}) = 0$ is contained in the outer space X_n if and only if no loop of G, i.e., a nontrivial element of $\pi_1(G)$, is contained in the union of these edges $e_{i_1},\dots,e_{i_r}$. (Note that if an edge length $\ell(e) = 0$, then the edge e is removed from the graph, and if the total length of a loop is equal to zero, the genus of the resulting graph will decrease.)

The outer space X_n is the union of open simplexes $\Sigma_{(G,h)}$, and its maximal dimensional simplices correspond to trivalent graphs (G,h). It can be proved by induction that such a trivalent graph G without vertices of valence 1 or 2 has $3n-3$ edges, and the normalization condition on the length gives a simplex $\Sigma_{(G,h)}$ of dimension $3n-4$.

There are several ways to put a topology on X_n. Probably the most direct one is to glue up the simplicial topology of these simplices. See [133] for more details and references on this and several other results on X_n recalled below.

An important result due to Culler-Vogtmann [42] is the following result.

Theorem 2.8. *The outer space X_n is contractible.*

The action of $\mathrm{Out}(F_n)$ is proper (see Proposition 2.15 below), but the quotient $\mathrm{Out}(F_n)\backslash X_n$ is noncompact. An explicit equivariant deformation S of X_n, a so-called spine, can be found such that S is a contractible simplicial complex of dimension $2n-3$ and the quotient $\mathrm{Out}(F_n)\backslash S$ is compact. One consequence of this result is that the virtual cohomology dimension of $\mathrm{Out}(F_n)$ is equal to $2n-3$ [42, p. 93].

Remark 2.9. In topology, different models of the universal space $\underline{E}\Gamma$ are unique up to equivariant homotopy equivalence. Usually people want to find models which are as explicit and small as possible. By Proposition 2.17 below, X_n is a model of $\underline{E}\Gamma$ for $\Gamma = \mathrm{Out}(F_n)$. This is the reason for defining outer space X_n by normalizing the total length of each metric graph to be 1. The space of all marked metric graphs of genus g without normalization, which will be identified with the tropical Teichmüller space $\mathcal{T}_n^{tr}$ of smooth compact tropical curves of genus n later, contains X_n as an equivariant deformation retract.

Simplicial metric on the outer space X_n

So far, we have been discussing topological properties of the outer space X_n. To understand its geometry, we need to define metrics on X_n. The most direct way to put a metric on X_n which is invariant under $\mathrm{Out}(F_n)$ is to glue the simplicial metrics of simplices $\Sigma_{(G,h)}$ together. Specifically, we can give each k-dimensional simplex $\Sigma_{(G,h)}$ the *standard simplicial metric* by identifying it with the standard simplex Δ_k in $\mathbb{R}^{k+1}$,

$$\Delta_k = \{(x_1,\dots,x_{k+1}) \in \mathbb{R}^{k+1} \mid x_1,\dots,x_k > 0, x_1+\dots+x_{k+1} = 1\},$$

whose edges have length $\sqrt{2}$. This metric extends and defines a metric on the closure of $\Sigma_{(G,h)}$ in X_n. It is also clear that for any collection of simplices with nonempty intersection, these metrics restrict to the same metric on their common simplicial faces. Therefore, it defines a *length function*, or a *length metric* on X_n, denoted by d_0.

Proposition 2.10. *The length metric d_0 on X_n is an incomplete metric on X_n which is invariant under* $\mathrm{Out}(F_n)$*, and its restriction to each simplex $\Sigma_{(G,h)}$ is the length metric induced from a Riemannian metric.*

Proof. Since all top dimensional simplices in X_n are isometric to the standard simplex Δ_n in $\mathbb{R}^{n+1}$, it is clear that the restriction to each simplex $\Sigma_{(G,h)}$ is the length metric of the flat Riemannian metric of the ambient Euclidean space $\mathbb{R}^k$, i.e., for every two points in $\Sigma_{(G,h)}$, there is no shortcut by going through neighboring simplices. Since the closure of each open simplex $\Sigma_{(G,h)}$ in X_n is noncompact and its diameter is equal to $\sqrt{2}$, d_0 is an incomplete metric. Since the action of $\mathrm{Out}(F_n)$ on X_n preserves the metrics of the simplices, the invariance of the length metric d_0 under $\mathrm{Out}(F_n)$ is clear. □

See [22, §1] for related discussion on how to glue up metrics on simplices.

We note that not every length space is a geodesic space, i.e., every two points are connected by a geodesic segment, which is an isometric embedding of an interval. Theorem 2.8 and the fact that all top dimensional simplices are isometric suggests the following conjecture.

Conjecture 2.11. *The length space (X_n, d_0) is a geodesic metric space.*

We also note that on a simply connected nonpositively curved complete Riemannian manifold, every two points are connected by a unique geodesic. This property is crucial for many applications, for example, for the validity of the Cartan fixed point theorem, which was used in the proof of Proposition 2.3. For an *incomplete* metric space, a replacement for this is the *geodesic convexity*: *a geodesic metric space is called geodesically convex if for every two points in the space, there exists a unique geodesic segment connecting them.* For a geodesically convex locally compact metric space, the Cartan fixed point theorem holds.

It also seems reasonable to make the following

Conjecture 2.12. *The length metric space (X_n, d_0) is geodesically convex.*

This is stronger than the result in Theorem 2.8 on the contractibility of X_n. An even stronger conjecture is

Conjecture 2.13. *The length metric space (X_n, d_0)* is a CAT(0)-space.

If this conjecture is true, then d_0 is a good analogue of the Weil-Petersson metric on Teichmüller space. Indeed, though the Weil-Petersson metric of Teichmüller space is incomplete, the Weil-Petersson metric is geodesically convex according to a result of Wolpert [136]. One might try to use this metric to understand elements of $\mathrm{Out}(F_n)$ as the Weil-Petersson metric was used in the paper [44]. On the other hand, the situation is more complicated since axes of fully irreducible outer automorphisms are not unique in general [11], [63], and [64].

Remark 2.14. Besides the simplicial metric defined above, each simplex $\Sigma_{(G,h)}$ of X_n also admits an analogue of Weil-Petersson metric [121]. It was motivated by the thermodynamical interpretation of the Weil-Petersson metric on the Teichmüller space in [98] and has the following properties: (1) for some graphs G, the metric on $\Sigma_{(G,h)}$ is incomplete, (2) and for some graphs, the curvature of the metric has both positive and negative signs. It is not obvious how these metrics can be glued into a metric on the whole outer space X_n and to get a complete metric on X_n.

2.3 Action of $\mathrm{Out}(F_n)$ on outer space X_n and applications

As mentioned before, the outer space X_n is an analogue of the Teichmüller space $\mathscr{T}_g$. Since the mapping class group Mod_g acts properly on $\mathscr{T}_g$, the following result is expected.

Proposition 2.15. *The natural action of* $\mathrm{Out}(F_n)$ *on the outer space* X_n *is proper.*

This follows from the fact that X_n is a locally finite simplicial complex, and for every simplex $\Sigma_{(G,h)}$ in X_n, its stabilizer in $\mathrm{Out}(F_n)$ is finite. In fact, the stabilizer of $\Sigma_{(G,h)}$ is isomorphic to a subgroup of the symmetry group S_N, where N is the number of vertices of the graph G.

Proposition 2.16. *Under the action of* $\mathrm{Out}(F_n)$ *on* X_n*, there are only finitely many orbits of open simplices* $\Sigma_{(G,h)}$.

Briefly, since the quotient $\mathrm{Out}(F_n)\backslash X_n$ is the moduli space of metric graphs of genus n, the $\mathrm{Out}(F_n)$-orbits of open simplices in X_n correspond to equivalence classes of finite graphs of genus n with only vertices of valence at least 3. Clearly, there are only finitely many such graphs of the fixed genus n up to equivalence.

The above results (Theorem 2.8, Proposition 2.15) suggest the following [135] and [85].

Proposition 2.17. *The outer space* X_n *is a model of* $\underline{E}\Gamma$ *for the group* $\Gamma = \mathrm{Out}(F_n)$, *i.e., a universal space for proper actions of* $\mathrm{Out}(F_n)$.

This is a direct analogue of Propositions 2.3 and 2.4, i.e., one important property shared by these three classes of group actions mentioned in §2.1. The analogy between them motivates us to suggest and prove properties of $\mathrm{Out}(F_n)$ based on those of arithmetic groups Γ and $\mathrm{Mod}_{g,n}$ in a systematic way. See [69] for a list of some properties of these groups.

On the other hand, as pointed out earlier, there are several differences between X_n, symmetric spaces X and $\mathscr{T}_{g,n}$. Besides the fact that the outer space X_n is not a manifold, there is *no known complete intrinsically defined geodesic metric on* X_n *which is invariant under* $\mathrm{Out}(F_n)$.

2.4 A natural approach to construct complete geodesic metrics on X_n

One natural method to construct an invariant metric on the outer space X_n is to find a metric space (Z,d) where $\mathrm{Out}(F_n)$ acts isometrically, and an embedding

$$i : X_n \hookrightarrow (Z,d)$$

equivariantly with respect to $\mathrm{Out}(F_n)$, and then to pull back the metric d.

Furthermore, if (Z,d) is a complete metric space and the image $i(X)$ is a complete metric subspace, then the pulled-back metric i^*d is an invariant complete metric on X_n.

On the other hand, even if (Z,d) is a geodesic metric, i^*d may not be a geodesic metric. One simple example which shows this is the embedding of the unit sphere S^{n-1} in the Euclidean space $\mathbb{R}^n$, where the pulled-back metric on S^{n-1} is the chord metric, which is not a geodesic metric on the sphere.

If X_n were a manifold, and (Z,d) is a Riemannian manifold, then an immersion $i : X_n \to Z$, which is not necessarily an embedding, will allow us to pull back the Riemannian metric tensor of Z to obtain a Riemannian metric tensor which defines a Riemannian distance on X_n.

One difficulty is that there is no obvious choice of *complete metric spaces* (Z,d) into which X_n can be *embedded equivariantly*.

Remark 2.18. We note that there is a natural embedding $X_n \to Y$ into a topological space Y, which is an infinite dimensional projective space, such that the closure is compact, but the ambient space Y is not a metric space. This embedding was one of the original ways to define the topology on X_n. See [43] and [133, §1.4] for details and references.

Instead, we will look for a complete metric space (Z,d) which admits an isometric action of $\mathrm{Out}(F_n)$ such that there exists an equivariant map $i : X_n \to (Z,d)$ satisfying the condition: *for any sequence of points $x_k \in X_n$ which converges to a boundary point as $k \to \infty$, i.e., a point in a missing simplex of X_n, the distance $d(i(x_k), i(x_0)) \to +\infty$ as $k \to \infty$, where $x_0 \in X_n$ is any fixed base point.*

In some sense, we give up the condition of being embedding and retain the completeness condition on the image. By combining it with the incomplete simplicial metric d_0 on X_n, we can construct complete geodesic metrics on X_n which are invariant under $\mathrm{Out}(F_n)$.

Given the above discussion on relations and analogy between the three classes of groups and their associated spaces: arithmetic groups Γ acting on symmetric spaces $X = G/K$, the mapping class group $\mathrm{Mod}_{g,n}$ acting on Teichmüller space $\mathscr{T}_g$, $\mathrm{Out}(F_n)$ acting on the outer space X_n, it is natural to explore metrics of symmetric spaces X and Teichmüller spaces $\mathscr{T}_g$ to construct complete metrics on X_n.

One purpose of this second part of the book is to show that such an equivariant map $i : X_n \to (Z,d)$ can be constructed by maps between the outer space X_n,

Teichmüller spaces, and symmetric spaces. Note that $\mathrm{GL}(n,\mathbb{R})$ is a reductive Lie group, $\mathrm{O}(n)$ is a maximal compact subgroup of $\mathrm{GL}(n,\mathbb{R})$, and the quotient space $\mathrm{GL}(n,\mathbb{R})/\mathrm{O}(n)$ with a $\mathrm{GL}(n,\mathbb{R})$-invariant Riemannian metric is a symmetric space of nonpositive curvature, though not of noncompact type.

Moduli spaces of flat tori

To identify $\mathrm{GL}(n,\mathbb{R})/\mathrm{O}(n)$ with a moduli space, we need to introduce the notion of *marked flat tori.* Similar to marked Riemann surfaces and marked graphs, we define

Definition 2.19. A marked flat torus of dimension n is a pair of a compact torus M^n with a flat Riemannian metric, and a diffeomorphism $\varphi : M^n \to \mathbb{Z}^n\backslash\mathbb{R}^n$, where φ is well-defined up to homotopy.

Proposition 2.20. *The symmetric space $X = \mathrm{GL}(n,\mathbb{R})/\mathrm{O}(n)$ can be identified with the space of positive definite quadratic forms in n variables, or equivalently, the space of marked compact flat tori of dimension n.*

Proof. First we note that $\mathrm{GL}(n,\mathbb{R})$ acts on the space of positive definite $n\times n$-matrices: for $g \in \mathrm{GL}(n,\mathbb{R})$ and a positive definite matrix A,

$$g\cdot A = gAg^t.$$

By the spectral decomposition of positive definite quadratic forms, we can prove that this action is transitive. Since the stabilizer of the identity matrix is equal to $\mathrm{O}(n)$, the first statement is proved.

To prove the second statement, we note that every positive definite quadratic form on $\mathbb{R}^n$ defines a flat Riemannian metric on $\mathbb{R}^n$ which descends to a flat Riemannian metric on the torus $\mathbb{Z}^n\backslash\mathbb{R}^n$. This torus $\mathbb{Z}^n\backslash\mathbb{R}^n$ has the marking giving by the identity map $\mathbb{Z}^n\backslash\mathbb{R}^n \to \mathbb{Z}^n\backslash\mathbb{R}^n$. Conversely, every marked flat torus (M^n,φ) of dimension n canonically induces a flat metric on $\mathbb{Z}^n\backslash\mathbb{R}^n$, which is lifted to a flat metric on $\mathbb{R}^n$, or equivalently a positive definite quadratic form on $\mathbb{R}^n$. Note that $\pi_1(M^n)\cong\mathbb{Z}^n$, and the marking φ induces an isomorphism $\mathbb{Z}^n\to\mathbb{Z}^n$. □

Corollary 2.21. *The group $\mathrm{GL}(n,\mathbb{Z})$ acts properly on the symmetric space X, and the quotient $\mathrm{GL}(n,\mathbb{Z})\backslash X$ is the isometry classes of compact flat tori of dimension n.*

Proof. Since $\mathrm{GL}(n,\mathbb{Z})$ is a discrete subgroup of $\mathrm{GL}(n,\mathbb{R})$, it acts properly on $X = \mathrm{GL}(n,\mathbb{R})/\mathrm{O}(n)$. To identify the quotient $\mathrm{GL}(n,\mathbb{Z})\backslash X$ with the set of isometry classes of flat metrics on $\mathbb{Z}^n\backslash\mathbb{R}^n$, we note that for any flat metric on $\mathbb{Z}^n\backslash\mathbb{R}^n$, a marking on $\mathbb{Z}^n\backslash\mathbb{R}^n$ corresponds to an isomorphism from $\mathbb{Z}^n$ to $\mathbb{Z}^n$, i.e., an element of $\mathrm{GL}(n,\mathbb{Z})$, and every marking arises in this way. When we identify the set of all flat metrics on $\mathbb{Z}^n\backslash\mathbb{R}^n$ and hence on $\mathbb{R}^n$ with X as in the previous proposition, then all different markings on a flat metric $\mathbb{Z}^n\backslash\mathbb{R}^n$ correspond to points in a $\mathrm{GL}(n,\mathbb{Z})$-orbit. □

Remark 2.22. In the above discussion of flat tori, we have fixed the underlying space $\mathbb{Z}^n\backslash\mathbb{R}^n$, and let flat metrics on it change. We can also fix the standard flat metric, i.e., the Euclidean metric, on $\mathbb{R}^n$, and change lattices $\Lambda \cong \mathbb{Z}^n$ so that we get flat tori $\Lambda\backslash\mathbb{R}^n$.

Periods of metric graphs

Given the above moduli space interpretation of $X = \mathrm{GL}(n,\mathbb{R})/\mathrm{O}(n)$, we can define a map

$$\Pi : X_n \to X, \tag{2.3}$$

which is equivariant with respect to the natural actions of $\mathrm{Out}(F_n)$ on X_n and $\mathrm{GL}(n,\mathbb{Z})$ on X, and the canonical homomorphism $\pi : \mathrm{Out}(F_n) \to \mathrm{GL}(n,\mathbb{Z})$ coming from the abelization of F_n.

For every marked metric graph (G,ℓ,h) in X_n, there is a canonical isomorphism

$$h_* : H_1(G,\mathbb{Z}) = F_n/[F_n,F_n] \cong \mathbb{Z}^n, \tag{2.4}$$

and hence an isomorphism $h_* : H_1(G,\mathbb{R}) \to \mathbb{R}^n$ by linear extension.

We note that this isomorphism $h_* : H_1(G,\mathbb{Z}) \cong \mathbb{Z}^n$ depends on the marking h of the graph. If we change the marking h by an element $\varphi \in \mathrm{Out}(F_n)$ to get an new marking φh, then φ projects to an element $\pi(\varphi) \in \mathrm{GL}(n,\mathbb{Z})$, which gives an automorphism of $\mathbb{Z}^n$, and the isomorphism $(\varphi h)_* : H_1(G,\mathbb{Z}) \cong \mathbb{Z}^n$ is obtained from h_* by composing with $\pi(\varphi)$.

The isomorphism $h_* : H_1(G,\mathbb{Z}) \cong \mathbb{Z}^n$ gives the compact torus $H_1(G,\mathbb{Z})\backslash H_1(G,\mathbb{R})$ a canonical marking. We will use the metric ℓ on the graph (G,ℓ,h) to define a flat metric on this torus, which is equivalent to a positive definite quadratic form Q on the real vector space $H_1(G,\mathbb{R})$, as expliained in the proof of Proposition 2.20.

Note that G is of dimension 1 and $H_1(G,\mathbb{R})$ is equal to the set of 1-cycles on G. Therefore, it suffices to define a quadratic form Q on the set $C_1(G,\mathbb{R})$ of all 1-chains on G and then to restrict it to the subspace of 1-cycles.

Fix an orientation for every edge e of G. Then every 1-chain σ on G can be written uniquely as $\sum_{e\in G} a_e e$, where $a_e \in \mathbb{R}$. For every two edges e,e' of G, define

$$Q(e,e') = \begin{cases} \ell(e), & \text{if } e = e', \\ 0, & \text{if } e \neq e'. \end{cases} \tag{2.5}$$

By extending it linearly, we obtain a quadratic form Q on $C_1(G,\mathbb{R})$ [34, p. 151].

Proposition 2.23. *The bilinear quadratic form Q on $C_1(G,\mathbb{R})$, and hence its restriction on the subspace $H_1(G,\mathbb{R})$, is positive definite.*

Proof. For any nontrivial 1-chain σ of G, write $\sigma = \sum_{e\in G} a_e e$, where $a_e \neq 0$ for at least one edge. It follows from the definition that

$$Q(\sum_{e\in G} a_e e, \sum_{e\in G} a_e e) = \sum_{e\in G} a_e^2 \ell(e) > 0.$$

□

In terms of the canonical identification $H_1(G,\mathbb{Z}) \cong \mathbb{Z}^n$ and $H_1(G,\mathbb{R}) \cong \mathbb{R}^n$, the quadratic form Q becomes a positive definite matrix of size $n \times n$. It is called the *period matrix* of the marked metric graph (G,ℓ,h). By Proposition 2.20, we obtain a *period map*

$$\Pi : X_n \to X = \mathrm{GL}(n,\mathbb{R})/\mathrm{O}(n), \quad (G,\ell,h) \mapsto Q. \tag{2.6}$$

By the above discussion on relation between the marking on the metric graph (G,ℓ,h) and the marking of the flat torus $H_1(G,\mathbb{Z}) \cong \mathbb{Z}^n$, it is clear that the period map Π is equivariant with respect to the canonical homomorphism $\pi : \mathrm{Out}(F_n) \to \mathrm{GL}(n,\mathbb{Z})$.

By Proposition 2.20 above again, this positive definite quadratic form Q induces a flat metric on the marked torus $H_1(G,\mathbb{Z})\backslash H_1(G,\mathbb{R})$. To be consistent with the Jacobian map of tropical curves below, we also denote the *period map* by

$$\Pi((G,\ell,h)) = (H_1(G,\mathbb{Z})\backslash H_1(G,\mathbb{R}), Q), \tag{2.7}$$

where the image is a marked compact flat torus.

We will explain below why it is called the period map and compare it with the period (or Jacobian) map of Riemann surfaces.

Remark 2.24. The definition of the quadratic form Q in Equation 2.5 might look puzzling, and may seem wrong. The reason is that the length of the chain e with respect to the metric defined by Q is equal to $\sqrt{\ell(e)}$, which is not a smooth function of the lengths of the metric graph (G,ℓ). From the above proof, it is clear that if we use $\ell(e)^2$ instead of $\ell(e)$ and define $\tilde{Q}(e,e) = \ell(e)^2$, then we also obtain a positive definite quadratic form on the torus $H_1(G,\mathbb{Z})\backslash H_1(G,\mathbb{R})$. This latter definition might look more natural, since the length of a chain e in this new metric would be equal to $\ell(e)$, which will depend smoothly on the metric ℓ of the metric graph (G,ℓ).

Remark 2.25. The proof of Proposition 2.23 shows that for every function $f(x)$ on $[0,+\infty)$ satisfying $f(x) > 0$ for $x > 0$, we can obtain a positive quadratic form Q_f on $H_1(G,\mathbb{R})$ by defining

$$Q_f(e,e) = f(\ell(e))$$

as in Equation 2.5 and hence an equivariant map

$$\Pi_f : X_n \to \mathrm{GL}(n,\mathbb{R})/\mathrm{O}(n).$$

To explain that the function $f(x) = x$ is the right choice, we need to go through tropical geometry. The definition of the quadratic form Q in Equation 2.5 turns out to be the right choice when $(H_1(G,\mathbb{Z})\backslash H_1(G,\mathbb{R}),Q)$ is viewed as the *tropical Jacobian variety*, a principally polarized tropical abelian variety, of the marked *compact smooth tropical curve* canonically associated with the marked metric graph (G,ℓ,h).

Remark 2.26. Another important point is that the quadratic form Q is a *linear function* when it is restricted to each simplex of the outer space X_n and hence it has the characteristic property of tropical maps in tropical geometry. This is one of the motivations for us to describe some basic notions in tropical geometry in this paper. We will see later this the geometry of the outer space X_n will be useful in studying the moduli space $\mathcal{M}_n^{tr}$ of compact smooth tropical curves. In fact, the quadratic form Q is linear on each simplex gives a natural local tropical structure on the tropical Teichmüller space $\mathcal{T}_n^{tr}$ and the structure of a tropical orbifold on $\mathcal{M}_n^{tr}$. See Proposition 8.25 below.

To solve Problems 1.1 and 1.3 on defining complete invariant metrics on X_n, there is one issue with using the period map Π in Equation 2.7 above. When $n \geq 3$, the homomorphism $\mathrm{Out}(F_n) \to \mathrm{GL}(n,\mathbb{Z})$ has an infinite kernel. Since the map Π is equivariant with respect to the homomorphism $\mathrm{Out}(F_n) \to \mathrm{GL}(n,\mathbb{Z})$, it cannot be injective. But as we will show below (Corollary 11.10), Π is not even injective locally when $n \geq 3$, i.e., the induced quotient map

$$\Pi : \mathrm{Out}(F_n)\backslash X_n \to \mathrm{GL}(n,\mathbb{Z})\backslash \mathrm{GL}(n,\mathbb{R})/\mathrm{O}(n)$$

is not injective, unlike the period map of Riemann surfaces in Equation 9.3 below. When it is restricted to each simplex in X_n, the period map Π is a differential map, but it is not injective or is an immersion in general. Though $\mathrm{GL}(n,\mathbb{R})/\mathrm{O}(n)$ is a symmetric space and has an invariant Riemannian metric, we cannot use this map to pull back and define a metric on X_n.

To overcome this difficulty, one simple and crucial idea is to make use of the simplicial metric d_0 on X_n in Proposition 2.10 as well and combine it with the pulled-back semi-metric by the period map Π. Several metrics are constructed in this way in §12.1 and §12.2. In §12.3, we also view d_0 as an analogue of the Weil-Petersson metric of the Teichmüller space $\mathcal{T}_g$ and follow the construction of the McMullen metric on Teichmüller spaces [97] in order to construct another invariant complete geodesic metric on X_n.

Remark 2.27. The above brief discussion on constructing complete invariant on the outer space X_n is one concrete example that shows the fruitful interaction among the three kind of spaces: symmetric space X, the Teichmüller space $\mathcal{T}_{g,n}$, and the outer space X_n.

3 Tropical semifields and tropical polynomials

Due to diverse applications of tropical geometry and unusual nature of tropical varieties (in comparison with algebraic varieties over the complex numbers and other fields), there have been many introductions and surveys of tropical geometry. Here are some [27], [28] [30], [124], [99], [100], [51], [94], [19], and [77]. There are also two recent books on tropical geometry [95], [106].

Briefly, the tropical geometry is the algebraic geometry over the tropical semifield $\mathbb{T}$, whose operations are rather different from the usual ones for real and complex numbers. We start with the definition and properties of $\mathbb{T}$.

3.1 The tropical semifield $\mathbb{T}$

The tropical semifield is $\mathbb{T} = \mathbb{R} \cup \{-\infty\}$ with two operations: It has an addition

$$\text{“}x + y\text{”} = \max\{x, y\},$$

and a multiplication

$$\text{“}x \cdot y\text{”} = x + y.$$

It is clear from the definition that $-\infty$ is the *additive zero*: for every $x \in \mathbb{T}$,

$$\text{“}x + (-\infty)\text{”} = x,$$

and 0 is the *multiplicative identity*: for every $x \in \mathbb{T}$,

$$\text{“}0 \cdot x\text{”} = x.$$

Note that $-\infty$ is also the *multiplicative zero*, since

$$\text{“}x \cdot (-\infty)\text{”} = -\infty.$$

But the tropical semifield does not have a subtraction, since we cannot recover x from “$x + y$” $= \max\{x, y\}$ and y when $x < y$. We also note that the *idempotent property*, which says that for every $x \in \mathbb{T}$,

$$\text{“}x + x\text{”} = x,$$

prevents the existence of subtraction.

The division

$$\text{“}x/y\text{”} = x - y$$

is defined when y is not equal to the multiplicative zero, i.e., $y \neq -\infty$. This corresponds to the usual rule that the division is defined when the denominator is not equal to zero.

It is also clear that these operations on $\mathbb{T}$ are commutative and satisfy the distribution law. Therefore, $\mathbb{T}$ is a semifield. If we ignore the division, then $\mathbb{T}$ is a semiring.

Remark 3.1. Such semirings have been studied in computer science and are often used to prove various finiteness results such as Burnside-type problems in groups, and semigroups and decidability problems in languages. See [119] for an introduction and references.

Though people are used to fields, semifields occur naturally. For example, $\mathbb{R}_{\geq 0}$ with the usual addition $+$ and multiplication $\cdot$ is probably the most familiar semifield. The restriction to nonnegative real numbers excludes the existence of substraction on $\mathbb{R}_{\geq 0}$.

It turns out that $(\mathbb{T},+,\cdot)$ is a degeneration of $(\mathbb{R}_{\geq 0},+,\cdot)$. Specifically, for any $t>1$, define a bijective map $\log_t$ by

$$\log_t : \mathbb{R}_{\geq 0} \to \mathbb{T}, \quad a \mapsto \frac{\log a}{\log t}.$$

Note that 1 is mapped to 0 and 0 is mapped to $-\infty$. The converse is given by

$$\mathbb{T} \to \mathbb{R}_{\geq 0}, \quad x \to t^x.$$

Using this map, the two operations on $\mathbb{R}_{\geq 0}$ are transferred to $+_t$ and $\cdot_t$ on $\mathbb{T}$: for any $x,y \in \mathbb{T}$,

$$x +_t y = \frac{\log(t^x + t^y)}{\log t}, \quad x \cdot_t y = \frac{\log(t^x \cdot t^y)}{\log t}.$$

Therefore, for each $t>1$, we have obtained a semifield $(\mathbb{T},+_t,\cdot_t)$ which is isomorphic to the classical semifield $(\mathbb{R}_{\geq 0},+,\cdot)$. It is clear that for all $x,y \in \mathbb{T}$,

$$x \cdot_t y = x + y = \text{“}x \cdot y\text{”},$$

and

$$\max\{x,y\} \leq x +_t y \leq \max\{x,y\} + \frac{\log 2}{\log t}.$$

Therefore, the multiplications in $(\mathbb{T},+,\cdot)$ and $(\mathbb{T},+_t,\cdot_t)$ are the same. When $t \to +\infty$, for $x,y \in \mathbb{R}$,

$$x +_t y \to \max\{x,y\} = \text{“}x + y\text{”}.$$

This implies that as $t \to +\infty$, the addition of $(\mathbb{T},+_t,\cdot_t)$ converges to the addition of $\mathbb{T}$ and hence $(\mathbb{T},+_t,\cdot_t)$ converges to the tropical semifield $(\mathbb{T},+,\cdot)$

Because of this, the classical semifield $(\mathbb{R}_{\geq 0},+,\cdot) \cong (\mathbb{T},+_t,\cdot_t)$ can be considered as a *deformation* (or *quantization*) of the tropical semifield $(\mathbb{T},+,\cdot)$, and the tropical semifield $(\mathbb{T},+,\cdot)$ is a *dequantization* of the classical semifield $(\mathbb{R}_{\geq 0},+,\cdot)$.

3.2 Tropical polynomials

We note that in defining polynomials, we only use two operations: a commutative addition and a commutative multiplication. Therefore, we can define polynomials over the tropical semifield $\mathbb{T}$.

It is clear that for any number of variables $x_1, \ldots, x_n$, the polynomials in $x_1, \ldots, x_n$ with coefficients in $\mathbb{T}$ form a semiring, denoted by $\mathbb{T}[x_1, \ldots, x_n]$. These tropical polynomials satisfy *the idempotent property*: for any tropical polynomial $p(x_1, \ldots, x_n)$,

$$\text{“}p(x_1, \ldots, x_n) + p(x_1, \ldots, x_n)\text{”} = p(x_1, \ldots, x_n).$$

For example,

$$\text{“}\frac{1}{2}x + \frac{1}{2}x\text{”} = (\text{“}\frac{1}{2} + \frac{1}{2}\text{”})x = \frac{1}{2}x,$$

instead of the familiar addition of polynomials over $\mathbb{C}$, for example,

$$\frac{1}{2}x + \frac{1}{2}x = x.$$

Polynomials over $\mathbb{C}$ and $\mathbb{R}$ define functions on $\mathbb{C}^n$ and $\mathbb{R}^n$. Tropical polynomials also define functions on $\mathbb{T}^n$. We denote the function defined by a tropical polynomial $p(x)$ also by $p(x)$. Usually, we restrict the domains of these polynomials to the subset $\mathbb{R}^n \subset \mathbb{T}^n$.

For example, for a tropical polynomial in two variables

$$p(x,y) = \text{“}ax^2 + bxy + cy^2\text{”},$$

when it is restricted to $\mathbb{R}^2 \subset \mathbb{T}^2$, the function defined by $p(x,y)$ is the maximum of linear functions,

$$p(x,y) = \max\{2x + a, x + y + b, 2y + c\},$$

and the graph of this function is a piecewise linear surface in $\mathbb{R}^3$.

For polynomials over $\mathbb{C}$, the usual convention is that the coefficient 1 is omitted from nontrivial powers of x and y. For example, the polynomial $x^2 + y^2 + 1$ is an abbreviation of $1 \cdot x^2 + 1 \cdot y^2 + 1$.

Similarly, since 0 is the multiplicative identity in $\mathbb{T}$, the tropical polynomial such as “$x + y + 0$” means “$0 \cdot x + 0 \cdot y + 0$”. We note that 0 cannot be dropped from the expression “$x + y + 0$” since the tropical polynomial “$x + y + 0$” is equal to

$$\max\{x, y, 0\},$$

while the tropical polynomial “$x + y$” is different and represents another function

$$\max\{x, y\}.$$

Over $\mathbb{C}$, we drop monimial terms with zero coefficients such as $0xy$. Similarly, for a tropical polynomial "$\sum_{ij} a_{ij}x^iy^j$", monimials "$(-\infty)x^iy^j$" are dropped, and all present coefficients satisfy the condition: $a_{ij} \neq -\infty$.

Therefore, the above discussions show that tropical polynomials are similar to usual polynomials over $\mathbb{R}$ and $\mathbb{C}$ as far as formal computations are concerned. On the other hand, there is one important point which is special about tropical polynomials: *two different tropical polynomials can represent the same function.*

This does not happen to polynomials over $\mathbb{R}$ or $\mathbb{C}$ since polynomials over $\mathbb{C}$ (or $\mathbb{R}$) are completely determined by their values over $\mathbb{C}$ (or $\mathbb{R}$), or even by their values at sufficiently finitely many points.

For example, take the tropical polynomials $p_1(x) =$ "x^2+0", and $p_2(x) =$ "x^2+x+0". They are clearly different tropical polynomials. But for all $x \in \mathbb{T}$, they take the same value:

$$p_2(x) = \max\{2x, x, 0\} = \max\{2x, 0\} = p_1(x).$$

Similarly, the tropical polynomials "$x^2+y^2+x+y+1$" and "x^2+y^2+1" represent the same function on $\mathbb{T}^2$.

Roots of polynomials are basic notions and invariants of polynomials. Before the 19th century, algebra basically meant solutions of polynomial equations. We next consider roots of tropical polynomials.

To start, we consider the root of a tropical linear polynomial "$x+b$". Since $-\infty$ is both the additive and multiplicative zero, the most obvious or direct definition for root of the polynomial $x+b$ is a solution of "$x+b$" $= -\infty$. But it does not have a solution when $b \neq -\infty$.

Remark 3.2. We note that for usual polynomials over $\mathbb{C}$, the root of $x+b$ is $x = -b$, which is obtained by subtracting b from both sides of $x+b=0$. The same procedure cannot be applied to $\mathbb{T}$ since there is no substraction in $\mathbb{T}$.

We will need to *use the geometry of the graph of a polynomial* p to define the roots of the polynomial p. We start with linear polynomials.

The graph of the function "$x+b$" $= \max\{x, b\}$ is a union of two line segments with a corner at $x = b$, or equivalently $x = b$ is the unique nonsmooth point of this piecewise linear function (Fig. 1).

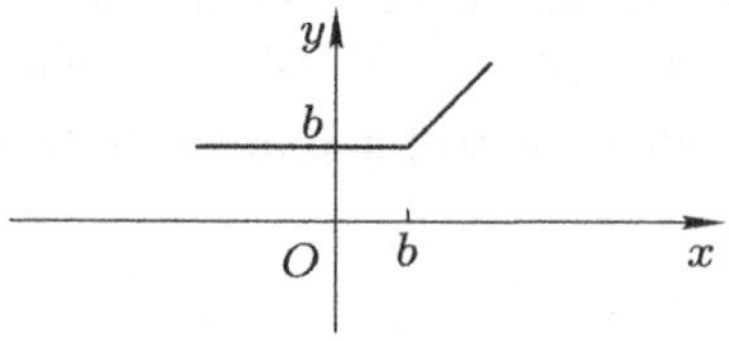

Fig. 1 a corner point

Clearly, this is the most special or important point of this function or its graph. Because of this, we can define the *root* of the tropical linear function "$x+b$" to be $x=b$.

Being special is probably not enough reason for defining it to be the root. How to justify this definition of the root of a tropical linear function?

Recall for any polynomial $p(x)$ over $\mathbb{C}$, x_0 is a root of $p(x)$ if and only if $1/p(x)$ is not regular (or smooth) at $x=x_0$. Similarly, over $\mathbb{T}$, it is reasonable to define that *x_0 is a root of a tropical polynomial $p(x)$ if and only if "$0/p(x)$" is not regular at $x=x_0$.* (Recall that 0 is the multiplicative identity in the tropical semifield.)

Now we need to define regular functions over $\mathbb{T}$. Certainly tropical polynomials are considered as regular functions over $\mathbb{T}$. They are piecewise linear convex functions with nonnegative integral coefficients in front of x. On the other hand, the constant term may not be an integer.

We note that *tropical Laurent polynomials* also give rise to convex piecewise linear functions. For example, since x^{-2} stands for "$0\cdot x^{-2}$" $=$ "$\frac{0}{x^2}$" $=0-2x=-2x$, the tropical Laurent polynomial

$$p(x) = \text{“}x^{-2}+2x^{-1}+3x+4\text{”}$$

gives rise to the function

$$\max\{-2x,-x+2,x+3,4\}.$$

This is a convex piecewise linear function.

The above discussion shows that *tropical Laurent polynomials* also give rise to piecewise linear convex functions with integral coefficients in front of x, and it is natural to consider them to be *regular functions* on $\mathbb{T}$ as well.

Definition 3.3. A tropical rational function is called regular or smooth at a point if its value is equal to the value of a Laurent polynomial in a neighborhood of the point.

For the tropical polynomial $p(x)=$ "$x+b$", we can check that "$0/p(x)$" $=-\max\{x,b\}$ is not regular at x_0 if and only if $x_0=b$. Note that b is the only strictly convex point of the function $\max\{x,b\}$, i.e., $-\max\{x,b\}$ is not locally convex at $x=b$. This explains the above definition of root of "$x+b$" to be $x=b$.

After we have defined the root of a tropical linear function, we can now give the general definition of roots of polynomials.

Definition 3.4. For any tropical polynomial $p(x)$, its roots are defined to be the nonsmooth points of the piecewise linear function defined by $p(x)$ over $\mathbb{R}$, i.e., the corner points of its graph, or equivalently the non-regular point of the function "$0/p(x)$".

We can also define the *multiplicity* of a root of a polynomial. It is clear that b should be a root of "$x+b$" of multiplicity (or order) 1, and for any positive integer

n, b should be a root of "$(x+b)^n$" with multiplicity n. We need to translate them into geometric conditions on the graphs of their associated functions.

For the function "$(x+b)^n$", in terms of piecewise linear functions, $x=b$ occurs at the corner point of $\max\{nx, nb\}$. The slopes of the graph of the piecewise linear function "$(x+b)^n$" jumps by n at the root $x=b$ (Fig. 2).

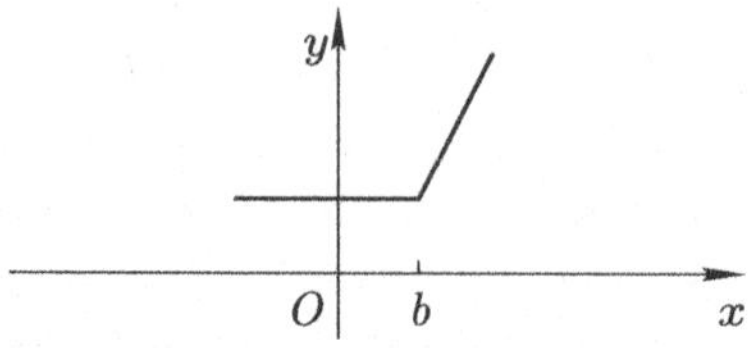

$b=1, n=2$

Fig. 2 Jump of slopes

This motivates the following definition.

Definition 3.5. Let b be a root of a tropical polynomal $p(x)$. It is a root of multiplicity $n \geq 1$, if for x in a neighborhood of b, the value of the polynomial $p(x)$ is equal to $\max\{nx, nb\} + \alpha x + \beta$ for some α, β, i.e., the slope of the piecewise linear function represented by $p(x)$ jumps by n at $x = b$.

There are some other differences with polynomials over $\mathbb{R}$ and $\mathbb{C}$. For example, any monomial $p(x) =$ "ax^n" has no root in $\mathbb{R}$. In this case, it seems to be natural to define that $-\infty$ is a root of multiplicity n, since $-\infty$ is the zero of $\mathbb{T}$. This is consistent with the situation over $\mathbb{R}$ and $\mathbb{C}$, where the only root of a monomial is 0. We also note that the quadratic tropical polynomial $p(x) =$ "$x^2 + x$" has only one root $x = 0$ and it has multiplicity 1.

The following result shows $\mathbb{T}$ is algebraically closed in some sense.

Proposition 3.6. *If a tropical polynomial $p(x)$ is not a monomial, then it has at least one root in $\mathbb{R}$. Furthermore, if x_0 is a root of multiplicity k, then $p(x)$ admits a factorization*

$$p(x) = \text{“}(x+x_0)^k q(x)\text{”},$$

where $q(x)$ is a polynomial which does not have x_0 as a tropical root. The tropical polynomial $p(x)$ admits a factorization

$$p(x) = \text{“}ax^{k_0}(x+x_1)^{k_1}\dots(x+x_m)^{k_m}\text{”},$$

where $k_0, \dots, k_m$ are positive integers, and $x_1, \dots, x_m$ are finite real numbers.

The proof follows from the fact that the graph of $p(x)$ is a piecewise linear convex curve with finitely many corner points, and each of the line segments has a nonnegative integral slope (Fig. 3). As we move from the left to the right, the slopes are given by increasing nonnegative integers. Let $x_1, \dots, x_m$ be x-coordinates of the corner points of the graph of $p(x)$. Assume that the slope of the line segments is increased by k_i at the point x_i. Then the graph can be built up by adding a linear function $k_i x$ starting at x_i until the next corner point x_{i+1}, i.e., by adding a function of the form $\max\{k_i x, k_i x_i\}$. The function $\max\{k_i x, k_i x_i\}$ corresponds to a tropical polynomial "$(x+x_i)^{k_i}$" and hence to a tropical root at x_i of multiplicity k_i. Note that the convexity of the curve implies that $k_i > 0$.

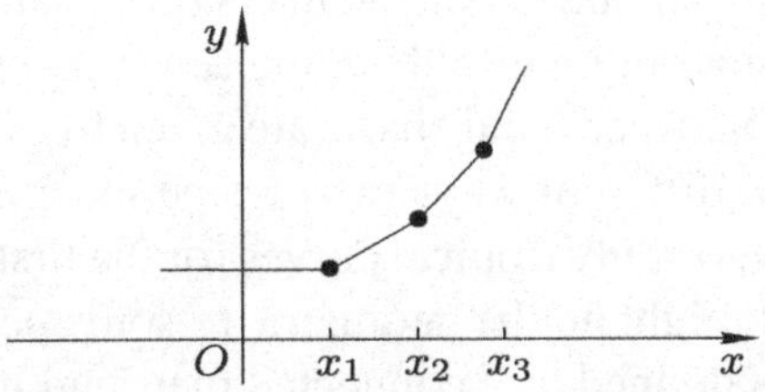

Fig. 3 A piecewise linear convex curve with jumps at the roots

Remark 3.7. As pointed out already, a tropical polynomial of degree n may not have n roots in $\mathbb{R}$ counted with multiplicity. As pointed out above, for a tropical monomial, it is natural to define $-\infty$ to be a root. If we include $-\infty$, then every tropical polynomial of degree n has n tropical roots in $\mathbb{T}$ counted with multiplicity. For example, the quadratic tropical polynomial $p(x) =$ "$x^2 + x$" has two roots $x = 0, -\infty$, each of multiplicity 1. Then this will be an exact analogue of the fundamental theorem of algebra, i.e., on complex roots of polynomials with complex coefficients.

On the other hand, the above proposition corresponds to the fact that every non-monomial polynomial over $\mathbb{C}$ has at least one *nonzero* root.

4 Tropical curves

Metric graphs have played a fundamental role in understanding properties of the outer automorphism group $\mathrm{Out}(F_n)$. Though metric graphs can be identified with smooth tropical curves, it takes some unwrapping to see this connection.

After explaining some basic definitions and properties of tropical curves, we will describe explicitly the 1-1 correspondence between the set of metric graphs without vertices of valence 1 or 2 and the set of *smooth* tropical curves. In the following, we will assume that all graphs have no vertices of valence 1 or 2 unless indicated otherwise.

This relation between metric graphs and tropical curves allows us to explain why the definition of the positive definite quadratic form Q in Equation 2.5 for a metric graph is natural. This also allows us to study the moduli space $\mathcal{M}_n^{tr}$ of compact smooth tropical curves of genus n using the outer space X_n, or rather the tropical Teichmüller space $\mathbb{T}_n^{tr}$. For example, this will be used in studying compactifications of $\mathcal{M}_n^{tr}$.

One reason why tropical curves are confusing to some people, in particular to the author of this second part of the book, is that one often uses plane tropical curves as examples of tropical curves. But these plane tropical curves are *singular in general* and do not correspond bijectively to metric graphs. As we shall see, only *smooth tropical curves* correspond canonically to metric graphs.

Definitions of abstract smooth tropical curves (and manifolds) basically follow the usual definition of smooth manifolds in the sense that they have local models and transition functions between local charts are given by $\mathbb{Z}$-affine transformations.

But there are several different ways to describe local models and they can be complicated to people who study tropical curves for the first time. For example, the local model in [107, §3] might not be geometric to some people. In [67, Definition 1.15], a tropical curve is defined as a connected metric graph.

We will follow the approach in [101] together with some ideas and statements in [99], [100]. There are other approaches, for example in [122], [51], [94], [95], [27], [106], some of which will be discussed below.

4.1 Plane tropical curves Γ_p

It is well-known that for any polynomial $p(x,y)$ with coefficients in $\mathbb{R}$, the equation $p(x,y) = 0$ defines a plane curve in $\mathbb{R}^2$, which could be empty or consist of only finitely many points. On the other hand, a nonzero complex polynomial defines an algebraic curve in $\mathbb{C}^2$, which is a subvariety of complex dimension 1. Naturally, we expect that tropical polynomials define tropical curves in $\mathbb{T}^2$. To avoid complication with the value $-\infty$, we will concentrate on points of the tropical curves inside $\mathbb{R}^2$.

Given a tropical polynomial $p(x,y)$, a direct generalization of a complex plane curve is to define the associated tropical curve by the locus of the equation

$$p(x,y) = -\infty.$$

(Recall that $-\infty$ is the tropical additive 0). But this has a serious problem of having no or too few points. For example, for a tropical linear polynomial $p(x,y) =$ "$x+y+0$", $p(x,y) = -\infty$ has no solution in $\mathbb{T}^2$, since

$$\text{“}x+y+0\text{”} = \max\{x,y,0\} \neq -\infty$$

for all x,y; and for the tropical polynomial $p(x,y) =$ "$x+y$", the equation $p(x,y) = -\infty$ has only one solution $(-\infty,-\infty)$ in $\mathbb{T}^2$, but no point in $\mathbb{R}^2$.

This difficulty can be overcome as in the case of defining roots of polynomials. We note that any tropical Laurent polynomial $p(x,y) =$ "$\sum_{ij} a_{ij}x^i y^j$" represents a convex piecewise linear function with integral coefficients of x and y, i.e., for any $x,y \in \mathbb{R}$,

$$\text{“}\sum_{ij} a_{ij}x^i y^j\text{”} = \max_{ij}\{ix + jy + a_{ij}\}.$$

The only difference between a tropical polynomial and a tropical Laurent polynomial is signs of the coefficients of x and y. Therefore, functions on $\mathbb{R}^2$ defined by tropical polynomials and tropical Laurent polynomials can both be regarded as regular tropical functions on $\mathbb{R}^2$.

Remark 4.1. It might also be helpful to remember that every Laurant polynomial over $\mathbb{C}$ is regular outside the zero, and hence it is reasonable to expect that a tropical Laurent polynomial is also regular outside the zero, which is equal to $-\infty$.

Similarly, a tropical rational function in x, y is called *regular* at a point if its value in a neighborhood of the point is equal to the value of a Laurent polynomial.

For a usual polynomial $p(x,y)$ over $\mathbb{C}$, the plane curve $p(x,y) = 0$ in $\mathbb{C}^2$ can also be defined to consist of points (x,y) where the function $1/p(x,y)$ is not regular.

For a tropical polynomial $p(x,y)$, we can similarly define *its plane tropical curve* in $\mathbb{R}^2$ to consist of points in $\mathbb{R}^2$ where the quotient

$$\text{“}0/p(x,y)\text{”} = -p(x,y)$$

is not regular. (Recall that 0 is the tropical multiplicative identity.) See [67, pp. 10-11] for more details.

Definition 4.2. Given any tropical polynomial $p(x,y)$ which is not a monomial, denote the plane tropical curve defined by p by Γ_p.

This means that points of the plane tropical curve Γ_p of $p(x,y)$ correspond to points where the value $p(x,y) = \max_{ij}\{ix + jy + a_{ij}\}$ is achieved by at least two linear functions.

It is helpful to note that in a small neighborhood of a point where two different linear functions f and g achieve the maximum value, the function $p(x,y)$ is strictly convex, and hence the function

$$-p(x,y) = -\max\{f,g\} = \inf\{-f,-g\}$$

is strictly concave down and not equal to any tropical Laurent polynomial.

Remark 4.3. The plane tropical curve Γ_p of a polymonial p can also directly be defined as the points (x,y) in $\mathbb{R}^2$ where the graph of p is not smooth.

Tropical lines

Since a line defined by a linear function in x, y with integral coefficients has a rational slope, the following result is clear from the definition.

Proposition 4.4. *For any tropical polynomial $p(x,y)$, its tropical plane curve Γ_p in $\mathbb{R}^2$ is a piecewise linear curve with rational slopes.*

Though the slope of each line segment is rational, its endpoints are not necessarily rational. For example, consider the tropical linear function

$$p(x,y) = \text{“}\sqrt{2}x + \sqrt{3}y + 0\text{”}.$$

Its plane curve Γ_p consists of three rays out of the vertex $(-\sqrt{2}, -\sqrt{3})$ (Fig. 4), one in each of the directions $(-1,0)$, $(0,-1)$, $(1,1)$.

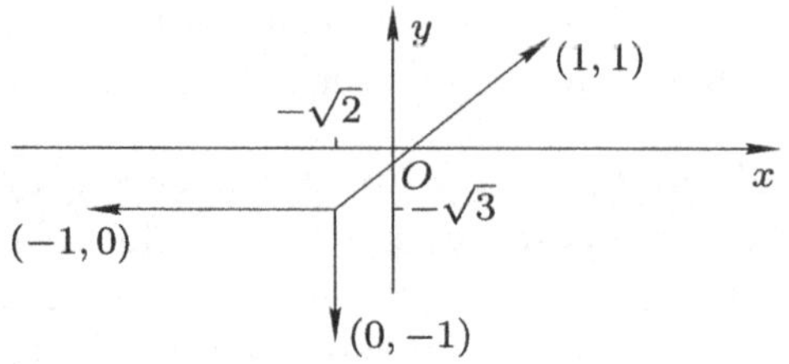

Fig. 4 a tropical line with vertex at $(-\sqrt{2}, -\sqrt{3})$

The analogy with complex curves suggests that when p is a linear tropical polynomial, the corresponding tropical curve Γ_p should be called a *tropical line.*

Clearly, lines with rational slopes are tropical lines defined by tropical linear functions without constant terms. Specifically, for $p(x,y) = \text{“}ax+by\text{”}$, the value $\max\{a+x, b+y\}$ is achieved by both of them when $a+x = b+y$, which gives the usual straight line $y = x + (a-b)$ in $\mathbb{R}^2$.

For a special linear tropical linear functions with nonzero constant terms, their tropical lines are also straight lines. For example, when $p(x,y) = \text{“}ax+c\text{”} = \max\{a+x, c\}$, its plane curve is the straight line $x = c-a$ in the usual sense.

But most tropical lines contain more than one line segments. For example, when $p = \text{“}x+y+0\text{”}$, the tropical curve consists of three rays from the origin: the half negative x-axis, the half negative axis, and the ray in the direction $(1,1)$. In some sense, this is a typical tropical curve (Fig. 5).

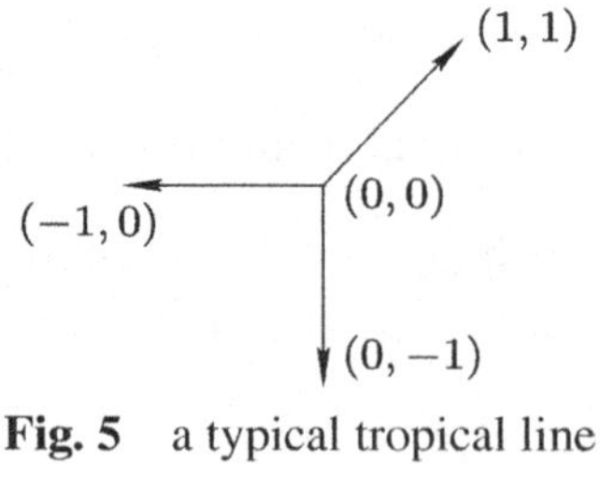

Fig. 5 a typical tropical line

In fact, the following proposition can be proved directly from definition.

Proposition 4.5. *For any linear tropical polynomial* $p(x,y) = \text{“}ax+by+c\text{”}$ *with none of* a, b, c *equal to* $-\infty$*, its plane curve* Γ_p *consists of three rays out of the point*

$(c-a,c-b)$ in the directions $(-1,0),(0,-1),(1,1)$, i.e., a shift of the tropical line defined by "$x+y+0$" by moving the origin $(0,0)$ to $(c-a,c-b)$.

Tropical quadratic curves

The situation for tropical quadratic polynomials is more complicated. We will consider some examples first.

Example 4.6. The quadratic tropical polynomial $p(x,y) =$ "$2x^2+2y^2+3xy+2x+2y+0$" is a generic one in certain sense, since all monomials are present. The plane tropical curve Γ_p defined by p contains two rays in the direction $(-1,0)$, whose y coordinates are the tropical roots of "$2y^2+2y+0$", i.e., $y=-2$, $y=0$. Similarly, Γ_p contains two rays in the direction $(0,-1)$, whose x coordinates are the tropical roots of "$2x^2+2x+0$", i.e., $x=-2,0$, and two more rays in the direction $(1,1)$ given by $2x+2=x+y$ and $2y+2=x+y$ (Fig. 6).

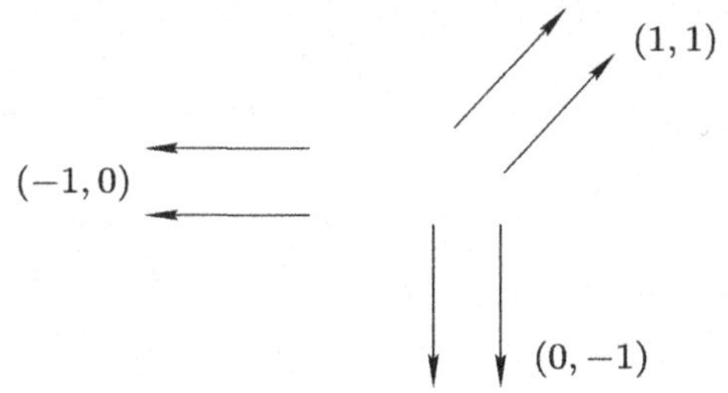

Fig. 6 two rays going to infinity in each direction

To see the above conclusions, we note that for any fixed y, when $x \ll 0$,

$$\text{"}2x^2+2y^2+xy+2x+2y+0\text{"} = \text{"}2y^2+2y+0\text{"} = \max\{2+2y, 2+y, 0\}.$$

This polynomial in y has two roots: $y=0$ from the equality

$$2+2y=2+y,$$

and $y=-2$ from the equality

$$2+y=0.$$

They correspond to the y-coordinates of the rays in the direction $(-1,0)$. (Note that the third case $2+2y=0$ does not occur, since for $y=-1$, the maximum is achieved by $2+(-1)=1>0$.) The rays in the direction $(0,-1)$ can also be obtained in the same way.

Since

$$\text{"}2x^2+2y^2+3xy+2x+2y+0\text{"} = \max\{2+2x, 2+2y, 3+x+y, 2+x, 2+y, 0\},$$

when $x \gg 0$ and y is bounded from above, the maximum cannot be realized by two terms, and hence (x,y) will not lie on the tropical curve. Similarly, no point (x,y) with $y \gg 0$ and x bounded from above will lie on the tropical curve.

When $x \gg 0, y \gg 0$, the value of $p(x,y)$ can potentially be achieved by any one of the three terms $2x+2, 2y+2, 3+x+y$. This gives the two rays in the direction $(1,1)$ determined by the equality

$$2x+2 = 3+x+y,$$

or the equality

$$2y+2 = 3+x+y.$$

The case $2x+2 = 2y+2$ can not happen since the maximum will be taken by $3+x+y$ under this condition.

Example 4.7. If we change the polynomial slightly to "$2x^2+2y^2+xy+2x+2y+0$", it will only have one ray in the direction $(1,1)$ of *multiplicity* 2. The reason is that when $x \gg 0, y \gg 0$, the maximum of $2x+2, 2y+2, x+y$ can only be achieved by $2x+2$ or $2y+2$, and their equality gives a line segment defined by the equation

$$2x = 2y.$$

The equation $x = y$ gives a ray of multiplicity 1, but the equation $2x = 2y$ gives a ray of multiplicity 2 (Fig. 7).

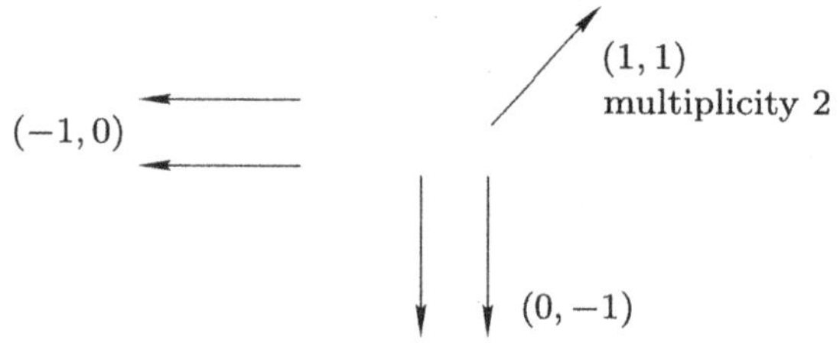

Fig. 7 only one ray in the direction $(1,1)$ with multiplicity 2

Example 4.8. The curve defined by the tropical quadratic polynomial $p(x,y) =$ "$2x^2 + 2y^2 + 3xy + x + y + 0$" will have only one ray in the direction $(-1,0)$, counted with multiplicity 2, one ray in the direction $(0,-1)$, counted with multiplicity 2, and two rays in the direction $(1,1)$ (Fig. 8).

By definition,

$$\text{“}2x^2+2y^2+3xy+x+y+0\text{”} = \max\{2+2x, 2+2y, 3+x+y, x, y, 0\}.$$

When $x \ll 0$,

$$\max\{2+2x, 2+2y, 3+x+y, x, y, 0\} = \max\{2+2y, y, 0\} = \max\{2+2y, 0\},$$

and the tropical polynomial "$2y^2+0$" $= \max\{2+2y, 0\}$ has a root $y = -1$ of multiplicity 2, which gives a ray in the direction $(-1,0)$ with multiplicity 2. The same reasoning gives a ray in the direction $(0,-1)$ with multiplicity 2. The ray in the direction $(1,1)$ can be determined as in the previous example.

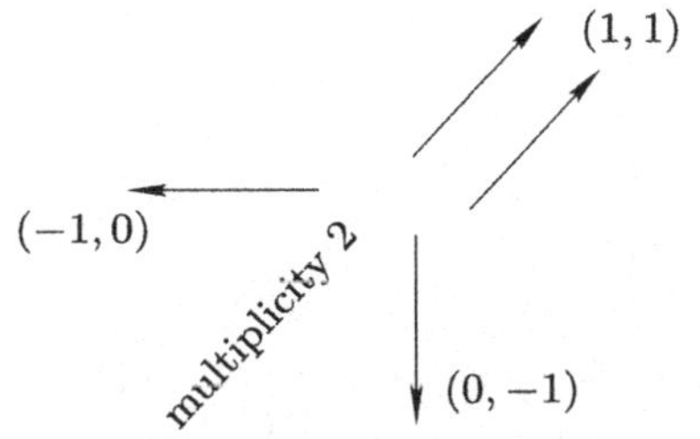

Fig. 8 only 1 ray in two directions but 2 rays in third direction

Example 4.9. On the other hand, the tropical curve of $p(x,y) =$ "$2x^2 + 2y^2 + x + y + 0$" has only one ray in each of the directions $(-1,0), (0,-1), (1,1)$ with multiplicity 2 (Fig. 9).

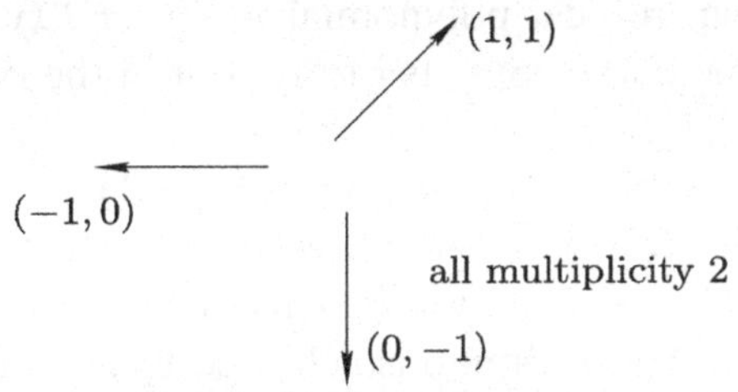

Fig. 9 only 1 ray in each of three directions with multiplicity 2

To see this, we note that

$$p(x,y) = \max\{2+2x, 2+2y, x, y, 0\}.$$

When $x \ll 0$ and y is bounded, $\max\{2+2x, 2+2y, x, y, 0\} = \max\{2+2y, 0\}$, and we get a ray in the direction $(-1,0)$ with multiplicity 2. The other two rays can be determined similarly.

Example 4.10. The above examples show that when coefficients of a quadratic tropical polynomials change, multiplicities of rays can change. Other changes can occur too. For example, there exist tropical curves Γ_p which *do not have rays in any of the directions* $(-1,0), (0,-1)$, and $(1,1)$.

Take the tropical polynomial $p(x,y) =$ "$x^2 + y + 0$" $= \max\{2x, y, 0\}$. It can be shown that Γ_p consists of three rays out of the origin in the directions $(-1,0)$, $(0,-1)$ and $(1,2)$ (Fig. 10).

The rays in the directions $(-1,0)$ and $(0,-1)$ can be computed as above. When $x \gg 0, y \gg 0$, the maximum value $\max\{2x, y, 0\}$ can be achieved by both $2x$ and y, and the equality $2x = y$ gives a ray in the direction $(1,2)$.

This is a degenerate tropical quadratic polynomial in a certain sense, and *there is no more ray in the direction* $(1,1)$.

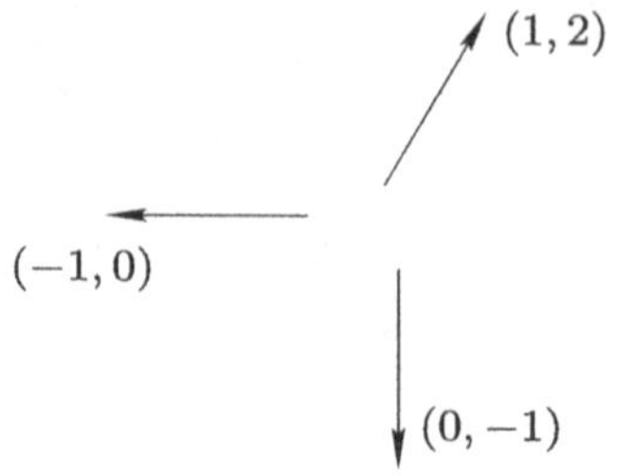

Fig. 10 no ray in direction $(1,1)$

Example 4.11. For the tropical polynomial $p(x,y) =$ "$xy+0$" $= \max\{x+y, 0\}$, the tropical plane curve Γ_p consists of a straight line $x+y=0$ in $\mathbb{R}^2$. Its direction is $(1,-1)$ or $(-1,1)$.

Example 4.12. For the tropical polynomial $p(x,y) =$ "$xy+x+0$" $= \max\{x+y, x, 0\}$, its tropical plane curve consists of rays out of the origin in the directions $(-1,1), (1,0)$, and $(0,-1)$.

Specifically, when $x<0$, the maximum is achieved by $x+y$ and 0 when $x+y=0$. This gives a ray in the direction $(-1,1)$. When $x>0$, the maximum is achieved by $x+y$ and x when $x+y=x$, i.e., $y=0$. This gives a ray in the direction $(1,0)$. When $x=0$, and $y<0$, the maximum is achieved by x and 0. This gives a ray in the direction $(0,-1)$.

This tropical plane curve does not have a ray in any of the directions $(-1,0)$, $(1,1)$ (Fig. 11).

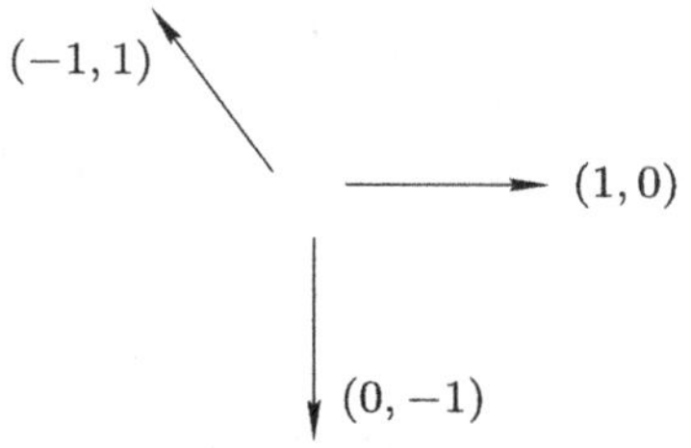

Fig. 11 no ray in any of the directions $(-1,0), (1,1)$

4.2 Newton polygons and tropical curves

The examples in the previous subsection show that directions of rays of a plane tropical curves Γ_p can be complicated and are not necessarily along $(-1,0)$, $(0,-1), (1,1)$. Instead, they depend on the indexes (i,j) such that $a_{ij} \neq -\infty$ in the polynomials $p(x,y) =$ "$\sum_{i,j} a_{ij}x^i y^j$", i.e., the exponents of nonzero monomials in p. It turns out that directions of rays can be described efficiently by Newton polygons associated with tropical polynomials.

Given a tropical polynomial $p(x,y) =$ "$\sum_{i,j} a_{ij}x^i y^j$", consider all indexes (i,j) with $a_{ij} \neq -\infty$, i.e., the indexes for all monomials with nonzero coefficients. Let Δ_p be the convex hull of these indexes (i,j). It is called the *Newton polygon* of the polynomial $p(x,y)$.

Proposition 4.13. *The plane tropical curve Γ_p has at least one ray in the direction of the outer normal vector of every edge of the Newton polygon Δ_p, and every ray of Γ_p is parallel to such an outer normal vector.*

Proof. Let $V = (a,b)$ be an outer normal vector of an edge e of Δ_p. Then there are at least two points of Δ_p lying on this edge e, i.e., there are at least two nonzero monomials $a_{i_1,j_1}x^{i_1}y^{j_1}$, $a_{i_2,j_2}x^{i_2}y^{j_2}$ such that $i_1 a + j_1 b = i_2 a + j_2 b$, since the vector $(i_1 - i_2, j_1 - j_2)$ is perpendicular to (a,b). For all other points (i,j) of Δ_p which do not lie on the edge e, $ia + jb < i_1 a + j_1 b$. Then for suitable rays of the form $x = x_0 + at, y = y_0 + bt$, $p(x,y)$ is achieved by at least two linear functions $a_{i_1 j_1} + i_1 x + j_1 x$, $a_{i_2 j_2} + i_2 x + j_2 x$, where $(i_1, j_1), (i_2, j_2)$ lie on the edge. Therefore such a ray belongs to Γ_p when $t \gg 0$. It is clear that the starting points of such rays depend on values of the coefficients a_{ij} too.

By reversing the above argument, we can show that every ray of Γ_p is of such a form, i.e., perpendicular to an edge of Δ_p. □

The number of rays perpendicular to each edge of Δ_p, counted with multiplicity, is less than or equal to the degree n of $p(x,y)$. The reason is that for any sufficiently large x_0, y_0, the function $p(x_0 - bs, y_0 + as)$ is a polynomial in s of degree at most n and has at most n tropical roots, which account for the nonsmooth rays of the graph of $p(x,y)$ in the direction (a,b).

The examples in the previous section on roots of tropical polynomials show that this number could be strictly less than n.

Remark 4.14. It is clear that it could happen that some indexes (i,j) will be contained in the interior of the Newton polygon Δ_p. All the indexes (i,j) and coefficients a_{ij} together will contribute to bounded line segments (i.e., the internal geometry) of the tropical curve Γ_p.

More precisely, the whole structure of a tropical curve Γ_p can be described in terms of a three dimensional generalization of the Newton polygon Δ_p.

Given a tropical polynomial $p(x,y) =$ "$\sum_{i,j} a_{ij}x^i y^j$", consider the convex hull of the points (i,j,a_{ij}) in $\mathbb{R}^3$. Project the upper envelope of this convex subset into $\mathbb{R}^2$ under the projection $(i,j,a_{ij}) \mapsto (i,j)$, and the image is the convex polygon Δ_p in $\mathbb{R}^2$ together with a distinguished subdivision into smaller polygons. (Note that the component $a_{i,j}$ is the height of the point (i,j,a_{ij}).) The dual graph to this subdivision of Δ_p is the plane tropical curve defined by p. See [124] for detail. Proposition 4.13 only describes its duality for the outer edges of Δ_p. See also [51], [94], [122].

Though the shape of each plane tropical curve Γ_p depends on the polynomial p, for a typical plane tropical curve, we have the following result:

Proposition 4.15. *For a generic tropical polynomial $p(x,y)$ of degree d, its plane curve in $\mathbb{R}^2$ contains d rays counted with multiplicity in each of the directions* $(-1,0),(0,-1),(1,1)$.

By a generic polynomial of degree d, we meant a polynomial

$$p(x,y) = \sum_{i,j:i+j\leq d} a_{ij}x^iy^j,$$

where for each pair (i,j) with $i+j\leq d$, the coefficient $a_{ij}\neq -\infty$. In other words, every allowable monomial appears. In this case, the Newton polygon Δ_p is the triangle with vertices $(d,0),(0,d),(0,0)$. Then Proposition 4.13 implies the above proposition. Of course, the internal geometry of the tropical curve Γ_p is determined by what happens inside Δ_p according to Remark 4.14.

As pointed out earlier in Proposition 3.6, the total number of roots counted with multiplicity of a tropical polynomial of degree d is less than d if a nontrivial power can be factored out. Similarly, for some particular plane tropical curves, the degree is not accurately reflected by the number of rays in each direction, even counted with multiplicity.

Remark 4.16. Though Newton polygons are basic in the study of singularities of curves and solutions of polynomial equations, they might not be so well-known to some people in geometry and topology. We give a basic summary here. For details, see [25, §8.3].

In his correspondences with Leibniz related to infinite series in 1676, Newton described a method to obtain asymptotic expansion in terms of fractional powers (called Puiseux series) at singular points of a plane curve defined by a polynomial equation $f(x,y)=0$. Suppose that $(0,0)$ is a singular point, and we are looking for the leading term of a solution $y=y(x)$ in an asymptotic expansion

$$y = tx^\mu + \dots.$$

Then comparing the leading terms of all monomials in $p(x,y)=\sum_{ij}a_{ij}x^iy^j$ will lead to an edge of the Newton polygon Δ_p with the slope $-\frac{1}{\mu}$, i.e., for such a solution of y exists, μ is determined by the slope $-\frac{1}{\mu}$ of the edge. An iteration of this procedure gives other terms of higher powers in x. In [25, §8.3], only the steepest edge of Δ_p is used, and this is enough when the germ of the singularity at $(0,0)$ is irreducible. In [35, Chapter 1], all edges of Δ_p are used and all roots of $p(x,y)$ or branches are found in [35, Corollary 1.5.5], and in this case, every edge corresponds to an irreducible factor of the germ at $(0,0)$.

A similar Newton polygon Δ_f for a polynomial equation $f(x)=1+a_1x+\ldots+a_nx^n=0$ can be used to describe its roots. Δ_f is well-suited to describe solutions of polynomial equations over fields with discrete valuations. More precisely, suppose that the polynomial is defined over the completion Ω of the algebraic closure of $\mathbb{Q}_p$, and ord_p is the evaluation of Ω which extends the p-adic evaluation of $\mathbb{Q}_p$. Then the Newton polygon Δ_f of f is the lower boundary of the convex hull of the points $(0,0),(1,\mathrm{ord}_p(a_1)),\ldots,(n,\mathrm{ord}_p(a_n))$, i.e., the highest convex polygonal line (i.e., a piecewise linear connected curve) connecting $(0,0),(1,\mathrm{ord}_p(a_1))$ and $(n,\mathrm{ord}_p(a_n))$ and passing through or below all other points. The slopes of edges of Δ_f are equal to $\mathrm{ord}_p(\frac{1}{\alpha})$, where α are roots of $f(x)$. See [79, pp. 97-98] for details.

For a polynomial equation over $\mathbb{C}$, $1+a_1x+\ldots+a_nx^n=0$, we can define its Newton polygon using the points $(0,0),(1,\log|a_1|),\ldots,(n,\log|a_n|)$ in the plane $\mathbb{R}^2$, and the slopes of its edges can be used to approximate the logarithms of absolute values of the roots of the equation. See [109, Theorem 1] for details. It might be helpful to keep in mind that an analogue of discrete valuations for $\mathbb{C}$ is $\log|z|$, $z\in\mathbb{C}$. This is related to the idea of amoebas of algebraic varieties below.

These comments show that the Newton polygon was motived and has been used to describe asymptotic behavior of solutions of algebraic curves and algebraic equations. This suggests that tropical curves are related to such asymptotic behaviors of algebraic curves and solutions of algebraic equations.

4.3 Balancing condition at vertices

In the previous section, examples of tropical plane curves show that the piecewise linear curves resulting from tropical polynomials can have very complicated patterns and directions. Then one natural question is what kind of piecewise linear curves can arise in this fashion. The crucial point is that for every tropical plane curve, there is a *balancing condition* (or the *zero tension* condition) at every vertex.

The tropical line Γ_p defined by the polynomial "$x+y+0$" has one vertex at $(0,0)$, and three rays going out of this vertex. The three integral primitive vectors along the rays add up to 0 (Fig. 12):

$$(-1,0)+(0,-1)+(1,1)=(0,0).$$

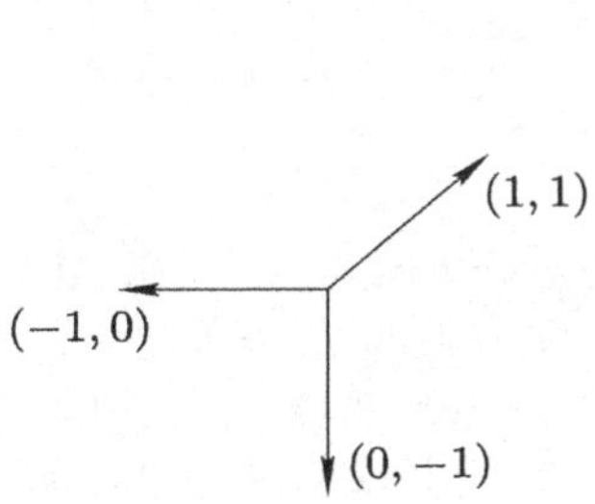

Fig. 12 balancing condition at $(0,0)$

On the other hand, the plane tropical curve defined by $p(x,y)=$ "x^2+y+0" also has a vertex at $(0,0)$ and the three rays from the vertex. The directions out of the vertex are $(-1,0),(0,-1)$, and $(1,2)$. The naive balancing condition *does not hold* (Fig. 13):

$$(-1,0)+(0,-1)+(1,2)\neq(0,0).$$

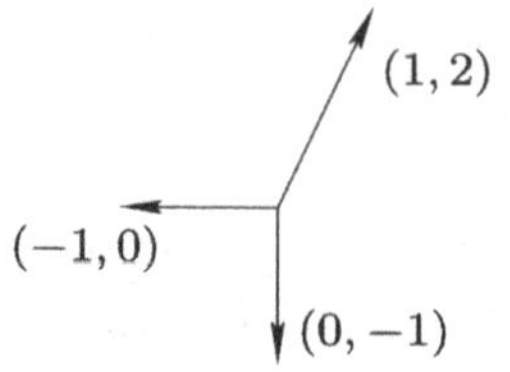

Fig. 13 naive balancing condition at $(0,0)$ does not hold

Instead, it should be replaced by

$$(-1,0)+2(0,-1)+(1,2)=(0,0).$$

We need to add a weight 2 to the ray in the direction $(0,-1)$. This can be explained the fact that this ray is defined by $2x=0$, or "x^2" $=0$. So it has multiplicity 2.

Now we explain the general balancing condition at vertices of a plane tropical curve Γ_p. Suppose $x_0\in\Gamma_p$ is a vertex of valence $n+1$. Let $e_0,\dots,e_n$ be all the edges out of x_0. Suppose that an edge e is defined by the equality of two linear functions "$a_1x^{j_1}y^{k_1}$" $=$ "$a_2x^{j_2}y^{k_2}$", i.e., the two linear functions $j_1x+k_1y+a_1$ and $j_2x+k_2y+a_2$ achieve the value of $p(x,y)$. Define $V(e)$ to be the *primitive integral vector* in the direction of the edge e (pointing away from the vertex), and the *weight* $W(e)$ of the edge e to be the greatest common denominator $GCD(j_1-j_2,k_1-k_2)$. The *weight* $W(e)$ can be considered as the multiplicity of the edge e, and is consistent with the multiplicity of roots of tropical polynomials.

Lemma 4.17 ([99, p. 512]). *Under the above notation, for every vertex x_0 of a plane tropical curve Γ_p, the following balancing condition is satisfied:*

$$\sum_{i=0}^{n}W(e_i)V(e_i)=0. \tag{4.1}$$

Proof. Suppose an edge e out of x_0 is defined by $f_1(x,y)=f_2(x,y)$, where $f_1(x,y)=j_1x+k_1y+a_1$ and $f_2(x,y)=j_2x+k_2y+a_2$. Assume that when we move counter-clockwise around the vertex, $f_1(x,y)-f_2(x,y)$ changes from positive values to negative values. Then it can be checked that $W(e)V(e)=(-(k_1-k_2),j_1-j_2)=(k_2-k_1,j_1-j_2)$. On the other hand, if $f_1(x,y)-f_2(x,y)$ changes from negative values to positive values, then $W(e)V(e)=(k_1-k_2,j_2-j_1)$. (Think of the simplest example of two linear functions x,y with the vertex at the origin.) For any vertex x of Γ_p, there are linear functions $f_1(x,y),\dots,f_k(x,y)$ such that all edges are defined by the equality of two linear functions which achieve the maximum value.

These edges divide a small disk with center at x into cone pieces, and for each cone Ω, one linear function f_i is the unique function which takes the maximum value on this cone. When we move counter-clockwise across the two edges e_1, e_2 of Ω defined by $f_i = f_{i_1}$ and $f_i = f_{i_2}$ for some f_{i_1} and f_{i_2}, both possibilities of sign changes of the values $f_i - f_{i_1}$ and $f_i - f_{i_2}$ can occur. This means the coefficient of x (and similarly the coefficient of y) of f_i will appear in $W(e_1)V(e_1)$ and $W(e_2)V(e_2)$ once as positive and another time as negative. By summing over all edges (or rather all cones), they all cancel out and the sum $\sum_{i=0}^{n} W(e_i)V(e_i)$ is zero. (Think of the simplest example of tropical curve defined by the tropical polynomial "$x+y+0$", which has only one vertex at the origin and 3 edges coming out of it.) □

Corollary 4.18. *For every tropical polynomial $p(x,y)$, its plane tropical curve Γ_p is a piecewise linear connected curve in $\mathbb{R}^2$ with rational slopes such that with respect to a suitable integral weight $W(e)$ for each edge e, it satisfies the balancing condition as in Equation 4.1 at every vertex.*

The converse of the above corollary is also true [99, p. 512].

Proposition 4.19. *Let C be a piecewise linear connected curve in $\mathbb{R}^2$ with rational slopes such that, with respect to a suitable integral weight for each edge, it satisfies the balancing condition in Equation 4.1 at every vertex. Then there exists a tropical polynomial p whose associated tropical curve Γ_p is equal to C, and the weight $W(e)$ of every edge e of Γ_p is equal to the given one.*

Briefly, this can be proved as in Lemma 4.17 by noting that if a finite sequence of numbers $a_1, \ldots, a_k$ sums to 0, then there exist numbers $b_1, \ldots, b_k$ such that $a_i = b_i - b_{i+1}$ for $i = 1, \ldots, k$.

Remark 4.20. We can also define piecewise linear connected curves in $\mathbb{R}^n$ with rational slopes satisfying the balancing condition at vertices with respect to suitable integral weights on edges, but a converse statement as in Proposition 4.19 does not hold in general. It is a complicated problem to decide when this is possible. See [123].

Remark 4.21. The above balancing condition in Corollary 4.17 also holds for tropical hypersurfaces in $\mathbb{R}^n$ and is called the *structure theorem of tropical varieties* [95, §3.3] [106, Theorem 2.5.7]. For a beautiful explanation of the balancing condition of plane tropical curves, see [51, §1.4]. For some pictures of plane tropical curves, see [99], [100].

4.4 Abstract smooth tropical curves

As it is known, plane curves over $\mathbb{C}$ of high degrees are not always smooth. This is also true for plane tropical curves. It will follow from our definition of smoothness of tropical curves below that plane tropical curves are not smooth whenever they contain a vertex of valence strictly greater than 3. In order to relate metric graphs to tropical curves, we need to introduce *smooth tropical curves* which are not necessarily embedded in some ambient tropical space $\mathbb{T}^n$, i.e., we need abstract tropical curves. In fact, compact tropical curves cannot be embedded in $\mathbb{T}^n$ (which can be seen by using the balancing condition at vertices). This is reasonable since tropical curves in $\mathbb{T}^n$ are affine tropical curves.

In this subsection, we introduce the notion of abstract smooth tropical curves. The basic reference is [101], and the papers [67], [124] are also helpful. More comprehensive discussions, including tropical spaces of higher dimension and tropical manifolds, are given in [106].

To be consistent with notation in the previous section, we use Γ to denote a tropical curve. *A smooth tropical curve* is a graph Γ endowed with a *smooth integral affine ($\mathbb{Z}$-affine) structure*, which consists of an open covering of Γ by compatible $\mathbb{Z}$-affine smooth charts satisfying the following conditions:

1. For each interior point x of an edge e, let U be a neighborhood of x homeomorphic to an interval. A $\mathbb{Z}$-affine structure on U is an embedding
$$\phi_U : U \longrightarrow \mathbb{R}.$$
Note that $\mathbb{R}$ has a canonical $\mathbb{Z}$-affine structure coming from $\mathbb{Z} \subset \mathbb{R}$, and this integral structure induces an integral structure on U.
2. If x is a vertex of valence $n+1$, let $e_0, \dots, e_n$ be the edges coming out of x. Let U be a neighborhood of x which is contained in the union of these edges and which consists of $n+1$ sub-intervals. A *smooth $\mathbb{Z}$-affine structure* on U is an embedding
$$\phi_U : U \longrightarrow \mathbb{R}^n$$
such that for every edge e_i, $\phi_U(U \cap e_i)$ is a line segment with a rational slope starting from $\phi_U(x)$, and $\phi_U(U)$ satisfies a *balancing and nondegeneracy condition* at the vertex $\phi_U(x)$. Specifically, let $v_i \in \mathbb{Z}^n$ be the *primitive vector* in the direction of $\phi_U(U \cap e_i)$ (pointing away from x). Then the *balancing condition* at $\phi_U(x)$ is the equation
$$\sum_{i=0}^{n} v_i = 0. \tag{4.2}$$
Note that in the above equation, each edge has multiplicity 1. And the *nondegeneracy condition* is that *any n vectors* of the $n+1$ vectors $v_0, \dots, v_n$ form a basis of $\mathbb{Z}^n$, which is equivalent to the condition that $v_0, \dots, v_{n-1}$ form a basis of $\mathbb{Z}^n$ when the balancing condition is satisfied. (One reason for imposing this non-degeneracy condition is that every two bases of $\mathbb{Z}^n$ can be interchanged by some element of $\mathrm{GL}(n, \mathbb{Z})$. This makes the edges out of x more symmetric.)
3. The above local charts are compatible as follows. Given two overlapping charts $\phi_1 : U_1 \longrightarrow \mathbb{R}^{n_1}$, $\phi_2 : U_2 \longrightarrow \mathbb{R}^{n_2}$, we can choose a common $\mathbb{R}^N$ and inclusions of $\mathbb{R}^{n_1}$, $\mathbb{R}^{n_2} \subset \mathbb{R}^N$ as linear subspaces

$$\phi_1 : U_1 \to \mathbb{R}^{n_1} \subset \mathbb{R}^N, \phi_2 : U_2 \to \mathbb{R}^{n_2} \subset \mathbb{R}^N,$$

and find an invertible $\mathbb{Z}$-affine linear map $\Phi_{12} : \mathbb{R}^N \to \mathbb{R}^N$ such that

$$\phi_1|_{U_1 \cap U_2} = \Phi_{12} \circ \phi_2.$$

(Recall that an *invertible $\mathbb{Z}$-affine linear map* is given by a map

$$v \in \mathbb{R}^N \mapsto Av + b,$$

where A is an *integrally invertible* $N \times N$-matrix, i.e., $A \in \mathrm{GL}(n, \mathbb{Z})$, and $b \in \mathbb{R}^N$.) Note that the overlap $U_1 \cap U_2$ can be complicated if U_1, U_2 are large. If they are sufficiently small, then $U_1 \cap U_2$ consists either of an open interval contained in an edge or a neighborhood of a vertex of Γ. The compatibility condition shows that on each edge and at every vertex of Γ, there is a unique choice of $\mathbb{Z}$-affine structure on its neighborhoods.

In the above definition, the non-degeneracy condition and the balancing condition with *multiplicity* 1 in Condition (2), not the weighted balancing condition, are an essential part of being a smooth tropical curve. One immediate corollary is the following.

Corollary 4.22. *Let $p(x,y)$ be a tropical polynomial and Γ_p the associated tropical plane curve. If Γ_p has a vertex of valence $k \geq 4$, i.e., with k edges coming out of this vertex, then Γ_p is not a smooth tropical curve.*

Proof. The plane tropical curve Γ_p is a connected piecewise-linear curve in $\mathbb{R}^2$ such that each edge has a rational slope. When Γ_p contains a vertex x_0 of valence $k > 3$, the primitive integral vectors along all edges connected to the vertex x_0 do not satisfy the nondegeneracy condition. Therefore, Γ_p is not a smooth tropical curve. □

On the other hand, we can always obtain a smooth tropical curve from a plane tropical curve by normalizing it as the next proposition shows.

Proposition 4.23. *Given any plane tropical curve Γ_p, there is a canonical smooth tropical curve $\tilde{\Gamma}$ which can be mapped bijectively to the plane tropical curve Γ_p by a locally $\mathbb{Z}$-affine map. The smooth tropical curve $\tilde{\Gamma}$ is called the normalization of the plane tropical curve Γ_p.*

Let Γ_p be a tropical plane curve. We explain how to put a smooth $\mathbb{Z}$-affine structure on Γ_p. Suppose that x_0 is an interior point of an edge e, which is defined by "$a_1 x^{j_1} y^{k_1}$" $=$ "$a_2 x^{j_2} y^{k_2}$", i.e., the equality of two linear functions $j_1 x + k_1 y + a_1 = j_2 x + k_2 y + a_2$. Define $V(e)$ to be a *primitive integral vector* along the edge e, and the *weight* $W(e)$ of the edge e to be the greatest common denominator $GCD(j_1 - j_2, k_1 - k_2)$ of $j_1 - j_2, k_1 - k_2$. Let U be a small neighborhood of x_0

homeomorphic to an interval. Define a $\mathbb{Z}$-affine map $\phi_U : U \longrightarrow \mathbb{R}$ such that it maps x_0 to 0, and under the natural extension, it maps the vector $W(e)V(e)$ to the unit vector 1 on $\mathbb{R}$,

$$\phi_U(x_0 + W(e)V(e)) = 0 + 1.$$

Suppose $x_0 \in \Gamma_p$ is a vertex of valence $n+1$. Let $e_0, \dots, e_n$ be all the edges from x_0. For each edge e, we have a *primitive integral vector* $V(e)$ and weight $W(e)$ as above.

Consider the following piecewise linear curve Γ_0 in $\mathbb{R}^n$ with one vertex of valence $n+1$ at the origin: the edges out of the origin are rays along the $n+1$ vectors: $v_0 = (1,\dots,1), v_1 = (-1,0,\dots,0), \dots, v_n = (0,\dots,0,-1)$. When $n = 2$, this is the plane tropical curve defined by the polynomial $p(x,y) =$ "$x+y+0$".

These edges of Γ_0 have rational slopes, and the primitive vectors along the edges are $v_0, \dots, v_n$ and satisfy the balancing and nondegeneracy conditions. Therefore, Γ_0 is a smooth tropical curve.

Proposition 4.24. *Suppose x is a vertex of a plane tropical curve Γ_p of valence $n+1$. Let $e_0, \dots, e_n$ be the edges out of x as above. For each edge e_i, let $W(e_i)$ be its weight and $V(e_i)$ be its primitive integral vector as defined above. Then there exists a unique $\mathbb{Z}$-affine linear map $\pi : \mathbb{R}^n \longrightarrow \mathbb{R}^2$ such that*

$$\pi(0) = x, \quad \pi(v_i) = W(e_i)V(e_i), \quad i = 0, \dots, n.$$

Proof. The nondegeneracy condition of $v_0, \dots, v_n$ implies that $v_1, \dots, v_n$ are linearly independent vectors of $\mathbb{R}^n$ and form a basis of $\mathbb{R}^n$. This implies that there exists a unique affine linear map $\pi : \mathbb{R}^n \longrightarrow \mathbb{R}^2$ such that $\pi(v_i) = W(e_i)V(e_i)$, $i = 0, \dots, n$, and $\pi(0) = x$. Since the vectors $v_0, \dots, v_n$ satisfy the balancing condition$\sum_{i=0}^n W(e_i)V(e_i) = 0$ by Lemma 4.17, and the vectors $e_0, \dots, e_n$ also satisfy $\sum_{i=0}^n v_i = 0$, it follows that

$$\pi(v_0) = W(e_0)V(e_0).$$

Since $v_1, \dots, v_n$ forms a basis of the lattice $\mathbb{Z}^n$ and $W(e_1)V(e_1), \dots, W(e_n)V(e_n)$ are integral vectors, the map π is a $\mathbb{Z}$-affine linear map. □

Let U_0 be a small neighborhood of the origin in Γ_0. When U_0 is small enough, $\pi(U_0)$ is a neighborhood of x in Γ_p, denoted by U, and the restricted map π_{U_0} : $U_0 \to U$ is a homeomorphism. Its inverse gives a smooth $\mathbb{Z}$-affine structure on U:

$$\phi_U = (\pi_{U_0})^{-1} : U \to U_0 \subset \mathbb{R}^n.$$

We note that if an interior point x of an edge is treated as a vertex of valence 2, then the above two constructions of a smooth $\mathbb{Z}$-affine structure at x agree.

Proposition 4.25. *For any plane tropical curve Γ_p, the $\mathbb{Z}$-affine structures on neighborhoods of points of Γ_p defined above give the structure of a smooth tropical curve on the graph associated with Γ_p. Denote this smooth tropical curve by $\tilde{\Gamma}$.*

Proof. We need to show that the local charts are compatible. Let U_1, U_2 be two charts of Γ_p. If they are neighborhoods of the same vertex, then they correspond to different choices of ordered integral bases of $\mathbb{Z}^n$, i.e., the vectors $v_1, \dots, v_n$ of $\{v_0, v_1, \dots, v_n\}$. Since every two bases of $\mathbb{Z}^n$ are related by invertible integral linear maps, they define compatible $\mathbb{Z}$-affine structures.

Otherwise, by our choice, $U_1 \cap U_2$ is an open interval of an edge e of C. Under the embeddings $\phi_1 : U_1 \longrightarrow \mathbb{R}^{n_1}, \phi_2 : U_2 \longrightarrow \mathbb{R}^{n_2}$, the integral vector $W(e)V(e)$ along the edge e is mapped to primitive integral vectors in $\mathbb{R}^{n_1}$ and $\mathbb{R}^{n_2}$ respectively. Embed $\mathbb{R}^{n_1}$, $\mathbb{R}^{n_2}$ into a common linear space $\mathbb{R}^N$ so that primitive integral vectors are mapped to primitive integral vectors. Since every two primitive integral vectors of $\mathbb{R}^N$ are related by an element $g \in \mathrm{GL}(n, \mathbb{Z})$, the $\mathbb{Z}$-affine structures defined by ϕ_1 and ϕ_2 are compatible. □

Proof of Proposition 4.23 on normalization of tropical curves

By definition, the underlying space of the tropical curve $\tilde{\Gamma}$ in Proposition 4.25 is the same as Γ_p, and the identity map from $\tilde{\Gamma}$ to Γ_p is locally a $\mathbb{Z}$-affine map. Roughly speaking, the normalization involves both making the edge vectors of every vertex nondenegerate and scaling $\mathbb{Z}$-affine maps on the edges according to their weights. This is achieved by opening them up in higher dimensions and scale them *down* suitably according to the primitive integral vectors and weights. □

Remark 4.26. From the above discussion, it is clear that the integral structure $\mathbb{Z}^n$ in $\mathbb{R}^n$ is used crucially. If all the weights $W(e)$ are equal to 1, then primitive integral vectors of edges are mapped to primitive integral vectors and the normalization amounts to making the primitive integral vectors of edges *nondegenerate* by embedding them into (or opening up in) a higher-dimensional linear space.

4.5 Identification between smooth tropical curves and metric graphs

By definition, a smooth tropical curve Γ is topologically a graph. The following results are well-known and are often used to define tropical curves in [101, Proposition 1.3], [107, Proposition 3.6], [67, Definition 1.15]. We fill in more detail for the convenience of the reader.

Proposition 4.27. *Every smooth tropical curve Γ admits a canonical metric and hence becomes a metric graph in a canonical way.*

Proof. We will define a metric on a smooth tropical curve Γ using a $\mathbb{Z}$-affine structure on charts. It suffices to define lengths of subintervals of the edges, and there are two cases to consider.

Let $\phi_U : U \longrightarrow \mathbb{R}$ be a $\mathbb{Z}$-affine chart of an interior point of an edge e. Let V be an open subinterval contained in U. Define the length of V to be equal to the length of the interval $\phi_U(V) \subset \mathbb{R}$ with respect to the standard metric of $\mathbb{R}$, i.e., the length of the interval $[0, 1]$ is equal to 1.

If x is a vertex of valence $n+1$ and $\phi_U : U \to \mathbb{R}^n$ is a $\mathbb{Z}$-affine chart, and $e_0,\dots,e_n$ are the edges from x, and $V(e_i)$ is an integral primitive vector on the ray $\phi_U(U \cap e_i)$, then we restrict the standard metric of $\mathbb{R}^n$ to $\phi_U(U \cap e_i)$ and scale down the metric on it so that the length of the vector $V(e_i)$ is equal to 1 (i.e., scale the metric on *each edge separately* so that the *length of a primitive integral vector* along it is equal to 1). Then for an interval V of an edge e_i contained in U, the length of V is equal to the length of the image $\phi_U(V)$ with respect to the *scaled metric* on the ray $\phi_U(U \cap e_i)$.

In summary, if a directed interval of a smooth tropical curve Γ is mapped to a primitive integral vector under a local $\mathbb{Z}$-affine chart, its length is equal to 1. Since these $\mathbb{Z}$-affine structures on overlapping charts are compatible, integral primitive vectors in one chart are also integral primitive vectors in another, and the metrics of these charts agree on the overlaps. By gluing these metrics together, we obtain a metric on the tropical curve Γ. □

In the above construction, primitive integral vectors are the basic *unit* vectors. They are closed under invertible integral affine transformations and hence form an intrinsic class.

The converse of the above proposition is also true.

Proposition 4.28. *Every metric graph admits a canonical structure of a smooth tropical curve.*

Proof. Let (G,ℓ) be a metric graph, and x is a vertex of valence $n+1$. As in the previous subsection, let Γ_0 be the tropical curve in $\mathbb{R}^n$ with a vertex at 0 of valence $n+1$, and the edges out of the origin are rays along the vectors

$$v_0 = (1,\dots,1), v_1 = (-1,0,\dots,0), \dots, v_n = (0,\dots,0,-1).$$

Note that they are primitive integral vectors satisfying the balancing condition. Put a metric on Γ_0 as in the previous proposition such that each of $v_0,\dots,v_n$ is of length 1. Map a small neighborhood U of x *isometrically* to a small neighborhood U_0 of 0. Different mappings give equivalent $\mathbb{Z}$-affine structures, and hence they define a $\mathbb{Z}$-affine structure on U.

If $x \in G$ is an interior point of an edge, and U is an interval of the edge and contains x, we can map U *isometrically* to a neighborhood of the origin of $\mathbb{R}$. This defines a $\mathbb{Z}$-affine structure on U.

According to the above definitions, for every large integer N, a vector along any edge of G of length $\frac{1}{N}$ is mapped to $\frac{1}{N}$-multiple of a primitive integral vector in $\mathbb{R}^n$, where $\mathbb{R}^n$ depends on the local chart. This implies that these $\mathbb{Z}$-affine structures on charts of G are compatible and hence define a structure of smooth tropical curve on the metric graph (G,ℓ). □

Remark 4.29. The basic point of the proof of the above proposition is that a $\mathbb{Z}$-affine structure on a one-dimensional space is equivalent to a length metric by viewing primitive integral vectors as unit vectors.

Remark 4.30. We can show easily that the two maps in Propositions 4.27 and 4.28 are the inverses of each other. Specifically, start with a *smooth* tropical curve Γ, obtain a metric graph (G,ℓ) by Proposition 4.27, and then obtain a new tropical curve Γ' from (G,ℓ) by Proposition 4.28. It can be shown that Γ' is isomorphic to Γ, i.e., two $\mathbb{Z}$-affine structures on the graph G are compatible. In the other way, starting with a metric graph (G,ℓ), we obtain a smooth tropical curve $\tilde{\Gamma}$, which induces a metric ℓ' on the graph G. It can be checked easily that these two metrics ℓ,ℓ' agree.

Every plane tropical curve Γ_p admits a *canonical metric* defined as follows. For every edge e of Γ_p, let $V(e)$ be a primitive integral vector on e, and $W(e)$ be the weight of e defined above. Then the metric on e is defined to be the scaling-down of the restriction of the standard metric of $\mathbb{R}^2$ such that the norm of $V(e)$ is equal to $1/W(e)$; in particular, the norm of $V(e)$ is equal to 1 if and only if the weight $W(e)=1$.

Corollary 4.31. *The metric on Γ_p defined above is the same as the metric defined by Proposition 4.28 on its normalization $\tilde{\Gamma}$ given in Proposition 4.23.*

Remark 4.32. In conclusion, for a plane tropical curve Γ_p, if it is smooth, then integral primitive vectors on Γ_p are of unit length with respect to the canonical metric. We also note that it is reasonable to require that the primitive integral vectors on edges of smooth tropical curves are normalized to have metric 1 since primitive integral vectors are preserved under invertible $\mathbb{Z}$-affine maps, and they are the basic building blocks in integral affine geometry.

Remark 4.33. The construction of a canonical metric in the above corollary also allows one to construct alternatively a normalization of the plane tropical curve Γ_p, i.e., by taking the tropical curve corresponding to the metric graph associated with Γ_p.

If we trace through the definitions involved here, it is clear that normalizing the metric on each edge e of a plane tropical curve Γ_p by dividing by its weight $W(e)$ is required in order to maintain the balancing condition, in view of the balancing conditions in Equations (4.1) and (4.2).

5 Definitions of tropical varieties and motivations

In the previous section, we defined tropical curves. Higher dimensional tropical varieties occur naturally. For example, motivated by the algebraic structure of the moduli space of algebraic curves over $\mathbb{C}$, one natural question is whether the moduli space of tropical curves is a tropical variety. We will also need tropical abelian varieties in §10 below in our study of $\mathrm{Out}(F_n)$ and of tropical curves.

There are several generalizations of tropical plane curves and of their characterizations. For affine tropical varieties in $\mathbb{R}^n$, we have at least two direct generalizations:

1. As subsets of $\mathbb{R}^n$ defined by tropical polynomials.
2. As rational (or integral) polyhedral complexes in $\mathbb{R}^n$ satisfying suitable balancing conditions.

Once we have local models of tropical varieties, we can define abstract tropical varieties which are not necessarily embedded into $\mathbb{R}^n$. See [106, Chap 6] for details. There are also related notions of affine manifolds and tropical (or tropical affine) manifolds. See [58, p. xiii]. The tropical abelian varieties are tropical affine manifolds.

5.1 Tropical hypersurfaces and higher codimension tropical subvarieties of $\mathbb{R}^n$

The first natural class of higher dimensional tropical varieties consists of tropical hypersurfaces in $\mathbb{R}^n$. The second class consists of tropical fans and will be treated later in §6.1.

There are two definitions of a tropical hyperplane in $\mathbb{R}^n$ by generalizing those of a plane tropical curve:

1. A tropical hyperplane in $\mathbb{R}^n$ is defined by a tropical polynomial.
2. A tropical hyperplane in $\mathbb{R}^n$ is defined as a codimension 1 rational polyhedral complex in $\mathbb{R}^n$ satisfying the balancing condition at every codimension 2 polyhedron.

More precisely, given any tropical polynomial $f(x_1,\dots,x_n)$ in n variables, its *tropical hyperhypersurface* $V(f)$ consists of points in $\mathbb{R}^n$ where two linear functions which appear in

$$f(x_1,\dots,x_n) = \text{“}\sum_{m\in I} a_m x^m\text{”} = \max_{m\in I}\{a_m + m_1x_1 + \dots + m_nx_n\}$$

take the maximum value, or equivalently, the points over which the piecewise linear function f or its graph is not regular.

As explained in the case of curves, $V(f)$ is the locus where the tropical function “$0/f$” fails to be regular.

Proposition 5.1. *The tropical hypersurface $V(f)$ is a rational polyhedron complex in $\mathbb{R}^n$ which satisfies the balancing condition at every codimension 1 polyhedron face in $V(f)$.*

Note that $V(f)$ is a polyhedral complex of dimension $n-1$, and the balancing condition is imposed only for all polyhedron faces of dimension $n-2$.

This is called the *structure theorem of tropical varieties* and also holds for more general tropical varieties defined below (Proposition 5.5). For a proof of Proposition 5.1, see [106, Theorem 2.5.7] or [95, Theorem 3.3.6].

We give some explanations of the statement. Recall that a polyhedron P in $\mathbb{R}^n$ is a convex subset defined by linear equations and linear inequalities. It is said to be of *rational slope* if the coefficients of the first order terms are integers. (Note that the constant terms of the linear equations and inequalities are not required to be integer. This phenomenon can be seen clearly in the case of $P \subset \mathbb{R}^2$, line segments with rational slope.)

Then a *polyhedral complex* $\mathscr{P}$ is a collection of polyhedra which satisfy two compatibility conditions:

1. If a polyhedron P is contained in $\mathscr{P}$, then every face of P is also contained in $\mathscr{P}$.
2. If $P_1, P_2 \in \mathscr{P}$ are two polyhedra in $\mathscr{P}$, then their intersection $P_1 \cap P_2$ is either empty or is a common face of both P_1 and P_2.

A polyhedral complex $\mathscr{P}$ is said to be of *rational slopes* if every polyhedron in it has rational slope.

From the definition of a tropical hypersurface $V(f)$, it is clear that $V(f)$ is a rational polyhedral complex.

To define the balancing condition at every codimension 1 polyhedron face Q of $V(f)$, we can proceed in two ways. Let $P_1, \dots, P_l$ be all the top dimensional polyhedra in $V(f)$ which contain Q as a face. For each P_i, there is an integral primitive vector n_i which is tangent to P_i and normal to Q, and points to the direction of P_i. Each P_i has a *canonical weight* (or multiplicity) w_i defined from the two linear functions which cut out P_i, or rather the $n-1$ dimensional affine subspace containing P_i, as in the case of plane tropical curves. Then the *balancing condition* at a codimension 1 face Q is (see [105, Property 3.2] [106, Theorem 2.5.7] [95, Theorem 3.3.6]):

$$\sum_i w_i n_i = 0.$$

Another method to define the balancing condition is to use the linear projection $\mathbb{R}^n \to \mathbb{R}^2$ to reduce to the case of tropical curves. Let $V(Q)$ be the linear subspace which is parallel to Q. Then

$$\mathbb{R}^n / V(Q) \cong \mathbb{R}^2,$$

and let $\pi : \mathbb{R}^n \to \mathbb{R}^n/V(Q)$ be the canonical projection. The image $\pi(Q)$ is the origin 0 of the vector space $\mathbb{R}^n/V(Q)$, and $\pi(P_i)$ is a line segment adjacent to 0. Let v_i be an integral primitive vector from 0 in the direction of $\pi(P_i)$, and let w_i be the weight of P_i as above. Then the *balancing condition* at the codimension 1 face Q is equivalent to the following condition in the plane $\mathbb{R}^n/V(Q)$ [105, Property 3.3]:

$$\sum_i w_i v_i = 0.$$

Remark 5.2. The converse of Proposition 5.1 is also true, i.e., if V is a balanced rational polyhedral complex of dimension $n-1$ in $\mathbb{R}^n$, then there exists a tropical polynomial f such that $V = V(f)$. See [106, Theorem 2.5.10] for detail.

After the definition of tropical hypersurfaces, one natural question is how to define tropical subvarieties of $\mathbb{R}^n$ of dimension smaller than $n-1$. If we follow the analogy with affine algebraic varieties over $\mathbb{C}$, then for any collection of tropical polynomials $f \in I$, we can define a *tropical variety* $V(I)$ by

$$V(I) = \cap_{f \in I} V(f).$$

Usually we take I to be an ideal in $\mathbb{R}[x_1, \ldots, x_n]$. Unlike the case for polynomials (or varieties) over $\mathbb{C}$, if $f_1, \ldots, f_k$ are generators of I, it is *not true in general* that

$$V(I) = V(f_1) \cap \ldots \cap V(f_k).$$

See [94, Remark 2.6] for more detail and references. See also [95, Chap 3].

5.2 Tropicalization of polynomials and tropical varieties as non-Archimedean amoebas

In the above discussion of tropical curves and tropical hypersurfaces, we have followed definitions and results of affine algebraic curves and varieties over $\mathbb{C}$. One natural question is, besides the formal analogy, whether there are some direct relations between varieties over the tropical semifield $\mathbb{T}$ and varieties over $\mathbb{C}$ and other fields.

The answer turns out to be positive, and one way is to use the notion of amoebas and tropicalization of polynomials and varieties. These relations can be best explained for fields with a nontrivial valuation.

Let K be a field. A *valuation* of K is a map

$$\text{val} : K \to \mathbb{R} \cup \{+\infty\}$$

which satisfies the following conditions:

1. $\text{val}(0) = +\infty$.
2. For all $a, b \in K^\times$, $\text{val}(ab) = \text{val}(a) + \text{val}(b)$.
3. For all $a, b \in K^\times$, $\text{val}(a+b) \geq \min\{\text{val}(a), \text{val}(b)\}$.

Then the valuation defines a *non-Archimedean norm* on K: for any $a \in K$, define

$$|a| = e^{-\text{val}(a)}.$$

Probably the most basic example of fields with valuation is the p-adic field $\mathbb{Q}_p$, where p is a prime number. The p-adic field $\mathbb{Q}_p$ is defined as follows: for every $x \in \mathbb{Q}$, write

$$a = p^n \frac{a}{b},$$

where a, b are integers not divisible by p. Then define the p-adic valuation of x by

$$\mathrm{val}_p(x) = n.$$

The p-adic field $\mathbb{Q}_p$ is the completion of $\mathbb{Q}$ with respect to the p-adic norm

$$|x|_p = p^{-\mathrm{val}_p(x)}.$$

We can also consider extensions of $\mathbb{Q}_p$, in particular its algebraic closure, and the valuation can be defined on such field extensions.

On the other hand, for our discussion on tropical varieties, the field of *Puiseux series* $\mathbb{C}\{\{t\}\}$ is most relevant. Recall that $\mathbb{C}\{\{t\}\}$ consists of *formal power series* of the form

$$f(t) = \sum_{i \geq i_0} c_i t^{\frac{i}{n}},$$

where $c_i \in \mathbb{C}$, n is a positive integer, i_0 is an integer and could be negative. There is no requirement on convergence of the series.

Let K be a field with a nontrivial valuation val. For any polynomial $f \in K[x_1, \ldots, x_n]$, write

$$f(x_1, \ldots, x_n) = \sum_u c_u x^u,$$

where u is a multi-index. Then define the *tropicalization* of f:

$$\mathrm{trop}(f)(x_1, \ldots, x_n) = \text{“}\sum_u \mathrm{val}(c_u) x^u\text{”}.$$

Note that if a coefficient $c_u = 0$, then the term $\mathrm{val}(c_u)x^u$ is not present.

Then for each polynomial f over a field K, we have two varieties:

1. The usual *affine subvariety* (or *affine hypersurface*) $V(f)$ contained in K^n defined over the field K, which is given by

$$V(f) = \{(x_1, \ldots, x_n) \in K^n \mid f(x_1, \ldots, x_n) = 0\}.$$

2. The *tropical hyperplane* $V(\mathrm{trop}(f))$ contained in $\mathbb{R}^n$ defined in the previous subsection.

More generally, for any ideal $I \subset K[x_1, \ldots, x_n]$, we can define its associated *tropical ideal* $\mathrm{trop}(I)$ in $\mathbb{R}[x_1, \ldots, x_n]$ by tropicalizing all polynomials in it,

$$\mathrm{trop}(I) = \{\mathrm{trop}(f) \mid f \in I\}.$$

Similarly, we obtain two varieties:

1. An *affine variety* $V(I) \subset K^n$,

$$V(I) = \cap_{f \in I} V(f).$$

2. A *tropical variety* $V(\text{trop}(I)) \subset \mathbb{R}^n$, defined by

$$V(\text{trop}(I)) = \cap_{f \in I} V(\text{trop}(f)) = \cap_{p \in \text{trop}(I)} V(p).$$

The valuation map $\text{val} : K^\times \to \mathbb{R}$ can be extended to a valuation map

$$\text{val} : (K^\times)^n \to (\mathbb{R})^n.$$

Then one relation between the algebraic variety $V(f)$ and its tropical counterpart $V(\text{trop}(f))$ is as follows.

Theorem 5.3. *Assume that K is an algebraically closed field with a nontrivial valuation. Then for any polynomial $f \in K[x_1, \ldots, x_n]$, the tropical hypersurface $V(trop(f)) \subset \mathbb{R}^n$ is equal to the closure with respect to the standard topology of $\mathbb{R}^n$ of the image* $\text{val}(V(f) \cap (K^\times)^n)$ *in $\mathbb{R}^n$ of the affine variety $V(f)$ under the map* val.

Similarly, for any ideal $I \subset K[x_1, \ldots, x_n]$, the tropical variety $V(trop(I))$ is the closure of the image $\text{val}(V(I) \cap (K^\times)^n)$ *in $\mathbb{R}^n$ of the affine variety $V(I)$.*

The closure of the image $\text{val}(V(f) \cap (K^\times)^n)$ is called a *non-Archimedean amoeba*, since the field K is a non-Archimedean field by assumption.

This is usually called the *fundamental theorem of tropical geometry*. See [45, Theorem 2.1.1], [95, Theorem 3.2.5], and [94, §4] for a history of and references on this result.

Probably one reason why this result is called the fundamental theorem of tropical geometry is that it relates the classical algebraic geometry over fields to the tropical geometry over the semifield $\mathbb{T}$. Consequently, the piecewise linear and combinatorial structures of tropical varieties can be used to study and understand their classical counterparts. Furthermore, this result also describes asymptotic properties of affine algebraic varieties over usual fields.

Remark 5.4. We note that in [45], a tropical hypersurface is defined via non-Archimedean amoeba of a hypersurface, i.e., the closure of the image under the valuation map val. The above theorem shows that this coincides with the definition used in this book.

As mentioned above in Proposition 5.1, a tropical hypersurface $V(\text{trop}(f))$ is a balanced rational polyhedron complex. The same conclusion holds for $V(\text{trop}(I))$.

Proposition 5.5. *Given any ideal $I \subset K[x_1, \ldots, x_n]$ such that its affine variety $V(I)$ is irreducible, let d denote the dimension of $V(I)$. Then its associated tropical variety $V(trop(I))$ is d-dimensional weighted rational polyhedral complex in $\mathbb{R}^n$ satisfying the balancing condition at every face of dimension $d-1$.*

The balancing condition was defined in the previous subsection. This result is usually called *structure theorem* of tropical geometry, and it is a generalization of the corresponding results for tropical plane curves and tropical hypersurfaces in Proposition 5.1. It follows from a combination of Theorem 5.3 and [15, Theorem A]. See also [94, §4] for more detail.

Remark 5.6. Recall from Proposition 4.19 that any piecewise linear plane curve with rational slopes satisfying the balancing condition at every vertex is a tropical plane curve. Though the same conclusion also holds for piecewise-linear hypersurfaces in $\mathbb{R}^n$, it does not hold for high codimensional piecewise linear subspaces. For example, there exist piecewise linear curves in $\mathbb{R}^3$ with rational slopes (or directions) which satisfy the balancing condition at every vertex but are not tropical varieties defined by ideals of tropical polynomials. See [30, §6.3]. This is one reason for difficulty in obtaining the most general and consistent definition of higher dimensional tropical varieties by imitating equivalent definitions for tropical plane curves. There are other approaches. For example, general tropical spaces of high dimension are defined in [106, Chap 6] differently.

5.3 Tropical varieties as logarithmic limit sets

One of the first instances where tropical geometry arose and was used crucially is the paper [10] on logarithmic limit sets of affine varieties. Several equivalent definitions related to the definitions of tropical varieties mentioned above were introduced in this paper.

As we have seen above, the crucial feature of tropical geometry is the combination of *integral* and *linear* (or *affine*) structures. The basic idea of [10] is to reduce a nonlinear affine variety to its tropical geometry at infinity and hence to exploit the linear structure.

The construction in [10] was motivated by the following problem in commutative algebra. Let k be a field, and $k[X_1^{\pm},\dots,X_n^{\pm}]$ the Laurent polynomial ring. Since this is the group algebra on the free abelian group of rank n, i.e., $\mathbb{Z}^n$, the group $\mathrm{GL}(n,\mathbb{Z})$ acts on it naturally. For every ideal $I \subset k[X_1^{\pm},\dots,X_n^{\pm}]$ which is of infinite dimension and infinite codimension as a k-vector space, let $H \subset \mathrm{GL}(n,\mathbb{Z})$ be its stabilizer. The problem is to prove that H contains a finite index subgroup H_0 which stabilizes an infinite subgroup of $\mathbb{Z}^n$ which is also of infinite index in $\mathbb{Z}^n$, i.e., elements of H_0 are of the block upper-triangular form $\begin{pmatrix} A & B \\ 0 & C \end{pmatrix}$.

The problem amounts to prove the existence of an infinite subgroup of $\mathbb{Z}^n$ which has finitely many orbits under H. The idea of the proof is as follows. Since H stabilizes the ideal I, it acts on its associated affine variety $V(I) \subset (\mathbb{C}^*)^n$. To pick out a "sublinear" object of $\mathbb{Z}^n$ as described above, consider the behavior at infinity of $V = V(I)$. Since $\mathrm{GL}(n,\mathbb{Z})$ acts linearly on the exponents of Laurent polynomials in $k[X_1^{\pm},\dots,X_n^{\pm}]$, it is more natural to consider the ratios $\log|x_1| : \dots : \log|x_n|$ when $x = (x_1,\dots,x_n)$ goes to infinity.

Assume that k has an absolute value $|\cdot|$, i.e., a function $|\cdot| : k \to \mathbb{R}_{\geq 0}$ satisfying the conditions:

1. $|0| = 0, |1| = 1$.
2. For $x, y \in k$, $|xy| = |x||y|$, and $|x+y| \leq |x| + |y|$.

Furthermore, assume that there exists $x \in k$ with $0 < |x| < 1$.

Let $B^n \subset \mathbb{R}^n$ be the unit ball, and $V(k) \subset k^n$ the set of k-points of V. Define a map

$$\log : V(k) \longrightarrow B^n$$

by

$$\log(x) = (1 + (\log|x_i|)^2)^{-\frac{1}{2}}(\log|x_1|, \ldots, \log|x_n|).$$

Then the limit points of $\log(V(k))$ on the unit sphere S^{n-1}, i.e., the intersection of the closure $\overline{\log(V(k))}$ with S^{n-1} is called the *logarithmic limit* set of V and denoted by $V_\infty^{(a)}$.

When k is algebrically closed, it is clear that $V_\infty^{(a)}$ is a nonempty subset, and the stabilizer H of I acts on $V_\infty^{(a)}$. When V is a hypersurface defined by a single equation, it is easy to see that $V_\infty^{(a)}$ is a proper subspace of S^{n-1}. Therefore, it is also a proper subspace in general. More precisely, it can be shown that $V_\infty^{(a)}$ is a proper subset of dimension equal to $\dim_k V - 1$ [10, Theorem 3, p. 467].

If $k = \mathbb{C}$ and I is a homogeneous ideal $k[x_1, \ldots, x_n]$, then limit points of ratios $x_1 : \ldots : x_n$ for $x \in V(k)$ will form a proper nonempty subvariaty of the complex projective space $\mathbb{C}P^{n-1}$.

On the other hand, one crucial result in [10] is that the logarithmic limit set $V_\infty^{(a)}$ is contained in a *finite union of great subspheres* in S^{n-1}, which are rational in certain sense and hence related to tropical varieties. Given this, one can show that there exists a nonempty unique such union U which is minimal with respect to the property of containing the limit set $V_\infty^{(a)}$. Since great subspheres of U correspond to subgroups of $\mathbb{Z}^n$ by duality, a solution to the above problem follows from the finiteness of great subspheres in the union U.

In order to prove the above result, two other related subsets $V_\infty^{(b)}$ and $V_\infty^{(c)}$ in S^{n-1} were introduced in [10].

The subset $V_\infty^{(b)}$ consists of points $(-v(x_1), \ldots, -v(x_n))$ as v runs over all *real-valued valuations* of $k[x_1, \ldots, x_n]/I$ satisfying $\sum_{j=1}^n v(X_j)^2 = 1$. Note that this is suggestive of the non-Archimedean amoeba of V defined in the previous subsection, since every point of $V(k)$ gives an evaluation (or specialization) map $k[x_1, \ldots, x_n]/I \to k$. But these maps are not real-valued evaluations.

The subset $V_\infty^{(c)}$ is defined to be the tropical variety $V(\text{trop}(I))$, when the valuation of K is the trivial one.

It was proved in [10, Theorem 2] that $V_\infty^{(b)} = V_\infty^{(c)}$ and contains $V_\infty^{(a)}$, and points in $V_\infty^{(c)}$ with rational coordinates are contained in $V_\infty^{(a)}$. The conjecture in [10, pp. 466-467] stating that $V_\infty^{(a)} = V_\infty^{(c)}$ was proved in [15, Theorem A].

5.4 Tropical varieties as limits of Archimedean amoebas

As explained above, tropical varieties can be realized as non-Archimedean amoebas. For affine varieties defined over $\mathbb{C}$, there is also a notion of amoebas, called Archimedean amoebas, and their limits give rise to tropical varieties. This gives another relation between classical varieties and tropical varieties.

The papers [104] and [51] give good introductions to Archimedean amoebas, and the papers [114] and [66] contain more comprehensive discussions. We will give a brief description of Archimedean amoebas.

Recall that for $t > 1$, we have defined a function

$$\log_t a = \log a / \log t \quad \text{for } a \in \mathbb{R}_{>0}.$$

For each positive integer n, we define a map

$$\mathrm{Log}_t : \mathbb{C}^n \to (\mathbb{T})^n, \quad (z_1, \dots, z_n) \mapsto (\log_t(|z_1|), \dots, \log_t(|z_n|)).$$

It restricts to

$$\mathrm{Log}_t : (\mathbb{C}^\times)^n \to \mathbb{R}^n.$$

In particular, when $n = 2$, we obtain a map

$$\mathrm{Log}_t : (\mathbb{C}^\times)^2 \to \mathbb{R}^2, \quad (z_1, z_2) \mapsto (\log_t(|z_1|), \log_t(|z_2|)).$$

For any algebraic variety V in $\mathbb{C}^n$, its image under the map Log_t,

$$\mathrm{Log}_t(V(\mathbb{C}) \cap (\mathbb{C}^\times)^n),$$

is called the *amoeba* of V. Since $\mathbb{C}$ is an Archimedean field, it is an *Archimedean amoeba.*

A basic result on Archimedean amoebas and tropical varieties is the following.

Proposition 5.7. *When $t \to +\infty$, $\mathrm{Log}_t(V(\mathbb{C}))$ converges to a tropical variety.*

We consider several examples of amoebas of plane algebraic curves. Let $P(z_1, z_2) = z_1 + z_2 + 1$, and C_P be its plane algebraic curve. The amoeba $\mathrm{Log}_t(C_P(\mathbb{C}))$ is a thin neighborhood of a curve in $\mathbb{R}^2$ which is the plane tropical curve Γ_p associated with the tropical polynomial $p(x,y) =$ "$x + y + 0$". As $t \to +\infty$, the amoeba $\mathrm{Log}_t(C_P(\mathbb{C}))$ converges to the plane tropical curve Γ_p (Fig. 14).

We can explain this result as follows. Suppose that z_1 is close to 0. Then z_2 is close to -1. This implies that $\mathrm{Log}_t(z_1, z_2)$ is close to $(\log_t(|z_1|, 0)$, one ray of the tropical curve Γ_p. The case of z_2 being small is similar. Suppose that z_1 is large, then z_2 is approximately $-z_1$, and hence $\mathrm{Log}_t(z_1, z_2)$ is close to the diagonal ray (x, x), $x \gg 1$. When z_1, z_2 are bounded, $\mathrm{Log}_t(z_1, z_2) \to (0,0)$ as $t \to +\infty$.

Similarly, for any nonzero numbers a, b, c, the amoeba of the plane algebraic curve defined by $P(z_1, z_2) = az_1 + bz_2 + c$ is also converging to the tropical plane curve Γ_p associated with the same tropical polynomial $p(x,y) =$ "$x + y + 0$".

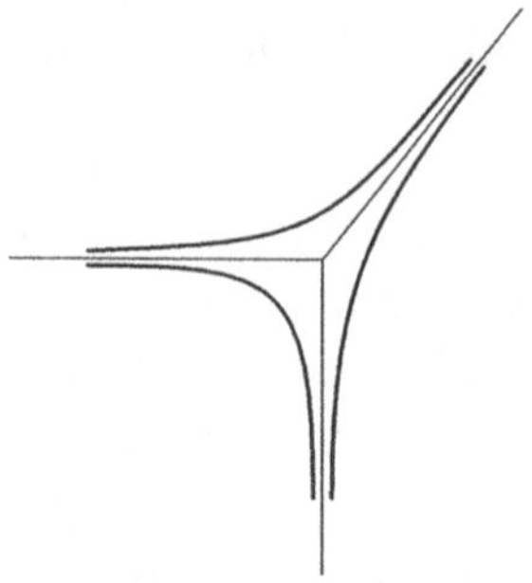

Fig. 14 amoeba of $z_1 + z_2 + 1 = 0$

On the other hand, if $c = 0$, the amoeba of the plane algebraic curve defined by $P_0(z_1, z_2) = az_1 + bz_2 = 0$ will converge to the straight line $x = y$, which is the tropical curve Γ_{p_0} for the tropical polynomial $p_0(x,y) =$ "$x+y$". (Note that p_0 is different from the tropical polynomial $p(x,y) =$ "$x+y+0$".)

As recalled earlier in §5.2, every polynomial over a field with valuation have a tropicalization. If we give $\mathbb{C}$ the trivial valuation val_0:

$$\mathrm{val}_0(z) = 0, \quad z \in \mathbb{C},$$

then for any polynomial $P(z_1, z_2)$ over $\mathbb{C}$, we can define its tropicalization

$$\mathrm{trop}_0(P) = \text{“}\sum_{ij} \mathrm{val}_0(a_{ij}) z_1^i z_2^j\text{”} = \text{“}\sum_{ij} z_1^i z_2^j\text{”},$$

where the sum is over the indexes such that $a_{ij} \neq 0$. By our discussion with tropical polynomial, the coefficient 0 is omitted from a monomial $0z_1^i z_2^j$.

Then Proposition 5.7 can be started more precisely in this special case.

Proposition 5.8. *For any polynomial $P(z_1, z_2)$ of two variables over $\mathbb{C}$, let $V(P)$ be the plane curve defined by P. When $t \to +\infty$, $\mathrm{Log}_t(V(P))$ converges to the tropical curve $V(\mathrm{trop}_0(P))$.*

Let us consider a few more examples. When $P(z_1, z_2) = az_1 + bz_2 + c$ with $a, b, c \neq 0$, its tropicalization $\mathrm{trop}_0(P)(x,y) =$ "$0x+0y+0$" $=$ "$x+y+0$", which is the limit of "$\log_t ax + \log_t by + \log_t c$". Similarly, the tropicalization of $P(z_1, z_2) = az_1 + bz_2$ is equal to "$x+y$".

Given any polynomial $P(z_1, \dots, z_n)$ over $\mathbb{C}$ in more variables, we can also define its tropicalization $\mathrm{trop}_0(P)$. Then the amoeba of the algebraic hypersurface $V(P)$ converges to the tropical hypersurface $V(\mathrm{trop}_0(P))$. More generally, Proposition 5.8 also holds all varieties in $\mathbb{C}^n$. See also [74].

It should be pointed out that tropical varieties obtained this way from subvariaries of $\mathbb{C}^n$ are rather special, and each is a union of polyhedral cones with the common vertex at the origin.

Remark 5.9. In the above discussion, we concentrated on the limit of amoebas of complex varieties. The topology of amoebas can be controlled also by the tropical varieties. For any complex hypersurface V defined by a polynomial P, the amoeba $\mathrm{Log}_t(V(\mathbb{C})$ can be deformation retracted to a spine given by the tropical hypersurface $\mathrm{trop}_0(P)$. See [114, p. 276] and [112].

5.5 Affine manifolds and tropical affine manifolds

In the above discussions, the underlying spaces of tropical curves, tropical hypersurfaces and tropical varieties are not topological manifolds in general. On other hand, some smooth manifolds are tropical varieties. Examples include lines in $\mathbb{R}^2$ with rational slopes and affine subspaces of $\mathbb{R}^n$ with rational slopes. No balancing conditions are needed or rather explicitly stated at any point since a neighborhood of every point is identically to an open subset of some $\mathbb{R}^k$, and the balancing condition is automatically satisfied for every point of $\mathbb{R}^k$.

If we use only these special tropical spaces as local models and patch them together, then we get *tropical manifolds*. Tropical abelian varieties discussed below are such tropical manifolds.

Specifically, a real smooth manifold M is called an *affine manifold* if it admits an atlas of local coordinate charts $\phi_i : U_i \subset\!\!\rightarrow V_i \subset \mathbb{R}^n$ such that for every pair of overlapping charts, the transition function $\phi_i\phi_j^{-1}$ is an affine function. An affine manifold M is called a *tropical manifold* if it admits an atlas of local coordinate charts as above such that the transition functions $\phi_i\phi_j^{-1}$ are *affine transformations whose linear parts have integral coefficients*, i.e., $\phi_i\phi_j^{-1} \in \mathrm{GL}(n,\mathbb{Z}) \rtimes \mathbb{R}^n$. These affine transformations are also called *integral affine transformations.*

It is helpful to emphasize that the translation parts of integral affine transformations are not necessarily integers. This implies that for any discrete subgroup Γ of $\mathbb{R}^n$, the quotient of $\Gamma\backslash\mathbb{R}^n$ under the translation action of Γ is a tropical manifold, and it inherits both the local linear and integral structures from $\mathbb{R}^n$. See [58, p. xiii].

6 Some alternative descriptions of tropical varieties and related spaces

In this section, we discuss some alternative descriptions of tropical varieties and variants. Though it is natural to expect that the moduli space of tropical curves is a tropical variety as in the case of complex curves, this is not be true in general. On the other hand, stacky fans seem to provide one appropriate framework to describe this tropical moduli space. The discussion of tropical fans also clarifies relations with fans used in toric varieties.

We will show later in §8.5 that another closely related framework to understand the moduli space $\mathcal{M}_n^{tr}$ is the notion of tropical orbifolds.

6.1 Fans and tropical fans

As pointed out earlier, for an algebraic variety $V \subset \mathbb{C}^n$ over $\mathbb{C}$, the limit of its amoeba $\mathrm{Log}_t(V(\mathbb{C}))$ is a tropical variety and consists of rational cones. They are related to fans for toric varieties, but there is an important difference between them due to the balancing conditions for tropical varieties.

We recall some basics of fans. The basic references used here are [52] and [38]. Let V be a finite dimensional real vector space, and $\Lambda \subset V$ a lattice, i.e., Λ is a discrete subgroup of $\mathbb{R}^n$ with a compact quotient $\Lambda\backslash\mathbb{R}^n$. Then the lattice Λ gives the vector space V an integral structure and consequently a rational structure. The most basic example is $V = \mathbb{R}^n$ and $\Lambda = \mathbb{Z}^n$.

Fix a lattice $\Lambda \subset V$. *A rational polyhedral cone* $\sigma \subset V$ is a convex cone defined by finitely many *integral linear* equalities and non-strict *integral linear* inequalities. Specifically, let Λ^* be the dual lattice in the dual vector space V^*. Then for any $v^* \in \Lambda^*$, a linear equation $v^*(x) = 0, x \in V$, is called an integral linear equation with respect to the integral structure induced by Λ. Integral linear inequalities are similarly defined.

Equivalently, there exist finitely many elements $u_1, \ldots, u_m \in \Lambda$ such that the cone σ can be written as

$$\sigma = \{\lambda_1 u_1 + \ldots + \lambda_m u_m \mid \lambda_1, \ldots, \lambda_m, \lambda \in \mathbb{R}_{\geq 0}\}.$$

In this case, we say that σ is spanned by $u_1, \ldots, u_m$. If σ can be generated by $\dim \sigma$ vectors, then σ is called a *simplicial cone*.

Let us consider a basic example. Take $V = \mathbb{R}^2$, $\Lambda = \mathbb{Z}^2$. Then

$$\sigma_0 = \{(x,y) \in \mathbb{R}^2 \mid x - y \geq 0, y \geq 0\}$$

is a rational polyhedral cone. It is clear that σ_0 is spanned by $u_1 = (1,1)$ and $u_2 = (1,0)$, and hence σ_0 is a simplicial cone (Fig. 15).

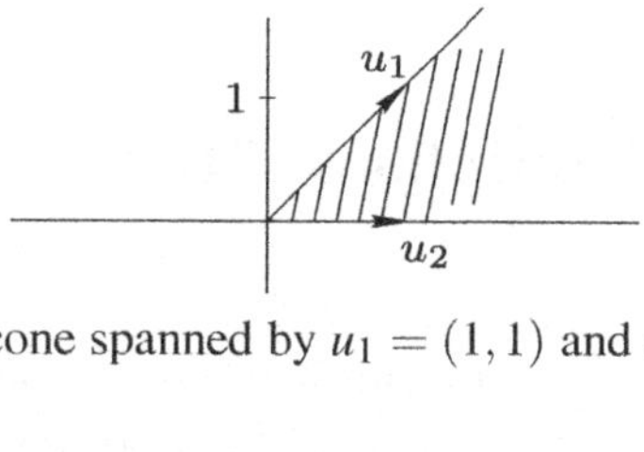

Fig. 15 cone spanned by $u_1 = (1,1)$ and $u_2 = (1,0)$

Definition 6.1. A rational polyhedral fan Σ in V is a collection of rational polyhedral cones in V satisfying the following conditions:

1. For any cone $\sigma \in \Sigma$, if τ is a face of σ, then τ also belongs to Σ.
2. The intersection of every two cones in Σ is either empty or a face of each of them.

If every cone σ of a fan Σ is a simplicial cone, then the fan Σ is called a *simplicial fan*. The dimension of Σ is defined to be the supremum of $\dim\sigma$, where $\sigma \in \Sigma$. The fan Σ is called *pure-dimensional* if each maximal cone under the partial order of face inclusion has dimension equal to $\dim\Sigma$, i.e., all maximal cones have the same dimension.

Unlike rational polyhedral fans which appear in toric geometry, each cone σ here is not required to be *strictly* convex, i.e., it can contain whole straight lines. For example, the whole space V and closed halfspaces are allowed.

Let $n = \dim\Sigma$, and $\Sigma^{(n)}$ denote the subset of Σ consisting of top dimensional cones in Σ. For any $\sigma \in \Sigma^{(n)}$ and each codimension 1 face τ of σ, there is a *primitive integral normal vector* of σ relative to τ which is normal to τ and points to σ, and is denoted by $u_{\sigma/\tau}$.

Definition 6.2. A weighted rational polyhedral fan in V is a pair (Σ, ω) of a rational polyhedral fan Σ and a positive integral weight function $\omega : \Sigma^{(n)} \to \mathbb{Z}_{>0}$, which assigns a weight to every top dimension cone σ. A tropical fan in V is a weighted rational polyhedral fan (Σ, ω) such that for every codimension 1 cone τ in Σ, the following balancing condition is satisfied:

$$\Sigma_{\sigma\in\Sigma^{(n)},\ \sigma\supset\tau} \quad \omega(\sigma)u_{\sigma/\tau} = 0.$$

When $\dim\Sigma = 1$ and $V \cong \mathbb{R}^2$, then a weighted rational polyhedral fan is a plane tropical curve discussed before.

When $\dim V \geq 2$ and $\dim\Sigma = 1$, then (Σ, ω) is a tropical curve in the real vector space V. Even if the weight function ω is identically equal to 1, Σ is not a *smooth* tropical curve in general.

Remark 6.3. Tropical fans are model spaces which can be used to define general tropical spaces in [106, Chap 6]. There is also a notion of smooth tropical varieties in higher dimension in [102].

One important reason for introducing the notion of tropical fans is the following result [102, Theorem 1] and [52, Theorem 3.7].

Proposition 6.4. *The moduli space $\mathcal{M}_{0,n}$ of rational tropical curves (i.e., tropical curves of genus 0) with n marked points is a tropical fan of dimension $n-3$ for $n \geq 3$. It is also a simplicial fan and a smooth tropical variety.*

For a definition of tropical curves with marked points, see §8.4 below.

6.2 Stacky fans

When $n \geq 1$, the moduli space $\mathcal{M}_n^{tr}$ of compact smooth tropical curves of genus n is not a tropical variety, or rather at least not known to be a tropical variety (see [37, p. 1136, p. 1163] and [100, §3.1. C]). To describe these moduli spaces in a proper framework, the notion of stacky fans and ideal stacky fans was introduced in [38]. The basic point is that $\mathcal{M}_n^{tr}$ does have structures induced from some fans, but does not satisfy balancing conditions such as those for tropical fans as in the previous subsection. We will propose the notion of tropical orbifolds later in §8.5.

To deal with noncompactness of $\mathcal{M}_n^{tr}$, ideal rational polyhedral cones were also introduced in [38]. The main difference with rational polyhedral cones is that they are not necessarily closed.

Ideal polyhedral cones and ideal fans

Let Λ be a lattice in a finite dimensional real vector space V as in the previous subsection. Rational polyhedral cones in V are closed subspaces and will be denoted by $\overline{\sigma}$.

Definition 6.5. An ideal rational polyhedral cone σ of V is an open convex subset of a rational polyhedral cone $\overline{\sigma}$ obtained by removing some of its proper faces. In particular, σ contains the interior of $\overline{\sigma}$.

For example, for any rational polyhedral cone $\overline{\sigma}$, its interior is an ideal rational polyhedral cone, and it has no proper face since all its faces have been removed.

Now we put together ideal rational polyhedral cones to form ideal rational polyhedral fans. One naive definition is that an ideal rational polyhedral fan Σ in V is a collection of ideal rational polyhedral cones $\{\sigma\}$ in V such that their closures $\{\overline{\sigma}\}$ form a rational polyhedral fan $\overline{\Sigma}$. But this has the problem of some missing rational polyhedral cones which appear as faces of some given cones and also contain some other given cones, and this makes it difficult to define a topology on the fan and understand face relations between the cones.

Instead, we can start with a rational polyhedral fan $\overline{\Sigma}$, and remove a subset Z of cones in $\overline{\Sigma}$. For any rational polyhedral fan $\overline{\Sigma}$, the union of its cone with the topology induced from the topology of each cone and the face relation between them is called the underlying topological space and denoted by $\mathrm{supp}(\overline{\Sigma})$. Similarly, any subset Z also has an underlying space $\mathrm{supp}(Z)$. Then the space $\mathrm{supp}(\overline{\Sigma}) - \mathrm{supp}(Z)$ can be written uniquely as a union of ideal rational polyhedral cones. Such a union of ideal rational polyhedral cones is called an *ideal rational polyhedral fan*, or an *ideal fan*, and is denoted by Σ.

Note that an ideal rational polyhedral fan Σ also satisfies the two basic properties:

1. A face τ of a cone σ in Σ is also a cone in Σ, though a face of the closure of σ may not belong to Σ.
2. And the intersection of two ideal cones is a face of each.

But we cannot define an ideal rational polyhedral fan by these two properties alone. The reason is that too many faces could be missing, and these properties alone do not impose enough restrictions between different cones.

Given an ideal fan Σ as above, we can also define its underlying space

$$\mathrm{supp}(\Sigma) = \mathrm{supp}(\overline{\Sigma}) - \mathrm{supp}(Z).$$

We note that a fan or an ideal fan Σ is called *locally finite* if every point x in $\mathrm{supp}(\Sigma)$ has some open neighborhood that meets only finitely ideal cones of Σ.

Note that when $\overline{\Sigma}$ is locally finite, then the support of any subset Z of cones in $\overline{\Sigma}$ is a closed subset. In applications to tropical Teichmüller spaces in this book, local finiteness might not hold and some subtlety occurs [38, §2, Lemma 2.1.5].

Lemma 6.6. *Let Σ be an ideal rational polyhedral fan in V. Let*

$$X_\Sigma = (\coprod \sigma_i)/\sim$$

be the space obtained by gluing the ideal cones $\sigma_i \in \Sigma$ on their interaction and endowed with the sum topology. Then Σ is locally finitely if and only if the natural map $X_\Sigma \longrightarrow \mathrm{supp}(\Sigma)$ is a homeomorphism.

In some sense, the abstract space X_Σ is more intrinsic. We can also glue ideal rational polyhedral cones and ideal rational polyhedral fans contained in different real vector spaces to get some abstract spaces which are not embedded in some concrete vector spaces.

Stacky fans are glued from ideal rational polyhedral cones and ideal rational polyhedral fans using integral linear maps.

Definition 6.7. Let V, V' be two finite dimensional linear spaces, $\Lambda \subset V$ and $\Lambda' \subset V'$ be lattices, and $\sigma \subset V$, $\sigma' \subset V'$ be ideal rational polyhedral cones. Let $L : \sigma \longrightarrow \sigma'$ be an inclusion of σ as a face of σ' and restricted from a linear map $L : V \longrightarrow V'$. Then L is called an integral (or lattice-preserving) inclusion if $L(\Lambda) \subset \Lambda'$ and $L(\sigma)$ is a face of σ'.

Let $\{\sigma_i \subset V_i\}$ be a collection of ideal rational polyhedral cones. Let $\{L_\alpha\}$ be a collection of integral inclusions of cones in $\{\sigma_i \subset V_i\}$, $L_\alpha : \sigma_i \longrightarrow \sigma_j$. Note that we allow $i = j$, and for each pair i, j, there could be multiple linear maps L_α.

Definition 6.8. Let $X = (\coprod_i \sigma_i)/\sim$, where $\sim$ is the equivalence relation generated by identifying each $x \in \sigma_i$ with $L_\alpha(x) \in \sigma_j$ for all possible maps $L_\alpha : \sigma_i \longrightarrow \sigma_j$. Then X is called an ideal stacky fan.

If every ideal rational polyhedral cone of the collection $\{\sigma_i \subset V_i\}$ is closed, i.e., it includes all its faces, then the ideal stacky fan is called a *stacky fan*.

When $\Sigma = \{\sigma_i\}$ is an ideal rational polyhedral fan in a finite dimensional vector space V, then the inclusions of faces as in Lemma 6.6 are integral, and induce an equivalence relation $\sim$ on $\coprod_i \sigma_i$. The quotient space $X_\Sigma = (\coprod_i \sigma_i)/\sim$ is an ideal stacky fan with cells $\{\sigma_i^0\}$.

More generally, we can also glue a collection of ideal rational polyhedral fans in different finite dimensional vector spaces via integral inclusions of cones and linear maps between vector spaces to get stacky fans.

There are basic two ingredients in constructing ideal stacky fans:

1. Gluing rational polyhedral cones.
2. Identifying polyhedral cones, for example via group actions.

Both steps are used in constructing the moduli space $\mathcal{M}_n^{tr}$ as an ideal stacky fan.

6.3 Tropical hypersurfaces as super currents

In complex analysis, a complex hypersurface can be represented by a positive current satisfying certain hypotheses. Positive currents in complex analysis are related to the convexity defined through plurisubharmonic functions.

Tropical varieties are special subsets of $\mathbb{R}^n$ related to two basic notions: *convexity* and *linear structures*. The linear structure, or rather piecewise linear structure, is clear, and the convexity comes from the fact that each tropical polynomial gives a real-valued conex function.

Motivated by convex functions, the paper [86] introduces a notion of positive currents on $\mathbb{R}^n$, and proves that tropical hypersurfaces in $\mathbb{R}^n$ correspond to closed, positive super current, i.e., $(1,1)$-currents, in $\mathbb{R}^n$ whose support is of dimension $n-1$, and whose normal vectors are integral [86, Proposition 5.3].

It also defines higher codimensional tropical varieties by the support of strongly positive, closed current [86, Definition 5.5].

The characterization of tropical hypersurfaces and the definition of tropical varieties are used to study intersection of tropical hypersurfaces in [86].

6.4 Tropical varieties versus Berkovich spaces

One application of the construction of tropical varieties via valuation maps of function fields is to understand relations between tropical varieties and Berkovich spaces [116].

For any algebraic variety X defined over an algebraically closed field K that is complete with respect to a *nontrivial non-Archimedean valuation*, there is an associated *Berkovich space*, which is its *Berkovich analytification* X^{an}.

As explained above in §5.2, when X is embedded into $(K^\times)^n$, its amoeba is a tropical variety $\mathrm{Trop}(X) \subset \mathbb{R}^n$. Similarly, when X is embedded into a toric variety over K, there is also an amoeba of X. The reason is that the K-torus $(K^\times)^n$ is an open dense subset of any toric variety of dimension n, and we can consider the intersection of a toric variety with $(K^\times)^n$ and take its image under val.

It is natural to expect that these two spaces are related. The paper [116] makes the relation explicit: the Berkovich analytification X^{an} is topologically the inverse limit of tropicalizations of X obtained from all toric embeddings of X and all equivariant toric maps between these toric embeddings.

Some brief explanations of the above summary are in order. For an affine variety U over the ground field K which is endowed with a non-Archimedean valuation, the *Berkovich analytification* U^{an} of U consists of all valuations of the ring of regular functions on U which extends the valuation on K. On the other hand, the tropical space $\mathrm{Trop}(U)$ associated with U is the image of these valuations on the coordinate functions. More specifically, such an image was shown to be a rational polyhedron complex in [15, Theorem A], and it was identified with the amoeba of U and hence the associated tropical variety in [45, Theorem 2.2.5]. Of course, we get a surjective map from U^{an} to the tropicalization $\mathrm{Trop}(U)$ of U. Now different choices of functions correspond to different embeddings of U into toric varieties, and the inverse limit over all these different choices exhaust the space U^{an}, i.e., there is no nontrivial valuation whose values on all functions are trivial.

7 Applications of tropical geometry

There are many applications of tropical geometry. This is the reason for the rapid development of tropical geometry in the recent years. In this section, we list some applications and make brief comments whenever it is possible for us to do. It is clear that we will have to miss many applications or do not explain them well. The interested reader can also look up some books and survey papers such as [96], [106], [58], [27], [28] [30], [124], [99], [100], [103], [104], [105], [51], [94], [19], and [77] to gain a bigger perspective on applications of tropical geometry.

7.1 Logarithmic compactification of algebraic varieties

Tropical varieties, or rather their intersections with spheres, appeared as the logarithmic limit sets V_∞ of algebraic varieties V in $(\mathbb{C}-\{0\})^m$ [10]. As explained above in §5.3, the paper [10] was motivated by a problem in commutative algebra.

The logarithmic limit sets V_∞ can also be used to understand the geometry at infinity of V, since it gives a description of asymptotic behavior (not directions) of the variety V in $(\mathbb{C}-\{0\})^m$. Together with the variety V, it gives a topological compactification

$$V \cup V_\infty.$$

It is called the *logarithmic compactification.* Since the logarithmic limit set V_∞ is contained in the unit sphere S^{n-1} in $\mathbb{R}^n$, while V is contained in $(\mathbb{C}-\{0\})^m$, the topology of the compactification $V \cup V_\infty$ is not so obvious.

To explain the topology of such compactifications, we start with some examples. The space $\mathbb{C}-\{0\}$ has two ends. Clearly, it has two logarithmic limit points, $-\infty$ and ∞. The variety $\mathbb{C}-\{0\}$ admits a natural compactification by adding these two points, so the compactified space is $\mathbb{C}P^1 = \mathbb{C}\cup\{\infty\}$.

When $m \geq 2$, $(\mathbb{C}-\{0\})^m$ has only one end. The logarithmic limit set is the whole unit circle S^1 in $\mathbb{R}^2$.

To understand this compactification, we map $(\mathbb{C}-\{0\})^m$ surjectively to $\mathbb{R}^2$ under the map

$$\mathrm{Log} : (\mathbb{C}-\{0\})^2 \to (\log|z_1|, \log|z_2|).$$

This gives $(\mathbb{C}-\{0\})^2$ the structure of a principle $(S^1)^2$-fiber bundle over $\mathbb{R}^2$. Since the fibers are $(S^1)^2$ and are compact, and the base $\mathbb{R}^2$ is compactified by adding the circle S^1, the circle at infinity, we obtain a compactification of $(\mathbb{C}-\{0\})^2$ by adding a circle S^1 to the base while collapsing the fibers,

$$(\mathbb{C}-\{0\})^2 \cup S^1.$$

In other words, the fibers $(S^1)^2$ over the interior base points of $\mathbb{R}^2$ collapse to single points when the base points move towards the boundary S^1 of the base space.

In some sense, it is similar to compactifications of $(\mathbb{C}-\{0\})^2$ via toric varieties. The difference is that in this toric variety case, we compactify the base $\mathbb{R}^2$ by a polyhedral complex dual to a fan which defines the toric variety, and the fibers $(S^1)^2$ partially or completely degenerate, i.e., converging to points or circles S^1, depending on where the base point lies on the boundary polyhedral complex.

Similar results hold for $(\mathbb{C}-\{0\})^n$. For example, the analogy of the map Log gives the following logarithmic compactification $(\mathbb{C}-\{0\})^n$,

$$(\mathbb{C}-\{0\})^n \cup S^{n-1}.$$

We note that toric varieties or toric compactifications of $(\mathbb{C}-\{0\})^n$ are often algebraic or projective varieties. On the other hand, the logarithmic compactification $(\mathbb{C}-\{0\})^n \cup S^{n-1}$ is not an algebraic space when $n \geq 2$, but only a topological space.

Using the above logarithmic compactification of $(\mathbb{C}-\{0\})^n$, we can construct a logarithmic compactification $V(\mathbb{C}) \cup V_\infty$ for general varieties V.

Proposition 7.1. *For any algebraic variety $V \subset (\mathbb{C}-\{0\})^n$, the closure of V in the logarithmic compactification $(\mathbb{C}-\{0\})^n \cup S^{n-1}$ is a compactification of V and is homeomorphic to $V \cup V_\infty$.*

Though the logarithmic compactification $V \cup V_\infty$ is not an algebraic space, the boundary V_∞ has a piecewise linear structure and is the intersection of the tropical variety $\lim_{t\to+\infty} \mathrm{Log}_t(V)$ in $\mathbb{R}^n$ with the unit sphere S^{n-1}. This fits well the description of the tropical variety as the limit of Archimedean amoebas $\lim_{t\to+\infty} \mathrm{Log}_t(V)$ in §5.4.

The above discussion also shows that the tropical variety $\lim_{t\to+\infty} \mathrm{Log}_t(V)$ is the polyhedral cone over the logarithmic limit set V_∞ with the vertex at the origin.

7.2 Topology of real algebraic curves

In algebraic geometry, we usually study algebraic varieties over algebraically closed fields such as $\mathbb{C}$.

On the other hand, the study of algebraic varieties over $\mathbb{R}$, in particular, the real locus of algebraic varieties defined over $\mathbb{R}$ is classical and important. For example, in elementary geometry, we consider ellipses (in particular circles), parabola, and hyperbola. They correspond to plane algebraic curves of degree 2. When extended to $\mathbb{C}$, ellipses and hyperbola become isomorphic.

Understanding the real locus of plane curves of high degree is a complicated problem. The structure of each connected component is relatively simple, and the problem is to understand the number of connected components and their relative positions.

A sharp upper bound on the number of connected components of real plane curves in terms of the degree of the defining polynomial was obtained by Harnack in 1876. Describing their relative positions is one part of Hilbert's Sixteenth Problem.

Viro developed a technique, *patchworking*, to complete the solution of this problem for degrees up to 7. His technique was one of the motivations of tropical plane curves. Briefly, start with a family of polynomials with real coefficients. We know that their complex loci have shadows described by the tropical plane curves which are defined by the tropicalization of the polynomials. Now a certain combinatorial procedure tells us how to modify the tropical plane curves, and the result describes well the mutual positions of the connected components of the real plane curves. See [27] for an introduction and [130], [131] for more details.

7.3 Geometric and combinatorial group theory

As mentioned in the introduction, $\mathrm{Out}(F_n)$ is a basic group in combinatorial and geometric group theory. The identification of smooth tropical curves with metrics graphs allows one to apply the tropical geometry to study $\mathrm{Out}(F_n)$. This was explained briefly earlier and will be discussed in more detail later.

But tropical varieties were used in combinatorial group theory before the tropical geometry appeared as a subject. One motivation was to find criteria for finite presentations of metabelian groups in [17]. We note that a group G is called *metabelian* if it is an extension of an abelian group by another abelian group, i.e., G fits into an exact sequence

$$1 \to A \to G \to Q \to \{1\}, \tag{7.1}$$

where A, Q are abelian groups. It is known that G is finitely generated if and only if Q is finitely generated and A is finitely generated as a Q-module. But the question about finite presentation of G is much more complicated.

Assume that Q is a finitely generated abelian group of torsion-free rank n. Identify $\mathrm{Hom}(Q,\mathbb{R})$ with $\mathbb{R}^n$. Two homomorphisms of $Q \to \mathbb{R}$ are defined to be equivalent if they differ only by a positive constant multiple. Let

$$V = \mathrm{Hom}(Q,\mathbb{R})/\sim$$

be the quotient under this equivalence relation. Then V can be identified with the unit sphere $S^{n-1} \subset \mathbb{R}^n$. For every $v \in V$, define the monoid

$$Q_v = \{q \in Q \mid v(q) \geq 0\}.$$

For any finitely generated Q-module A, define a subspace

$$\Sigma_A = \{v \in V \mid A \text{ is finitely generated over } Q_v\}$$

of the topological space $V = S^{n-1}$. Two main results of [17] are the following:

1. G is polycyclic if and only if $\Sigma_A = S^{n-1}$.
2. G is finitely presented if and only if $\Sigma_A \cup -\Sigma_A = S^{n-1}$.

One consequence is that whether G is finitely presented depends only on the Q-module A, but not on the extension class given by Equation (7.1).

The subset Σ_A is related to the logarithmic limit set of [10] and was also extended and studied in [16]. In some cases [16, Corollary F], such sets are disjoint unions of finitely many open convex rational polyhedra in the unit sphere, and cones over them give rational polyhedral cones.

7.4 Three-dimensional topology

There are several applications of tropical geometry to 3-dimensional manifolds. Here is the basic way in which tropical varieties are used. For a given 3-dimensional manifold M, consider the character variety X which consists of representations of $\pi_1(M)$ into $\mathrm{SL}(2,\mathbb{C})$. This is a noncompact affine variety and we can consider its logarithmic limit set X_∞. Some topological invariants of M can be obtained in this way. They include relations between ideal points in the logarithmic limit set X_∞, and essential surfaces in M and slopes of the boundary of essential surfaces. (By definition, an essential surface of M is an immersion of a surface $i : S \to M$ such that i_* is injective on $\pi_1(S)$ and the image $i_*\pi_1(S)$ is not conjugate to a subgroup of $\pi_1(\partial M)$, where ∂M is the boundary of M.)

Since these results are technical and take substantial space to explain them, we only list some papers here without explaining their results: [41], [108], [40], and [127].

For example, in the roughest sketch, one result of the paper [41] can explained as follows: From a cusp of an algebraic curve in the character variety X, we get a valuation of the function field of the variety X, which gives an action on a tree. By the general result of Stalling, we get an essential surface in the 3-manifold M using this action on a tree.

The paper [108] generalizes [41] to obtain more actions on trees in order to compactify the character variety X, which generalizes Thurston's compactification of Teichmüller space.

In [128], logarithmic sets (or tropical geometry) were used to compactify character variety and also to construct essential surfaces explicitly. Tropical curves also occur in another study of knots via Jones polynomials and Newton polygons in [50].

7.5 Boundary of Teichmüller space $\mathscr{T}_g$

It is known that Teichmüller space $\mathscr{T}_g$ of compact Riemann surfaces of genus g was introduced by Teichmüller to study the moduli space $\mathscr{M}_g$ of compact Riemann surfaces of genus g. It is known that $\mathscr{M}_g$ is a quasi-projective variety. Since $\mathscr{T}_g$ is an infinite orbitfold covering of $\mathscr{M}_g$, it might be a bit surprising that $\mathscr{T}_g$ has the structure of a real algebraic space. This can indeed happen when we identify $\mathscr{T}_g$ with a subspace of the character (or representation) variety of the surface fundamental group $\pi_1(S_g)$, where S_g is a compact oriented surface of genus g.

Viewing $\mathscr{T}_g$ as a real algebraic space, we can use the idea of understanding the structure at infinity of algebraic varieties via the logarithmic limit set or tropical geometry to study $\mathscr{T}_g$ and some related spaces. For example, the Thurston compactification of $\mathscr{T}_g$ can be recovered in this way in [108]. (We note that when $g \geq 2$, the Teichmüller space $\mathscr{T}_g$ is differmorphic to $\mathbb{R}^{6g-6}$ and can intuitively be compactified by adding a sphere S^{6g-7} at infinity, though there are many different ways to carry this out. Since Mod_g acts continuously and properly on $\mathscr{T}_g$, it is natural to expect that this action of Mod_g extends continuously to such compactifications of $\mathscr{T}_g$. It turns out that among several known compactifications of $\mathscr{T}_g$, only on the Thurston compactification the mapping class group Mod_g acts continuously.) For some related related papers on compactifications of $\mathscr{T}_g$ from the above point of view, see [41], [4].

It is also known that the boundary of the Thurston compactification of $\mathscr{T}_g$ has a piecewise linear structure. See for example [118]. This piecewise linear structure can be well explained by the piecewise linear structure of the logarithmic limit sets, or tropical varieties. See [4] and also [117].

7.6 Enumerative algebraic geometry

This is a major direction of applications of tropical geometry and has been actively studied by many people. In some sense, these works brought the tropical geometry into prominence.

The basic idea is that for algebraic curves over $\mathbb{C}$, their Archimedian amoebas and associated tropical curves (i.e., limits of the amoebas) are effective approximations to the curves as far as counting of curves is concerned. The piecewise linear and combinatorial properties of tropical curves make the counting easier, and explicit combinatorial formulas for numbers of some specified classes of curves can be obtained. For example, fix two integers d, g, and a finite set $\mathscr{X}$ of points in the plane $\mathbb{C}^2$, a basic problem in enumerative algebraic geometry is to count the number of plane algebraic curves of degree d and genus g passing through $\mathscr{X}$. When $g = 0$, there is a recursive formula for this number via the associativity of the quantum cohomology [80], and for $g \geq 0$, there is also an algorithm to compute this number by using more classical algebraic geometry [33]. Using tropical geometry, explicit combinatorial formulas for this number of curves have been derived for all in [103, pp. 313-314].

There are many papers and survey papers on this, for example, [103], [105], [94], [51], and [28]. For more recent results, see [53], [54], [29], and [47].

These applications also have motivated people to develop intersection theories of tropical varieties and other general results too. See [122] for example.

7.7 Mirror symmetry of Calabi-Yau manifolds

In string theory, a certain type of compact Kähler-Einstein manifolds, called *Calabi-Yau manifolds*, is an essential ingredient. Briefly, a Calabi-Yau manifold is a Kähler manifold whose first Chern class is trivial. By Yau's celebrated solution to Calabi conjecture, such a manifold admits a Ricci flat Kähler-Einstein metric and hence its canonical line bundle it trivial. Usually the fundamental group of Calabi-Yau manifolds is assumed to be finite or even trivial. Calabi-Yau manifolds are higher dimensional generalizations of K3 algebraic surfaces, which are simply connected projective algebraic surfaces with trivial first Chern class.

There is an observed *mirror symmetry* between Calabi-Yau manifolds, for example, in terms of Hodge numbers. More importantly, computation of periods on one Calabi-Yau manifold predicts number of certain rational curves on the mirror Calabi-Yau manifold. This provides a surprising and effective method to count the number of rational curves, which are Gromov-Witten invariants. As mentioned in the previous subsection, counting such numbers is a fundamental problem in enumerative algebraic geometry.

One challenging problem is to understand this mirror symmetry and explain the reason behind it. There have been several conjectures which were made to explain this phenomenon. The most prominent ones are Kontsevich's *Homological Mirror Symmetry Conjecture* and the more geometric *SYZ conjecture* by Strominger, Yau, and Zaslow. The basic idea of the SYZ conjecture is to explain the mirror symmetry by a pair of torus fibrations of the Calabi-Yau manifolds whose fibers are dual to each other.

Both conjectures are still widely open. Gross and Siebert (see the book [58]) have a program to do this using tropical geometry, or the related log geometry. (Note that it is the log function which relates the tropical semifield to the semifield $\mathbb{R}_{\geq 0}$.)

Their approach focuses on the base of the special Lagrangian fibrations in the SYZ conjecture. The bases should be *tropical affine manifolds*, i.e., smooth real manifolds with a $\mathbb{Z}$-affine structure as defined in §5.5, or even more general singular spaces whose smooth loci support tropical affine structures. Briefly, given a tropical affine manifold B, it has a canonical integral basis $\frac{\partial}{\partial x_1}, \dots, \frac{\partial}{\partial x_d}$ of fibers of the tangent bundle TB and a canonical integral basis $dx_1, \dots, dx_d$ of the cotangent bundle T^*B from the local affine coordinates $x_1, \dots, x_d$ of B. They produce families of lattices Λ in TB and Λ^* in T^*B. Then the quotients $\Lambda \backslash TB$ and $\Lambda^* \backslash T^*B$ are fibers of the dual fibrations for a mirror pair.

One point of their program is to relate symplectic geometry and complex geometries of total spaces of the fibrations above to tropical geometrical properties of the base spaces. Then the common root in tropical geometry explains the mirror symmetry between the symplectic world (Gromov-Witten invariants) and the complex world (period integrals). Both the Gromov-Witten invariants and the mirror about period integrals are very complicated. If the program can be carried through, then, according to Gross [58, p. xv], "the geometry that now seems to underpin mirror symmetry, namely tropical geometry, is very simple and requires no particular background to understand".

The book [58] gives a detailed introduction to this approach through tropical geometry to mirror symmetry and many related concepts. Tropical geometry and combinatorial objects are also useful in constructing degenerations of Calabi-Yau manifolds and other problems in mathematical physics. See for example [81].

7.8 Arithmetic algebraic geometry

Another important application of tropical geometry is concerned with the generalized Bogomolov's conjecture for totally degenerate abelian varieties over function fields in [59] and [60].

This is an analogue of the generalized Bogomolov's conjecture for number fields which was proved in [137], and the complex analytic methods used in [137] were replaced by methods of tropical geometry and convex geometry in [60] to prove this conjecture over function fields.

Briefly, let K be an algebraic number field and $\overline{K}$ its algebraic closure. For an abelian variety A defined over K, we denote by $h(x)$ the Néron-Tate height of $x \in A(\overline{K})$ with respect to a symmetric and ample line bundle on A. A subvariety X of A is called *torsion* if X is a translate of an abelian subvariety by a torsion point of A. Then the generalized Bogomolov conjecture states that if a subvariety X of A is not torsion, then for a sufficiently small $\varepsilon > 0$, the set $\{x \in X(\overline{K}) \mid h(x) \leq \varepsilon\}$ is not Zariski dense in X. The statement for function fields is similar.

Let K be an algebraically closed field endowed with a nontrivial non-archimedean complete absolute value $|\ |$. For any algebraic subvariety X of the torus $\mathbb{G}_m^n \cong (\overline{K}^\times)^n$ defined over K, the valuation defines a map

$$\mathrm{val} : \mathbb{G}_m^n(K) \to \mathbb{R}^n, \quad x \mapsto (-\log|x_1|, \dots, -\log|x_n|).$$

One of the main results in [59], which is used in [60], says that *when X is an irreducible closed algebraic subvariety of $\mathbb{G}_m^n$ defined over K of dimension d, the tropical variety associated with X, i.e., the closure of $\mathrm{val}(X(K))$ in $\mathbb{R}^n$ (or the non-Archemedean amoeba of X), is a connected totally concave locally finite union of d-dimensional Γ-rational polytopes, where $\Gamma = \mathrm{val}(K^\times)$.*

Tropical varieties are also used in [59] to describe measures arising from Arakelov theory [39].

8 Moduli spaces of tropical curves and tropical Teichmüller spaces

As it is known, the moduli space $\mathscr{M}_g$ of compact Riemann surfaces of genus g is one of the most basic and extensively studied algebraic varieties. In fact, since every smooth algebraic curve over $\mathbb{C}$ is a Riemann surface, and every Riemann surface of conformally finite type (i.e., conformal to the complement of finitely many points of a compact Riemann surface) comes from an algebraic curve, $\mathscr{M}_g$ is also the moduli space of smooth projective algebraic curves over $\mathbb{C}$ of genus g.

The moduli space $\mathscr{M}_g$ was first introduced by Riemann in 1857 in his famous paper titled *the theory of Abelian functions*. Riemann counted the number of complex parameters needed to parametrize points, which is equal to $3g-3$ when $g \geq 2$. The problem of putting structures on and understanding properties of $\mathscr{M}_g$ became Riemann's moduli problem. After efforts by many people, it was only proved in the early 1960s that $\mathscr{M}_g$ is a quasi-projective algebraic variety [110]. For a history of $\mathscr{M}_g$, see [71].

In this section, we study the moduli space $\mathscr{M}_n^{tr}$ of compact smooth tropical curves of genus n. One natural question is whether $\mathscr{M}_n^{tr}$ *is a tropical variety.* The answer seems to be no in general. (See [37, p. 1136, p. 1163], [100, §3.1. C].)

The next question is what other kinds of properties, which are similar to properties of tropical varieties, are visible on the moduli space $\mathscr{M}_n^{tr}$.

The analogy between algebraic curves over $\mathbb{C}$ and tropical curves has motivated many problems and results for tropical curves. Similarly, the analogy between the two moduli spaces $\mathscr{M}_g$ and $\mathscr{M}_n^{tr}$ should also be fruitful.

In the study of $\mathscr{M}_g$, Teichmüller space $\mathscr{T}_g$ plays an important role. For example, the fact that $\mathscr{M}_g$ is a complex orbifold follows from the results that $\mathscr{T}_g$ is a complex manifold of complex dimension $3g-3$ when $g \geq 2$, and $\mathscr{M}_g$ is a quotient of $\mathscr{T}_g$ under the proper action of mapping class group Mod_g. This was the first time that Riemann's count of number of moduli of $\mathscr{M}_g$ is made rigorously by Teichmüller.

As it is known, later developments in topology and geometric group theory show that this action of Mod_g on $\mathscr{T}_g$ has played an essential role in understanding many properties of the group Mod_g.

Similarly, we will introduce tropical Teichmüller space $\mathscr{T}_g^{tr}$ and use it to study $\mathscr{M}_n^{tr}$. As we will see, $\mathscr{T}_g^{tr}$ can be identified with the *unnormalized* outer space $\tilde{X}_n$, and $\mathscr{M}_n^{tr}$ can be realized as the quotient $\mathrm{Out}(F_n)\backslash\tilde{X}_n$. This identification gives $\mathscr{M}_n^{tr}$ a natural stacky fan structure. Furthermore, when $\mathscr{T}_n^{tr}$ is given a natural local structure of tropical variety, the moduli space $\mathscr{M}_n^{tr}$ inherits the structure of a tropical orbifold.

As mentioned before, the action of $\mathrm{Out}(F_n)$ on $\tilde{X}_n$ or the normalized outer space X_n is also essential in the study of $\mathrm{Out}(F_n)$.

8.1 Moduli space $\mathcal{M}_n^{tr}$ of tropical curves and a general philosophy on moduli spaces

Fix any positive integer n. Consider all compact smooth tropical curves of genus n. In this section, we will study moduli spaces of tropical curves.

Two compact smooth tropical curves Γ_n and Γ_n' of genus n are *isomorphic* if their underlying graphs are homeomorphic and they induce the same *smooth* $\mathbb{Z}$-*affine structure* on the graph (see §4.4).

Two metric graphs (G,ℓ) and (G',ℓ) are *isometric* if there is a isometry φ : $(G,\ell) \to (G',\ell)$ between them as metric spaces. Clearly, φ maps edges and vertices of G to edges and vertices of G', and preserves the lengths of edges and valences of vertices.

By the identification of smooth tripical curves with metric graphs without vertices of valences 1 or 2 in §4.5, we have the following result.

Proposition 8.1. *Two compact smooth tropical curves* Γ_n *and* Γ_n' *are isomorphic if and only if they are isometric as metric graphs.*

The *moduli space* $\mathcal{M}_n^{tr}$ of compact smooth tropical curves of genus n is the set of isomorphism classes of compact smooth tropical curves of genus n, or equivalently the set of isometry classes of compact metric graphs of genus g.

One natural and important problem is to put additional structures on $\mathcal{M}_n^{tr}$ and understand its properties.

We start with **a general philosophy on moduli spaces.** Suppose that $\{V\}$ is a collection of spaces under classification with respect to an equivalence relation $\sim$ on them, and let $\mathcal{M} = \{V\}/\sim$ be the set of isomorphism classes of V, i.e., $\mathcal{M}$ is the moduli space of the spaces $\{V\}$. Assume that the equivalence relation $\sim$ is *compatible* with structures on V. Then it is *expected that* $\mathcal{M}$ *will inherit properties of the spaces* V *under classification.*

The best example to illustrate this philosophy is the moduli space $\mathcal{M}_g$ of compact Riemann surfaces Σ_g of genus g.

Each Riemann surface Σ_g is a 1-dimensional complex manifold. A natural isomorphism between complex manifolds is given by biholomorphic maps between. Since $\mathcal{M}_g$ is the set of biholomorphic classes of compact Riemann surfaces of genus g, $\mathcal{M}_g$ is expected to have a complex structure. The fact that this is true is the culmination of the works of Teichmüller, Rauch, Ahlfors, and Bers. Briefly, it was shown that the Teichmüller space $\mathcal{T}_g$ has a natural complex structure which is invariant the mapping class group Mod_g, and hence $\mathcal{M}_g = \mathrm{Mod}_g \backslash \mathcal{T}_g$ is a complex orbifold. See [71] for a summary.

One of the motivations of Riemann to introduce the concept of Riemann surfaces is that to each algebraic curve, there is a Riemann surface. (In fact, an algebraic equation $P(z,w) = 0$ gives a multi-valued function of $w = w(z)$, which becomes a single-valued holomorphic function on the associated Riemann surface.)

Conversely, Riemann also showed that every compact Riemann surface can be realized as an algebraic curve, i.e., comes from an algebraic curve. Then a biholomorphic map between two compact Riemann surfaces becomes an isomorphic morphism between two algebraic curves. Therefore, $\mathcal{M}_g$ is also the set of isomorphism classes of smooth projective algebraic curves of genus g over $\mathbb{C}$. It was proved by Mumford that $\mathcal{M}_g$ is a quasi-projective variety in [110].

When $g \geq 2$, by the uniformization theorem, each compact Riemann surface Σ_g has a canonical hyperbolic metric together with an orientation, and the complex structure is uniquely determined by the pair of hyperbolic metric and orientation. Therefore, $\mathcal{M}_g$ is the set of *oriented isometry classes* of *oriented hyperbolic surfaces* of genus g. One natural question concerns what are corresponding structures on $\mathcal{M}_g$. This question has not been studied as much as the previous two, but we should expect *natural metrics* on $\mathcal{M}_g$ defined purely in terms of hyperbolic metrics and orientations of Riemann surfaces Σ_g. This problem will be treated in more detail in another publication.

Motivated by the above general philosophy and the analogy between algebraic curves and tropical curves, we expect that $\mathcal{M}_n^{tr}$ should inherit properties of compact smooth tropical curves Γ_n. Specifically, the following conclusions sound reasonable:

1. Each tropical curve Γ is a 1-dimensional simplicial complex with metric. Therefore, $\mathcal{M}_n^{tr}$ should have a natural simplicial structure and a compatible metric.
2. Each tropical curve Γ_n is a 1-dimensional tropical curve. Therefore, $\mathcal{M}_n^{tr}$ should be a tropical variety, or have some structures related to tropical varieties.
3. Each *smooth* tropical curve can be identified with a metric graph without vertices of valences 1 or 2. Therefore, $\mathcal{M}_n^{tr}$ should admit metrics which are defined in solely terms of the metrics on graphs.

We will show below that expectations (1), (3) are true. Though expectation (2) is not true, i.e., $\mathcal{M}_n^{tr}$ is not a tropical variety, it does have the structure of a *tropical orbifold.*

Before explaining them, we note that any of the above three properties implies that $\mathcal{M}_n^{tr}$ should have natural topologies. Therefore, we start with constructing natural topologies on moduli spaces.

Recall that the complex structure of the moduli space $\mathcal{M}_g$ satisfies the following condition: Given a holomorphic family of Riemann surfaces

$$\pi : Y \to B,$$

where Y and B are complex manifolds, and π is a holomorphic map such that for every $b \in B$, the fiber $\pi^{-1}(b)$ is a compact Riemann surface of genus g, there a natural map

$$f : B \to \mathcal{M}_g,$$

where $b \in B$ is mapped to the isomorphism class of the Riemann surface $\pi^{-1}(b)$. Then the condition on the complex structure on $\mathcal{M}_g$ is that the map $f : B \to \mathcal{M}_g$ is a holomorphic map. (Strictly speaking, this is only one part of the requirement. There is also a requirement that any other complex structure on $\mathcal{M}_g$ should factor through it. See [71] for more detail.)

Similarly, if we consider a continuous family of Riemann surfaces $\pi : Y \to B$, i.e., Y and B are *topological spaces,* and π is a *continuous map* such that each fiber $\pi^{-1}(b)$ is a compact Riemann surface of genus g, then the induced map $f : B \to \mathcal{M}_g$ is a continuous map. This condition uniquely determines the natural topology of $\mathcal{M}_g$.

Since a holomorphic family of Riemann surfaces is clearly a continuous family of Riemann surfaces, the above two conditions are compatible.

The above consideration works for any moduli space. Therefore, the topology of the moduli space of tropical curves $\mathcal{M}_n^{tr}$ should be determined by similar conditions.

As we explained above, a tropical curve can be identified with a metric graph. To show that the topology determined by tropical curves is the same as the topology determined by the metric graphs, we note that the following result.

Proposition 8.2. *Let $p(x,y,t)$ be a tropical polynomial in three variables $x,y,t \in \mathbb{R}$. For each $t \in \mathbb{R}$, let*

$$p_t(x,y) = p(x,y,t),$$

a polynomial in two variables x,y which depends on the parameter t. Let Γ_{p_t} be the plane tropical curve defined by p_t, and Γ_t the associated metric graph. Then Γ_t is a continuous family of metric graphs.

Proof. The function defined by the tropical polynomial $p_t(x,y)$ is a piecewise linear function which is the maximum of several linear functions. It is clear that each linear function has constants which are linear functions of t. From this, it is clear that the plane tropical curve Γ_{p_t} and its associated metric graph Γ_t depends continuously on t and hence forms a continuous family of tropical curves and a continuous family of metric graphs. □

Remark 8.3. Though a general notion of tropical families of tropical curves is not easy to define, the above proposition and its proof suggests that tropical families of tropical curves are continuous families of metric graphs. Therefore, they induce the same topology on $\mathcal{M}_n^{tr}$.

In view of the above proposition and remark, in the following, we will define a natural topology on $\mathcal{M}_n^{tr}$ in terms of metric graphs.

8.2 Gromov-Hausdorff distance and topology on $\mathcal{M}_n^{tr}$

The moduli space $\mathcal{M}_n^{tr}$ has only been defined as a set. As we pointed out in the previous subsection, the most basic structure on $\mathcal{M}_n^{tr}$ is a topology which should be natural and reflect the metric (or tropical) properties of tropical curves. Naturally, the topology should be Hausdorff. As in a standard course in general topology, the most direct way to define a topology on a space is to use a metric on the space. We will construct a natural metric on $\mathcal{M}_n^{tr}$.

By definition, each point of $\mathcal{M}_n^{tr}$ is an isometry class of compact metric graphs. By the discussion in the previous subsection, for a continuous family of compact metric graphs of genus g, two metric graphs in the family in nearby fibers, as metric graphs, should correspond to nearby points in $\mathcal{M}_n^{tr}$. On the other hand, these two metric graphs should be close to each other as metric spaces.

Therefore, one way to measure distance between two points in $\mathcal{M}_n^{tr}$ represented by metric graphs Γ_n and Γ_n' is to measure the difference or rather similarity between the metric graphs Γ_n and Γ_n'.

In the theory of compact metric spaces, there is such a distance called *Gromov-Hausdorff distance* which measures the closeness or difference between two metric spaces.

We recall its definition and basic properties. Given a metric space (Z,d), there is a Hausdorff distance on the set $\mathscr{C}(Z)$ of closed subspaces of Z. For any two compact subsets $X,Y \subset Z$, the Hausdorff distance between them is

$$d_Z^H(X,Y) = \inf\{\varepsilon \mid X \subset N_\varepsilon(Y), Y \subset N_\varepsilon(X)\},$$

where $N_\varepsilon(Y)$ is the ε-neighborhood of X with respect to the metric d. It can be shown that this defines a metric on $\mathscr{C}(Z)$.

For any two compact metric spaces (X,d) and (Y,d), consider all metric spaces (Z,d) in which (X,d), (Y,d) can be isometrically embedded:

$$i\colon (X,d) \to (Z,d), \quad i\colon (Y,d) \to (Z,d).$$

Consequently, we get the Hausdorff distance between the two subsets $i(X)$ and $i(Y)$ in (Z,d). Then the *Gromov-Hausdorff distance* between (X,d) and (Y,d) is defined to be

$$d^{GH}(X,Y) = \inf_Z (d_Z^H(i(X), i(Y)),$$

where Z runs over all metric spaces admitting isometric embeddings of X,Y.

It can be shown that $d^{GH}(\cdot,\cdot)$ defines a metric on the set of isometry classes of compact metric spaces. In particular, two compact metric spaces (X,d) and (Y,d) are isometric if and only if the Gromov-Hausdorff distance $d^{GH}(X,Y) = 0$.

Roughly speaking, the Gromov-Hausdorff distance gives a precise measurement of how closely two metric spaces look alike. See [9, Chap. 7] for details about the above general properties.

Proposition 8.4. *The Gromov-Hausdorff distance defines a metric on the moduli space $\mathcal{M}_n^{tr}$, denoted by d^{GH}. Its induced topology on $\mathcal{M}_n^{tr}$ is called the Gromov-Hausdorff topology.*

Proof. We only need to note that every point of $\mathcal{M}_n^{tr}$ represents an isometry class of compact metric graphs. □

Though the definition of Gromov-Hausdorff distance is intrinsic and can be defined easily, it is not easy to compute it. On the other hand, it is easier to determine whether a sequence of metric spaces converges to a limiting metric space.

To explain this, we need to introduce several related notions. Let (X, d_X) and (Y, d_Y) be two metric spaces, and $f : X \to Y$ an arbitrary map. Then the *distortion* of f is defined by

$$\mathrm{dis}(f) = \sup_{x_1, x_2 \in X} |d_Y(f(x_1), f(x_2)) - d_X(x_1, x_2)|.$$

For any $\varepsilon > 0$, $f : X \to Y$ is called an *ε-isometry* if

1. $\mathrm{dis}(f) \leq \varepsilon$.
2. And every point of Y is contained in the ε-neighborhood of $f(X)$.

Then a basic result in the theory of Gromov-Hausdorff distance is [9, Corollary 7.3.28].

Proposition 8.5. *Let (X, d_X) and (Y, d_Y) be two metric spaces. For any $\varepsilon > 0$, we have:*

1. *If $d^{GH}(X, Y) < \varepsilon$, then there exists a 2ε-isometry from X to Y.*
2. *Conversely, if there exists an ε-isometry from X to Y, then $d^{GH}(X, Y) < 2\varepsilon$.*

Therefore, the Gromov-Hausdorff convergence of a sequence of metric spaces, i.e., with respect to the Gromov-Hausdorff distance, can be determined by the existence of ε-isometries for small values of ε.

Proposition 8.6. *Let (G_j, ℓ_j) be a sequence of metric graphs in $\mathcal{M}_n^{tr}$ converging to a metric graph (G_∞, ℓ_∞) in the Gromov-Hausdorff topology. Then for $j \gg 1$, there is a continuous surjective map $\pi : G_j \to G_\infty$ such that:*

1. *Every vertex of G_j is mapped to an vertex of G_∞.*
2. *Every edge of G_j is either mapped to a vertex or an edge of G_∞.*
3. *If an edge e of G_j is mapped to a vertex, then $\ell_j(e) \to 0$. If an edge e is mapped to an edge e_∞ of G_∞, then $\ell_j(e) \to \ell_\infty(e_\infty)$, as $j \to \infty$.*

In other words, G_∞ is obtained from G_j by shrinking and collapsing a collection of subgraphs, and the lengths of other edges converge. The converse is also true.

Proof. For each vertex of G_∞, there should be a portion of G_j which approximately looks like it when $j \gg 1$. Therefore, there can be only shrinking of edges. Outside these shrinking regions, the metric graphs G_j looks more and more similar to G_∞. Then the result is clear. □

Corollary 8.7. *Under the assumption in the previous proposition, if G_∞ has the same number of vertices as G_j, and hence of the same number of edges as G_j for $j \geq 1$, then there is a bijective correspondence between edges of G_j and G_∞ such that the lengths of the edges of G_j converge to the lengths of the corresponding edges of G_∞.*

Remark 8.8. In [32], the moduli space $\mathcal{M}_n^{tr}$ was constructed as a topological space, and in [38], $\mathcal{M}_n^{tr}$ was constructed as a stacky fan. In both cases, the topology on $\mathcal{M}_n^{tr}$ was defined in terms of convergence of the lengths of corresponding edges and collapsing edges. One point of using the Gromov-Hausdorff distance is that it is an *intrinsically defined metric* and can be defined solely in terms of the metrics on the graphs, without refereeing to the graph structures.

Remark 8.9. When defining and discussing structures on moduli spaces, we need to explain why they are natural from the perspective of moduli spaces. From Proposition 8.2 and Remark 8.3, it is clear that for any tropical family of smooth compact tropical curves, the induced family of metric graphs is a continuous family in the Gromov-Hausdorff topology. Therefore, the Gromov-Hausdorff topology is natural. This remark also clarifies that the topology on $\mathcal{M}_n^{tr}$ constructed in [32] and [38] mentioned in the previous remark is natural.

In metric theory, especially in the theory of Riemannian manifolds, one natural question is whether a metric is complete.

Proposition 8.10. *The Gromov-Hausdorff metric d^{GH} on $\mathcal{M}_n^{tr}$ is incomplete.*

Proof. Take any compact metric graph Γ_n of genus n with one loop given by one edge e. Let its length $t = \ell(e)$ go to 0 while fixing the lengths of other edges of Γ_n. Then we get a divergent family of compact metric graphs $\Gamma_{n,t}$, i.e., points in M_n^{tr}. In fact, as $t \to 0$, the Gromov-Hausdorff limit of $\Gamma_{n,t}$ is a metric graph obtained from Γ_n by collapsing the edge e. Since e is a loop, the limit has genus $n-1$ and hence does not belong to $\mathcal{M}_n^{tr}$. □

We will discuss the Gromov-Hausdorff completion of $\mathcal{M}_n^{tr}$ in Proposition 8.28 later.

8.3 The moduli space $\mathcal{M}_n^{tr}$ and tropical Teichmüller space $\mathcal{T}_g^{tr}$ as stacky fans

As mentioned before, since each tropical curve is a 1-dimensional simplial complex, it is natural to expect that the moduli space $\mathcal{M}_n^{tr}$ should have some simplicial or polyhedral complex structures.

By definition, a metric graph is a graph G together with an assignment of edge lengths,

$$\ell : E(G) \to \mathbb{R} > 0.$$

If we fix a graph G of genus n, all metrics on G are parametrized by the polyhedral cone

$$\sigma_G = \mathbb{R}_{>0}^{E(G)}.$$

Specifically, each metric graph (G,ℓ) is mapped to the vector $(\ell(e)), e \in E(G)$.

On the other hand, two metrics graphs corresponding to different points in $\sigma_G = \mathbb{R}_{>0}^{E(G)}$ may be isometric, for example, this occurs when the metric graphs admit nontrivial isometries and they are obtained from each other by isometries. Therefore, the natural map

$$\sigma_G \to \mathscr{M}_n^{tr},$$

which is obtained by mapping each metric graph to its isometry class, is not injective in general.

As in the general theory of moduli spaces, to avoid problems caused by nontrivial automorphisms, we rigidify metric graphs or tropical curves by additional data, i.e., markings in this case. Consequently, we will introduce the tropical Teichmüller space, which is essentially the outer space $\tilde{X}_n$ defined in §8.

For other more refined structures on $\mathscr{M}_n^{tr}$, we also need to use the tropical Teicumüller space $\mathscr{T}_n^{tr}$. This is consistent with the results for the moduli space $\mathscr{M}_g$ of Riemann surfaces and the Teichmüller space $\mathscr{T}_g$, since the Teichmüller space $\mathscr{T}_g$ was introduced to study the moduli space $\mathscr{M}_g$. For example, both the topology and complex structure on $\mathscr{M}_g$ were first introduced through $\mathscr{T}_g$.

To define marking on metric graphs (or smooth tropical curves), let R_n be a rose with n petals, i.e., a bouquet of n circles. Then a *marked graph* of genus n is a graph G together with an homotopy equivalence $h : G \to R_n$, denoted by (G,h). We always assume that graphs do not have vertices of valence 1 or 2. Then a *marked metric graph* of genus n is a metric graph (G,ℓ) together with a marking $h : G \to R_n$, denoted by (G,ℓ,h).

By definition, two homotopy equivalent maps $h,h' : G \to R_n$ define the same marking on a graph G. Similarly, two marked metric graphs (G,ℓ,h) and (G',ℓ',h') are *equivalent* if there is an isometry $i : (G,\ell) \to (G',\ell')$ which commutes with the markings, i.e., $h, h' \circ i : G \to R_n$ are homotopy equivalent.

By the correspondence between compact smooth tropical curves and metric graphs, a marked metric graph (G,ℓ,h) corresponds to a marked compact smooth tropical curve.

Definition 8.11. The set of marked compact smooth tropical curves of genus n, i.e., compact metric graphs (G,ℓ,h) of genus n, is called the tropical Teichmüller space and denoted by $\mathscr{T}_n^{tr}$.

We want to put a topology on $\mathscr{T}_n^{tr}$ and also more refined structures on it.

Remark 8.12. The tropical Teichmüller space is the extended outer space $\tilde{X}_n$ defined in §8, and strictly contains the outer space X_n. The difference is that in the definition of X_n, each metric graph has a normalized total edge length 1, while for $\mathcal{T}_n^{tr}$, there is no such normalization on the edge lengths.

To put a stacky fan structure on $\mathcal{T}_n^{tr}$, we need to identify polyhedral cones in it.

Lemma 8.13. *Fix a marked graph (G,h). The set of isometry classes of marked metric graphs supported by (G,h) is paramatrized by the simplicial cone*

$$\sigma_{(G,h)} = \mathbb{R}_{>0}^{E(G)}.$$

Proof. Suppose that G has k edges. Fix any order on them and denote them by $e_1,\dots,e_k$. Then each marked metric graph (G,ℓ,h) has the well-defined ordered edges $e_1,\dots,e_k$, and their lengths $\ell(e_1),\dots,\ell(e_k)$ give a point in the cone $\sigma_{(G,h)}$. By definition, two marked metric graphs are equivalent if and only if their corresponding edges have the same lengths. □

For every fixed n, there are only finitely many graphs G of genus n which have no vertices of valences 1 or 2. On the other hand, there are infinitely many marked graphs (G,h) of genus n. As a set, we have the following decomposition:

$$\mathcal{T}_n^{tr} = \coprod_{(G,h)} \sigma_{(G,h)}, \tag{8.1}$$

where (G,h) runs over all marked graphs of genus n. We need to glue together all these simplicial cones $\sigma_{(G,h)}$ in a suitable way.

Given a marked graph (G,h). If an edge e of G is not a loop, then collapsing the edge e will produce a graph $G' = G/e$ of genus n, and the marking $h : G \to R_n$ will also induce an homotopy equivalence $h' : G' \to R_n$. Therefore, we obtain a new marked graph (G',h'). The simplicial cone $\sigma_{(G',h')}$ appears naturally as a codimension 1 face of the cone $\sigma_{(G,h)}$.

More generally, for any disjoint union of subtrees Ω of G, collapsing Ω in G will produce a graph $G' = G/\Omega$ of genus g, and a marking of G will also induce a marking $h' : G' \to R_n$. Similarly, the cone $\sigma_{(G',h')}$ also appears as a face of the cone $\sigma_{(G,h)}$, and the codimension is determined by the number of edges in Ω collapsed.

Definition 8.14. A marked graph (G',h') is called a specialization or collapsing of another marked graph (G,h) if (G',h') is obtained from (G,h) by collapsing a disjoint union Ω of subtrees as above.

We note that for every marked graph (G,h), the cone $\sigma_{(G,h)}$ is an open simplicial cone. Define a *partial compactification* $\hat{\sigma}_{(G,h)}$ of the cone $\sigma_{(G,h)}$ by adding only the missing faces corresponding to specializations (G',h') of (G,h):

$$\hat{\sigma}_{(G,h)} = \sigma_{(G,h)} \cup \coprod_{(G',h')} \sigma_{(G',h')}, \tag{8.2}$$

where (G',h') runs over all specializations of (G,h).

Clearly, for each specialization (G',h), the cone $\hat{\sigma}_{(G',h')}$ is included in $\hat{\sigma}_{(G,h)}$ as a face.

Lemma 8.15. *For every marked graph (G,h), with respect to the standard lattice $\mathbb{Z}^{E(G)} \subset \mathbb{R}^{E(G)}$, the simplicial cone $\sigma_{(G,h)}$ is a rational polyhedral cone. For every every specialization (G',h') of (G,h), the inclusion of $\hat{\sigma}_{(G',h')} \hookrightarrow \hat{\sigma}_{(G,h)}$ is an integral affine (i.e. $\mathbb{Z}$-affine) map.*

Proposition 8.16. *The tropical Teichmüller space $\mathcal{T}_n^{tr}$ can be canonically identified with the underlying space of the stacky fan*

$$\coprod_{(G,h)} \hat{\sigma}_{(G,h)}/\sim,$$

where two points $p \in \hat{\sigma}_{(G,h)}$ and $p' \in \hat{\sigma}_{(G',h')}$ are identified if (G',h') is a specialization of (G,h), and p' is identified with p under the inclusion of $\hat{\sigma}_{(G',h')}$ as a face of $\hat{\sigma}_{(G,h)}$.

Proof. By the previous lemma and the definition of stacky fans, the space $\coprod_{(G,h)} \hat{\sigma}_{(G,h)}/\sim$ has a the canonical structure of a stacky fan. Since

$$\coprod_{(G,h)} \hat{\sigma}_{(G,h)}/\sim = \coprod_{(G,h)} \sigma_{(G,h)}/\sim,$$

Equation 8.1 implies that $\mathcal{T}_n^{tr}$ can be identified with the stacky fan here. □

Since we are mainly interested in the moduli space $\mathcal{M}_n^{tr}$, we need to undo markings on metric graphs.

Every homotopy equivalence $f: R_n \to R_n$ induces a change of marking of metric graphs by

$$(G,h) \mapsto (G, f \circ h),$$

and hence f induces a map on $\mathcal{T}_n^{tr}$. By Lemma 2.6 and Equation 2.2, the action of the space of all homotopy equivalences $f: R_n \to R_n$ induces an action of $\mathrm{Out}(F_n)$ on $\mathcal{T}_n^{tr}$.

It is clear from definition that as a set,

$$\mathcal{M}_n^{tr} = \mathrm{Out}(F_n)\backslash \mathcal{T}_n^{tr}. \tag{8.3}$$

Proposition 8.17. *The action of* $\mathrm{Out}(F_n)$ *on $\mathcal{T}_n^{tr}$ is given by integral affine maps, and the quotient* $\mathrm{Out}(F_n)\backslash \mathcal{T}_n^{tr}$ *is a stacky fan. Therefore, $\mathcal{M}_n^{tr}$ admits the structure of a stacky fan. As a topological space with respect to the Gromov-Hausdorff topology, M_n^{tr} is homeomorphic to the underlying space of the stacky fan* $\mathrm{Out}(F_n)\backslash \mathcal{T}_n^{tr}$.

Proof. The action of $\mathrm{Out}(F_n)$ on the cones $\sigma_{(G,h)}$ amounts to permutations of coordinates in $\mathbb{R}^{E(G)}$, and it is clearly given by integral affine maps. Therefore, the quotient $\mathrm{Out}(F_n)\backslash\mathscr{T}_n^{tr}$ inherits the stacky fan structure from $\mathscr{T}_n^{tr}$. The last statement follows from Proposition 8.6. □

8.4 The tropical Teichmüller space $\mathscr{T}_g^{tr}$ as a tropical space

As explained before, since $\mathscr{M}_n^{tr}$ is the moduli space of compact smooth tropical curves of genus n, one would hope that $\mathscr{M}_n^{tr}$ is a tropical variety. It seems that the answer is no in general, see [37, p. 1136, p. 1163] and [100, §3.1. C].

In this subsection, we will follow [100] and explain that the tropical Teichmüller space $\mathscr{T}_n^{tr}$ is *locally a tropical variety*, and hence $\mathscr{M}_n^{tr}$ is a *tropical orbifold.* Furthermore, some *finite covers of* $\mathscr{M}_n^{tr}$ *are tropical varieties*. Some special cases of the latter result when $n = 2,3$ were proved in [37, §7].

This result shows the importance of studying the tropical Teichmüller space $\mathscr{T}_n^{tr}$. It is consistent with the results for the usual moduli space $\mathscr{M}_g$ and Teichmüller space $\mathscr{T}_g$, or rather inspired by the importance of $\mathscr{T}_g$ in studying $\mathscr{M}_g$.

To explain this result, we need to introduce tropical curves with marked points. In the study of the moduli spaces $\mathscr{M}_g$ and $\mathscr{T}_g$, Riemann surfaces with punctures are crucial. For example, the universal curve over the Teichmüller space $\mathscr{T}_g$ can be constructed from the Teichmüller space $\mathscr{T}_{g,1}$ with compact Riemann surfaces of genus g with one point removed.

If we remove n points from a compact Riemann surface Σ_g of genus g, we obtain a Riemann surface of genus g with n punctures, denoted by $\Sigma_{g,n}$.

When the Euler number of $\Sigma_{g,n}$ is negative, it admits a complete hyperbolic metric of finite area. Such surfaces occur naturally in many different contexts. For example, for any finite index subgroup $\Gamma \subseteq \mathrm{SL}(2,\mathbb{Z})$, the quotient of the upper halfplane $\mathbb{H}^2$ by Γ is such a hyperbolic surface.

In spite of the analogy between algebraic curves and tropical curves, we cannot remove points from a compact tropical curve to get another smooth tropical curve which has desirable geometry.

Instead, we can think of $\Sigma_{g,n}$ as a compact Riemann surface Σ_g with n distinct marked points. For a smooth tropical curve Γ_n, we can also specify finitely many distinct marked points which are *disjoint* from the vertices of Γ_n as a graph.

When we add an interior point of an edge to a graph as a marked point, we essentially add a new vertex. Since we assume that graphs have no vertices of valence 2, people usually add a closed interval of infinite length $[0,+\infty]$ at each marked point in order to keep this condition. Clearly such a new edge only depends only on the location of the basepoint. This gives a definition of *compact smooth tropical curves with marked points.* (Note on metric graphs with marked points, we have vertices of valence 1. The edges of infinite length connected to them are called leaves in [32].)

Once we have the notion of smooth tropical curves $\Gamma_{n,k}$ of genus n with k marked points, we can similarly define isomorphisms between them and hence define the *moduli space* $\mathcal{M}_{n,k}^{tr}$ *of smooth tropical curves* $\Gamma_{n,k}$ *of genus n with k distinct marked points.*

Similarly, we can define marking on $\Gamma_{n,k}$ by requiring that homoropy equivalences preserve the marked points. Then we can define the *tropical Teichmüller space* $\mathcal{T}_{n,k}^{tr}$ *of marked smooth tropical curves* $\Gamma_{n,k}$ *of genus n with k distinct marked points.*

The discussion and proofs of the previous subsection work and prove the following more general result.

Proposition 8.18. *For any* $j,k \geq 0$*, the tropical moduli spaces* $\mathcal{M}_{n,k}^{tr}$ *and* $\mathcal{T}_{n,k}^{tr}$ *are stacky fans.*

When $n = 0$ and $k \geq 3$, each tropical curve $\Gamma_{0,k}$ has only trivial automorphism which fixes the marked points (or the infinite length edges). Therefore in this case,

$$\mathcal{M}_{0,k}^{tr} = \mathcal{T}_{0,k}^{tr}.$$

One basic result in [102, Theorem 1] and [52, Theorem 3.7] is

Proposition 8.19. *For any* $k \geq 3$*, the moduli space* $\mathcal{M}_{0,k}^{tr}$ *is a tropical variety.*

Briefly, $\mathcal{M}_{0,k}^{tr}$ is embedded into a vector space $\mathbb{R}^N$ for some N such that the image is a rational polyhedral fan satisfying the balancing condition at every codimension face in Definition 6.2. The embedding

$$\mathcal{M}_{0,k}^{tr} \hookrightarrow \mathbb{R}^N$$

is achieved through functions which are linear combinations of length of edges of tropical curves $\Gamma_{0,k}$. Therefore, we have the following corollary.

Corollary 8.20. *The tropical variety structure on* $\mathcal{M}_{0,k}^{tr}$ *in the above proposition is natural as a moduli space of tropical curves.*

Proof. We note that when we have a family of plane tropical curves, lengths of edges are linear functions of the coefficients of the tropical polynomials. This means that the edge length functions of metric graphs give the natural tropical structure on $\mathcal{M}_{0,k}^{tr}$. □

Remark 8.21. By Remark 2.26, and also Equations 2.7 and 10.2, periods of tropical curves are also linear functions of the edge lengths of the underlying metric graphs. This also implies that the edge length functions of metric graphs give the natural tropical structure on $\mathcal{M}_{0,k}^{tr}$.

As explained in [100, §3.1. D], the above result can be used to prove

Theorem 8.22. *The tropical Teichmüller space $\mathscr{T}_n^{tr}$ has a natural local structure of a tropical variety of dimension $3n-3$, i.e., every point of $\mathscr{T}_n^{tr}$ has a neighborhood which is isomorphic to a rational polyhedal complex in a real vector space $\mathbb{R}^N$ which satisfies the balancing condition on every face of codimension 1. Consequently, $\mathscr{T}_n^{tr}$ is locally a tropical variety.*

Proof. By the above discussion, $\mathscr{T}_n^{tr}$ is a stacky fan and can be identified with

$$\mathscr{T}_n^{tr} = \coprod_{(G,h)} \sigma_{(G,h)}.$$

If we have a rational polyhedral complex in an Euclidean space $\mathbb{R}^n$, to show that it is a tropical variety, we need to check the balancing condition at every codimension 1 cone (see Proposition 6.2 and Definition 6.2). Since $\mathscr{T}_n^{tr}$ is an abstract polyhedral cone complex (or stacky fan), we need to embed a neighborhood of every point of $\mathscr{T}_n^{tr}$ into some Euclidean space and to identify it with an open subset of a tropical fan (or variety). Then the balancing condition is automatically satisfied.

Top dimensional cones of $\mathscr{T}_n^{tr}$ are associated with marked graphs whose vertices all have valence 3. For every point in a top dimensional cone $\sigma_{(G,h)} = \mathbb{R}_{>0}^{3n-3}$, it has a neighborhood equal to an open subset of $\mathbb{R}_{>0}^{3n-3}$. Such a neighborhood has a natural tropical structure.

If a point is contained in a cone $\sigma_{(G,h)}$ of non-top dimension, then the graph G contains vertices of valence strictly less than 3.

Suppose $\sigma_{(G,h)}$ is of codimension 1. Then the graph G contains exactly one vertex v of valence 4. Top dimensional cones $\sigma_{(\hat{G},h)}$ are associated with marked graphs $(\hat{G},h)$ which realizes (G,h) as a specialization. We need to show that for any point in $\sigma_{(G,h)}$, there is a neighborhood of p in

$$\sigma_{(G,h)} \cup \coprod_{(\hat{G},h)\to(G,h)} \sigma_{(\hat{G},h)},$$

which can be identified with a neighborhood in a tropical fan, i.e., the balancing condition in Definition 6.2 is satisfied. (Note that each $\sigma_{(\hat{G},h)}$ contains $\sigma_{(G,h)}$ as a codimension 1 face.)

The *key point* is to observe that such a neighborhood of p is essentially given by a neighborhood of a point in a codimension 1 cone of the tropical variety $\mathscr{M}_{0,4}^{tr}$, the moduli space of rational tropical curves with 4 marked points.

Specifically, to move from the point p to points in the top dimensional cones

$$\coprod_{(\hat{G},h)\to(G,h)} \sigma_{(\hat{G},h)},$$

we need to expand a neighborhood of the vertex v in the graph G into a graph which has one closed and finite length edge e together with four one-side-open edges.

We note that there are 3 different ways to expand the vertex v. For example, number the edges coming out this vertex of valence 4 by e_1, e_2, e_3, e_3. Each expansion separates these 4 edges into 2 groups: $\{e_1, e_2; e_3, e_4\}$, $\{e_1, e_3; e_2, e_4\}$, $\{e_1, e_4; e_2, e_3\}$. All possibilities of such nearby expansion of the vertex v is described by a neighborhood of a point in $\mathcal{M}_{0,4}$.

For any graph $\hat{G}$ which specializes to G, e is an edge of $\hat{G}$. (In fact, there are exactly three types of $\hat{G}$.) At the same time, we also need to change lengths of other edges of $\hat{G}$.

Note that all other edges of one $\hat{G}$ are in a canonical correspondence with the other edges of another graph which specializes to G. There are $3n-4$ such edges, and their lengths are parametrized by points in $\mathbb{R}^{3n-4}$.

Therefore, there exists a neighborhood of p in

$$\sigma_{(\hat{G},h)} \cup \coprod_{(\hat{G},h)\to(G,h)} \sigma_{(\hat{G},h)}$$

which is the product $U \times V$, where U is a neighborhood of $\mathcal{M}_{0,4}$, and V is a neighborhood of $\mathbb{R}^{3n-4}$.

Since both U and V are open subsets of tropical fans, $U \times V$ is also an open subset of a tropical fan.

Suppose that a cone $\sigma_{(G,h)}$ is of codimension strictly greater than 1. Then the graph G has either more than 1 vertices of valence 4 or at least one vertex of valance at least 5.

For each vertex of G of valence 4, we can expand it into an edge with two vertices of valence 3 as above, and a small neighborhood of it is expanded into a neighborhood of a tropical curve of genus 0 with 4 marked points.

Suppose that G has a vertex of valence $k \geq 4$. Then a small neighborhood of the vertex can be expanded into open subset of a tropical curve $\Gamma_{0,k}$ of genus 0 with k marked points such that the valences of vertices of $\Gamma_{0,k}$ are strictly less than k.

These expansions of vertices of G can be done independently of each other, and lengths of edges of G can also be deformed independently.

Let $v_1, \ldots, v_m$ be the set of vertices of G with valence $k_1, \ldots, k_m$. Then for every point in the cone $\sigma_{(G,h)}$, there exists a neighborhood of p in $\mathcal{T}_n^{tr}$ which is a product

$$U_1 \times \ldots \times U_m \times U,$$

where, for $i = 1, \ldots, k$, U_i is an open subset of the moduli space $\mathcal{M}_{0,k_i}$, and U is an open subset of $\mathbb{R}^{E(G)}$. Since $\mathcal{M}_{0,k_i}$ is a tropical variety, each of these sets U_i is an open subset of a tropical variety, and consequently the neighborhood of p in $\mathcal{T}_n^{tr}$ is identified with an open subset of a tropical variety. This proves that $\mathcal{T}_n^{tr}$ has a local structure of a tropical variety.

We also note that this local tropical variety structure on $\mathcal{T}_n^{tr}$ is natural. For complex spaces, the complex structure is determined by holomorphic functions on it. For a tropical variety, the tropical structure is also determined by regular functions on it.

When tropical curves are identified with metric graphs, given a tropical family of tropical curves, the lengths of edges of the underlying metric graphs are regular functions in the parameter. See Remark 8.3 and Proposition 8.2. In the above description of the tropical structure on $\mathcal{T}_n^{tr}$, the tropical variety structures on the open subsets U and V are both induced by lengths of edges of the underlying metric graphs. Therefore, the tropical variety structure on $\mathcal{T}_n^{tr}$ is natural.

It is helpful to note that the edge length functions are linear functions on each simplicial cones in $\mathcal{T}_n^{tr}$ when it is considered as a stacky fan. Therefore, they are natural from the perspective of the stack fan structure on $\mathcal{T}_n^{tr}$. Probably the converse is more important: the local tropical variety structure on $\mathcal{T}_n^{tr}$ shows that the stacky fan structure on $\mathcal{T}_n^{tr}$ is natural.

The above reasoning on the naturalness of the local tropical structure on $\mathcal{T}_n^{tr}$ is also supported by the result in Remark 2.26 that periods of compact smooth tropical curves are also linear functions of the edge lengths of metric graphs, i.e., they are linear functions on each simplicial cones in $\mathcal{T}_n^{tr}$. See also Equations 2.7 and 10.2 about the periods of metric graphs. □

Remark 8.23. If we recall that smooth tropical curves defined by local charts in §4.4, it is natural to say that $\mathcal{T}_n^{tr}$ is a *higher dimensional tropical manifold*. This will be helpful when we talk about $\mathcal{M}_n^{tr}$ as a tropical orbifold.

Remark 8.24. *Universal tropical curve over the tropical Teichmüller space* $\mathcal{T}_n^{tr}$. We have defined the Teichmüller space $\mathcal{T}_{n,k}^{tr}$ of compact smooth tropical curves of genus n with k marked points. The same argument as in the proof of the the previous proposition shows that $\mathcal{T}_{n,k}^{tr}$ has a natural local tropical structure. It can be shown that the forgetful map by erasing the last marked point

$$\mathcal{T}_{n,k+1}^{tr} \to \mathcal{T}_{n,k}^{tr}$$

is given an integral affine map with respect to the local tropical structures, and $\mathcal{T}_{n,k+1}^{tr}$ is a family of tropical curves over $\mathcal{T}_{n,k}^{tr}$ such that the fiber over each point b in $\mathcal{T}_{n,k}^{tr}$ is the tropical curve of genus n with k marked points which is represented by the point b. In particular,

$$\mathcal{T}_{n,1}^{tr} \to \mathcal{T}_n^{tr}$$

is an analogy of the universal algebraic curve of genus g,

$$\mathcal{C}_g = \mathcal{T}_{g,1} \to \mathcal{T}_g,$$

over the Teichmüller space $\mathcal{T}_g$ of compact Riemann surfaces of genus g.

When $n = 0, k \geq 3$,

$$\mathcal{T}_{0,k}^{tr} = \mathcal{M}_{0,k}^{tr},$$

and the universal tropical curve over $\mathcal{M}_{0,k}^{tr}$ and its functorial property was constructed and proved in [49].

8.5 The moduli space $\mathscr{M}_n^{tr}$ as a tropical orbifold

In the previous subsection, we showed in Theorem 8.22 that the tropical Teichmüller space $\mathscr{T}_n^{tr}$ has a natural local structure of tropical variety. On the other hand, the moduli space $\mathscr{M}_n^{tr}$ is in general not a tropical variety, as pointed out before (see [37, p. 1136, p. 1163], [100, §3.1. C]). After explaining this obvious incompatibility, we will explain that $\mathscr{M}_n^{tr}$ should be viewed as a tropical orbifold, and $\mathscr{M}_n^{tr}$ admits finite covers which are tropical varieties.

Quotients of locally tropical spaces are not tropical spaces

As we pointed out before, we have the identification

$$\mathscr{M}_n^{tr} = \mathrm{Out}(F_n)\backslash\mathscr{T}_n^{tr}.$$

To explain that $\mathscr{M}_n^{tr}$ may not inherit the local tropical structure from $\mathscr{T}_n^{tr}$, we illustrate by simple examples that quotients of spaces which are locally tropical varieties are not necessarily locally tropical varieties.

Let $\{\pm 1\}$ act on $\mathbb{Z}$ by multiplication, and $\mathbb{Z} \rtimes \{\pm 1\}$ be the associated semidirect product. Then the group $\mathbb{Z} \rtimes \{\pm 1\}$ acts on $\mathbb{R}$ by affine transformations: for any $(a,b) \in \mathbb{Z} \rtimes \{\pm 1\}$ and $x \in \mathbb{R}$,

$$(a,b)\cdot x = 2a + bx.$$

Then $\mathbb{Z} \rtimes \{\pm 1\}$ acts on $\mathbb{R}$ with a fundamental domain equal to $[0,1]$.

Let $(\mathbb{R},\mathbb{Z})$ be the metric graph whose underlying space is $\mathbb{R}$ and vertices correspond to the integers of $\mathbb{Z}$. Clearly $(\mathbb{R},\mathbb{Z})$ has a local structure of tropical variety. The group $\mathbb{Z} \rtimes \{\pm 1\}$ acts on $(\mathbb{R},\mathbb{Z})$ by integral affine transformations. But the quotient $\mathbb{Z} \rtimes \{\pm 1\}\backslash(\mathbb{R},\mathbb{Z})$ is a metric graph and is equal to the unit interval $[0,1]$. Clearly, $[0,1]$ does not satisfy any balancing condition at its end points, and hence it is not tropical curve.

In this case, if we ignore the vertices $\mathbb{Z}$, then $\mathbb{R}$ is a smooth manifold, and the quotient $\mathbb{Z} \rtimes \{\pm 1\}\backslash\mathbb{R}$ is an orbifold. If we keep the metric graph (or tropical curve) structure on $\mathbb{R}$, then the quotient $\mathbb{Z} \rtimes \{\pm 1\}\backslash(\mathbb{R},\mathbb{Z})$ should be called a tropical orbifold of dimension 1.

This example might be objected since every vertex in $\mathbb{Z}$ has valence 2 and we have usually excluded vertices of valence 2 in considering metric graphs. To avoid this, we can consider the product $(\mathbb{R},\mathbb{Z}) \times (\mathbb{R},\mathbb{Z})$. This has a natural local tropical manifold structure of dimension 2. The group $(\mathbb{Z} \rtimes \{\pm 1\})^2$ acts on it by integral affine transformations, and the quotient $(\mathbb{Z} \rtimes \{\pm 1\})^2\backslash(\mathbb{R},\mathbb{Z}) \times (\mathbb{R},\mathbb{Z})$ is the unit square $[0,1] \times [0,1]$, which clearly does not satisfy the balancing condition on any codimension 1 face and hence is not a tropical variety. On the other hand, this should also be called a tropical orbifold of dimension 2.

If we prefer a metric graph with valences of all vertices at least 3, we can take the metric graph Γ in $(\mathbb{R},\mathbb{Z}) \times (\mathbb{R},\mathbb{Z})$ whose vertices are integral points $\mathbb{Z}^2$, edges consist of all horizontal and vertical line segments connecting them. Then every vertex has valence 4. It clearly has a local structure of a tropical curve. The group $(\mathbb{Z}^2 \rtimes \{\pm 1\})^2$ acts on Γ by integral affine transformations, and the quotient $(\mathbb{Z}^2 \rtimes \{\pm 1\})^2 \backslash \Gamma$ is a metric graph and equal to the boundary of the unit square $[0,1] \times [0,1]$, which clearly does not satisfy the balancing condition at any corner point.

Definition of a tropical orbifold

Before explaining that $\mathcal{M}_n^{tr}$ should be viewed as a tropical orbifold, we need to define the notion of tropical orbifolds.

Recall that an orbifold in differential topology is a Hausdorff topological space M satisfying the conditions:

1. Every point $p \in M$ admits a neighborhood which is the quotient of an open subset of $\mathbb{R}^n$ (or a smooth manifold) under a finite group action by diffeomorphisms.
2. These local finite covers are compatible.

For a precise description of these conditions, see [3, Definition 1.1, p. 2]. Motivated by the definition of orbifolds in differential topology, we define a *tropical orbifold* to be a Hausdorff topological space M such that the following conditions are satisfied:

1. Every point p of M admits a neighborhood which is the quotient of an open subset V of a tropical variety under a finite group F of integral affine maps, i.e., there is an embedding of $F \backslash V \to M$, and the image is a neighborhood of p in M. Denote the map $V \to M$ by φ, and call (V, F, φ) an orbifold chart over $\varphi(V)$.
2. An embedding $\lambda : (V, F, \varphi) \to (V', F', \varphi')$ between two such orbifold charts is a injective integral affine map $\lambda : V \to V'$ such that $\varphi' \lambda = \varphi$.
3. If two orbifold charts (V_1, F_1, φ_1), (V_2, F_2, φ_2) overlap, i.e., $\varphi_1(V_1) \cap \varphi_2(V_2) \neq \emptyset$, then for every point $p \in \varphi_1(V_1) \cap \varphi_2(V_2)$, there exists a neighborhood U of p and a orbifold chart (V, F, φ) over U which can be embedded into both charts (V_1, F_1, φ_1), (V_2, F_2, φ_2).

We note that the moduli space $\mathcal{M}_g$ of Riemann surfaces is a complex orbifold. This follows from the fact that the mapping class group Mod_g acts on the Teichmüller space $\mathcal{T}_g$ by holomorphic maps and the identification

$$\mathcal{M}_g = \mathrm{Mod}_g \backslash \mathcal{T}_g.$$

Therefore, the analogy between the complex geometry and the tropical geometry suggests that the following result is reasonable.

Proposition 8.25. *With respect to the natural local structure of tropical variety on* $\mathcal{T}_n^{tr}$ *defined in Theorem* 8.22, *the action* $\mathrm{Out}(F_n)$ *on* $\mathcal{T}_n^{tr}$ *preserves this tropical structure, and the quotient* $M_n^{tr} = \mathrm{Out}(F_n) \backslash \mathcal{T}_n^{tr}$ *is a tropical orbifold.*

Proof. Since the local tropical structure on $\mathscr{T}_n^{tr}$ is determined by integral linear combinations of lengths of edges of metric graphs, and $\mathrm{Out}(F_n)$ maps integral linear combinations of edge lengths to integral linear combinations of edge lengths, it is clear that the action of $\mathrm{Out}(F_n)$ preserves this local tropical structure on $\mathscr{T}_n^{tr}$.

Since $\mathrm{Out}(F_n)$ acts properly on $\mathscr{T}_{tr}$, it is clear that the local tropical structure on $\mathscr{T}_n^{tr}$ induces a structure of tropical orbifold on the quotient $\mathrm{Out}(F_n) \setminus \mathscr{T}_n^{tr}$. More specifically, for every point p in $\mathscr{T}_n^{tr}$, a small neighborhood of p which is stable under the stablilizer of p in $\mathrm{Out}(F_n)$ gives an orbifold cover of the image of p in $\mathscr{M}_n^{tr}$. The compatibility between these orbifold charts comes from the fact that $\mathrm{Out}(F_n)$ preserves the local tropical variety structure on $\mathscr{T}_n^{tr}$. □

To construct tropical orbifolds which are covers of $\mathscr{M}_n^{tr}$, we need to get rid of torsion elements of $\mathrm{Out}(F_n)$.

First we explain that the mapping class group Mod_n admits torsion-free subgroups of finite index. As mentioned above, there is a natural surjective homomorphism

$$\mathrm{Mod}_g \to \mathrm{Sp}(2g, \mathbb{Z}),$$

which is induced from the action of Mod_g on the first homology group of the base surface S_g used in the definition of $\mathscr{T}_g$.

For each integer N, consider the composed homomorphism

$$\mathrm{Mod}_g \to \mathrm{Sp}(2g, \mathbb{Z}) \to \mathrm{Sp}(2g, \mathbb{Z}/N\mathbb{Z}).$$

The kernel of this homomorphism is called the *congruence subgroup* of Mod_g of level N, denoted by $\mathrm{Mod}_g(N)$. It is known that when $N \geq 3$, $\mathrm{Mod}_g(N)$ is torsion-free. See [68] for example.

Recall that a *level N structure* on a compact Riemann surface Σ_g is a choice of basis of $H_1(\Sigma_g, \mathbb{Z}/N\mathbb{Z})$. Then the above discussion implies the following result.

Proposition 8.26. *For $N \geq 3$, the quotient $Mod_g(N)\backslash\mathscr{T}_g$ is a complex manifold, and is the moduli space of compact Riemann surfaces of genus g with level N structure. It is a finite cover of the moduli space $\mathscr{M}_g$ as orbifolds.*

Similarly, there is a surjective homomorphism

$$\mathrm{Out}(F_n) \to GL(n, \mathbb{Z})$$

induced from the action of $\mathrm{Out}(F_n)$ on the first homology group of the rose R_n with n petals used in defining markings of metric graphs.

For every integer N, the kernel of the composed homomorphism

$$\mathrm{Out}(F_n) \to GL(n, \mathbb{Z}) \to GL(n, \mathbb{Z}/N\mathbb{Z})$$

is called the *congruence subgroup* of $\mathrm{Out}(F_n)$ of level N, and denoted by $\mathrm{Out}(F_n)(N)$.

It is also known that when $N \geq 3$, $\mathrm{Out}(F_n)(N)$ is torsion-free subgroup of $\mathrm{Out}(F_n)$ of finite index. See for example [69] and references there.

Given a smooth tropical curve Γ_n of genus n, a *level N-structure* on Γ_n is a choice of a basis of $H_1(\Gamma_n, \mathbb{Z}/N\mathbb{Z})$.

Theorem 8.27. *For any positive integer $N \geq 3$, the moduli space of compact smooth tropical curves of genus n with a level N-structure is a tropical variety given by*

$$\mathrm{Out}(F_n)(N)\backslash \mathscr{T}_n^{tr}.$$

It is a finite cover of the moduli space $\mathscr{M}_n^{tr} = \mathrm{Out}(F_n)\backslash \mathscr{T}_n^{tr}$ as tropical orbifolds.

Proof. Since $\mathscr{T}_n^{tr}$ has a local structure of tropical variety and $\mathrm{Out}(F_n)(N)$ acts on $\mathscr{T}_n^{tr}$ fixed point freely by integral affine maps, every point p in the quotient $\mathrm{Out}(F_n)(N)\backslash \mathscr{T}_n^{tr}$ has a neighborhood which is equal to the neighborhood of a lift of p in $\mathscr{T}_n^{tr}$. Therefore, the local affine structure on $\mathscr{T}_n^{tr}$ passes to and defines a local affine on the quotient space $\mathrm{Out}(F_n)(N)\backslash \mathscr{T}_n^{tr}$. Since $\mathrm{Out}(F_n)(N)$ is a normal subgroup of $\mathrm{Out}(F_n)$ of finite index, the second statement is clear. □

8.6 Compactifications of the moduli space $\mathscr{M}_n^{tr}$

When $g \geq 1$, the moduli space $\mathscr{M}_g$ of Riemann surfaces of genus g is noncompact since Riemann surfaces Σ_g can degenerate. For example, when $g \geq 2$, each Σ_g has a canonical hyperbolic metric. By pinching along a simple closed geodesic, we get a diverging sequence of points in $\mathscr{M}_g$. Pinching along several disjoint simple closed geodesics also works, and this is essentially the only way for Riemann surfaces in $\mathscr{M}_g$ to diverge to infinity.

As an algebraic variety, $\mathscr{M}_g$ admits the well-known Deligne-Mumford compactification $\overline{\mathscr{M}_g}^{DM}$, which is a projective variety and obtained by adding stable nodal Riemann surfaces of Euler characteristic $2-2g$.

Similarly, the tropical moduli space $\mathscr{M}_n^{tr}$ is also noncompact. There are two sources of noncompactness of $\mathscr{M}_n^{tr}$:

1. There exist sequences of metric graphs $\Gamma_{n,j}$, $j = 1, 2, \ldots$, such that, when $j \to +\infty$, the length of some loops of $\Gamma_{n,j}$ goes to 0, while all other edge lengths stay bounded.
2. There exist sequences of metric graphs $\Gamma_{n,j}$, $j = 1, 2, \ldots$, such that, when $j \to +\infty$, the length of some edges of $\Gamma_{n,j}$ goes to ∞.

The above two cases can be combined. For example, there exist sequences of metric graphs $\Gamma_{n,j}$, $j = 1, 2, \ldots$, such that, when $j \to +\infty$, the length of some loops of $\Gamma_{n,j}$ goes to 0, and the length of some edges of $\Gamma_{n,j}$ goes to ∞.

In the first case, by passing to a subsequence, we can assume that there exists a metric graph $\Gamma_{n_\infty,\infty}$ of lower genus $n_\infty < n$ such that $\Gamma_{n,j}$ converges to $\Gamma_{n_\infty,\infty}$. The convergence is defined by convergence of lengths of corresponding edges of

$\Gamma_{n,j}$ and $\Gamma_{n_\infty,\infty}$. This convergence implies the condition that the Gromov-Hausdorff distance between $\Gamma_{n,j}$ and $\Gamma_{n_\infty,\infty}$ goes to zero.

The divergence in the first case and the resulting completion of $\mathscr{M}_n^{tr}$ can be described more precisely. By Proposition 8.10, the Gromov-Hausdorff metric d^{GH} on $\mathscr{M}_n^{tr}$ is incomplete.

Proposition 8.28. *The metric completion of $\mathscr{M}_n^{tr}$ with respect to the Gromov-Hausdorff metric d^{GH} consists of all compact metric graphs of genus less than n. Denote the completion by $\overline{\mathscr{M}_n^{tr}}^{GH}$.*

Proof. First we note that every compact metric graph of genus less than n is the Gromov-Hausdorff limit of a sequence of compact metric graphs of genus n. For example, suppose that Γ_k is a compact metric graph of genus $k < n$. For any positive t, by attaching $n-k$ loops of length t to Γ_k, we get a family of compact metric graphs of genus n, denoted by $\Gamma_{n,t}$. Clearly, when $t \to 0$, $\Gamma_{n,t}$ converges to Γ_k with respect to the Gromov-Hausdorff distance. Therefore, Γ_k belongs to the completion of $\mathscr{M}_n^{tr}$ with respect to d^{GH}.

Conversely, suppose $\Gamma_{n,j}$ is a Cauchy sequence in $\mathscr{M}_n^{tr}$ with respect to the metric d^{GH}. Then all edge lengths of $\Gamma_{n,j}$ are bounded. If there does not exist a loop in $\Gamma_{n,j}$ whose length goes to 0, then the sequence of metric graphs $\Gamma_{n,j}$ will converge to a compact metric graph of genus n as $j \to +\infty$. Otherwise, the sequence of compact metric graphs $\Gamma_{n,j}$ will converge to a compact metric graph of genus which is strictly less than n. □

We note that in the above proposition, the most degenerate metric graph consists of only one point (i.e. vertex), when all edge lengths go to 0.

In the theory of moduli spaces, we usually want to interpret ideal or boundary points of compactifications or bordifications of moduli spaces as some degenerate spaces. In Proposition 8.28, each boundary point in the Gromov-Hausdorff completion of $\mathscr{M}_n^{tr}$ can be interpreted as a compact metric graph, but its genus or Euler number can be different from that of the interior metric graphs. This is different from the situation of the Deligne-Mumford compactification $\overline{\mathscr{M}_g}^{DM}$ of the moduli space $\mathscr{M}_g$ of compact Riemann surfaces where stable nodal Riemann surfaces in the boundary have also Euler characteristic $2-2g$.

To overcome this problem, the notion of stable weighted metric graphs was introduced in [21, §3.3] and [32]. Briefly, a *stable weighted metric graph* is a graph G, an assignment of a positive length to each edge ℓ, together with a weight function on vertices $V(G)$,

$$w : V(G) \to \mathbb{Z}_{\geq 0}$$

such that any vertex of weight 0 has valence 3.

A metric graph of genus g can be identified with a weighted graph of zero weight. (Recall that by the convention in this book, all vertices of a metric graph have valence at least 3.)

The genus of a weighted metric graph (G,ℓ,w) is equal to the sum of the genus G and all the weights of vertices.

A bordification of $\mathscr{M}_n^{tr}$ was constructed in [32] and [21] by adding compact weighted metric graphs of genus n. Denote it by $\overline{\mathscr{M}_n^{tr}}^{W}$.

Proposition 8.29. *There exists a surjective continuous map from the bordification $\overline{\mathscr{M}_n^{tr}}^{W}$ of $\mathscr{M}_n^{tr}$ to the Gromov-Hausdorff completion $\overline{\mathscr{M}_n^{tr}}^{GH}$ of $\mathscr{M}_n^{tr}$ which is not injective when $n \geq 3$.*

Proof. By definition, each compact weighted metric graph (G,ℓ,w) has an underlying compact metric graph (G,ℓ). This gives the map $\pi : \overline{\mathscr{M}_n^{tr}}^{W} \to \overline{\mathscr{M}_n^{tr}}^{GH}$ in the proposition. It is clear that this map is surjective.

When $n \geq 3$ and a metric graph (G,ℓ) has genus strictly less than n but contains at least 2 vertices of valence at least 4, then there are different ways to assign weights to the vertices. Since these different weights do not change the metric on the graph, this shows that the map π is not injective. □

Since lengths of metric graphs in the Gromov-Hausdorff completion $\overline{\mathscr{M}_n^{tr}}^{GH}$ of $\mathscr{M}_n^{tr}$ are not bounded, this completion is not compact. This is related to the second type of divergence mentioned at the beginning of this subsection.

To obtain a compactification of $\overline{\mathscr{M}_n^{tr}}^{GH}$, we need to extend the definition of metric graphs to allowing some edges to have infinite lengths and some other edges to have lengths 0 as in [32].

Given a generalized compact metric graph (G,ℓ) of genus n, let $e_1,\dots,e_i$ be the set of its edges whose lengths are equal to ∞. A sequence of compact metric graphs (G_j,ℓ_j) of genus n is defined to converge to (G,ℓ) if the following conditions are satisfied:

1. When $j \gg 1$, there is a correspondence between the edges of G_j and G. Denote the edges of G_j corresponding to $e_1,\dots,e_i$ by $e_{j,1},\dots,e_{j,i}$.
2. The lengths of $e_{j,1},\dots,e_{j,i}$ go to ∞ when $j \to \infty$.
3. By collapsing the edges $e_1,\dots,e_i$, we obtain a new metric graph from G, denoted by $(G,\ell)/\{e_1,\dots,e_i\}$. Similarly, there is a collapsed graph $(G_j,\ell_j)/\{e_{1,j},\dots,e_{i,j}\}$. Then the Gromov-Hausdorff distance between $(G,\ell)/\{e_1,\dots,e_i\}$ and $(G_j,\ell_j)/\{e_{1,j},\dots,e_{i,j}\}$ goes to 0.

This convergence of metric graphs gives a topology to the union of the Gromov-Hausdorff completion $\overline{\mathscr{M}_n^{tr}}^{GH}$ and the space of generalized compact metric graphs of genus n. Denoted the union by $\overline{\mathscr{M}_n^{tr}}$.

Proposition 8.30. *The topological space $\overline{\mathscr{M}_n^{tr}}$ is a compact space.*

Proof. Since a compact metric graph of genus g has at most $3n-3$ edges, we can show that for any sequence of compact metric graphs, there is a convergent subsequence. Specifically, there are two cases. If all edge lengths are bounded, then a subsequence will converge to a point in the Gromov-Hausdorff completion $\overline{\mathcal{M}_n^{tr}}^{GH}$. On the other hand, if some edge lengths are not bounded, by passing to a subsequence, we can separate the edges of each metric graph into two groups depending on whether they are bounded or not, and a subsequence of the metric graphs will converge to a generalized metric graph. □

Remark 8.31. Instead of the tropical Teichmüller space $\mathcal{T}_n^{tr}$ which consists of all marked metric graphs, we can consider the outer space X_n which consist of only metric graphs whose total edge length is equal to 1. Then some results about compactifications can be simplified.

For example, the Gromov-Hausdorff completion of the moduli space $\mathrm{Out}(F_n)\backslash X_n$ is compact, since in this case, lengths of edges of metric graphs in X_n are always bounded from above by 1.

Remark 8.32. *Bordification by weighted metric graphs.* The bordification $\overline{\mathcal{M}_n^{tr}}^{w}$ of $\mathcal{M}_n^{tr}$ by weighted tropical curves in [32] and [21] can also be constructed using the ideal stacky fan structure on $\mathcal{T}_n^{tr}$. Each ideal polyhedral (or rather simplicial) cone in $\mathcal{T}_n^{tr}$ can be completed or compactified by taking its closure in the vector space which contains it. Then we get a stacky fan $\hat{\mathcal{T}}_n^{tr}$ in [21]. The action of $\mathrm{Out}(F_n)$ on $\mathcal{T}_n^{tr}$ extends to $\hat{\mathcal{T}}_n^{tr}$ by integral affine maps, and the quotient $\mathrm{Out}(F_n)\backslash\hat{\mathcal{T}}_n^{tr}$ is the bordification by weighted tropical curves and has the structure of stacky fan. We note that the Deligne-Mumford compactification $\overline{\mathcal{M}_g}^{DM}$ is a compact orbifold. It seems unlikely that $\overline{\mathcal{M}_n^{tr}}^{w}$ is a tropical orbifold.

Remark 8.33. As mentioned in the previous remark, a bordification of the tropical Teichmüller space $\mathcal{T}_n^{tr}$ was constructed by adding marked weighted metric graphs of genus g in [21]. This restricts to a bordification $\overline{X_n}^{w}$ of the outer space X_n.

In [14], another bordification $\overline{X_n}^{BS}$ of X_n was constructed, motivated by the Borel-Serre bordification of symmetric spaces of noncompact type, which induces the Borel-Serre compactification of locally symmetric spaces.

The action of $\mathrm{Out}(F_n)$ on X_n extends to a continuous action on the bordification $\overline{X_n}^{w}$ and also on $\overline{X_n}^{BS}$. It can be shown that the action of $\mathrm{Out}(F_n)$ on $\overline{X_n}^{w}$ is not proper, but $\mathrm{Out}(F_n)$ acts properly on $\overline{X_n}^{BS}$.

Furthermore, it can be shown that the identity map on X_n extends to a continuous surjective map

$$\overline{X_n}^{BS} \to \overline{X_n}^{w}.$$

The reason is that the stacky fan $\hat{\mathcal{T}}_n^{tr}$ is a partial compactification of $\mathcal{T}_n^{tr}$ and is an analogue of the Satake partial compactification of a symmetric space of noncompact type by adding rational boundary components endowed with the Satake topology. It is known that for a symmetric space X of noncompact type, the Borel-Serre partial compactification $\overline{X}^{BS}$ dominates any Satake partial compactification

$\overline{X}^S$ of the symmetric space X, i.e., the identity map on X extends to a continuous map

$$\overline{X}^{BS} \to \overline{X}^S,$$

which is then automatically surjective. See [20] for example. Similar arguments can be applied to the tropical counterparts and to prove the above result on $\overline{X_n}^{BS} \to \overline{X_n}^W$.

Remark 8.34. There are close relations between the Deligne-Mumford compactification $\overline{\mathcal{M}}_g^{DM}$ of the moduli space $\mathcal{M}_g$ of Riemann surfaces and the bordification $\overline{\mathcal{M}_n^{tr}}^W$ of the tropical moduli space $\mathcal{M}_n^{tr}$ by adding weighted metric graphs above. See [1] for detail.

9 Jacobian variety of a Riemann surface and the Siegel upper-half space

In the rest of the second part of this book, we want to construct an *intrinsic complete metric* on the outer space X_n which is invariant $\mathrm{Out}(F_n)$. We will do this by using the tropical analogy of the Jacobian variety of a compact Riemann surface.

To motivate the construction, we recall in this section the notion of polarized varieties, abelian varieties, periods of compact Riemann surfaces, and the Jacobian variety of a compact Riemann surface.

It is also worthwhile to mention that though one basic object of study in this book is the moduli space of compact Riemann surfaces, another closely related moduli space is the moduli space of abelian varieties. Consequently, we also mention the Siegel upper half-space, and the moduli space of principally polarized abelian varieties, and the Torelli theorem for Riemann surfaces. Their tropical counterparts will be discussed in the next subsection and will play a crucial role in the rest of this book.

9.1 Polarized algebraic varieties

Let M be a compact complex manifold. Recall that a complex line bundle $L \to M$ over the manifold M is called *very ample* if its global sections, i.e., elements of $H^0(M, \mathcal{O}(L))$, give an embedding

$$M \to \mathbb{C}P^N.$$

Equivalently, if there exists a holomorphic embedding $\pi : M \to \mathbb{C}P^N$, then $L = \pi^* H$ is a very ample line bundle over M, where H is the line bundle over the complex projective $\mathbb{C}P^N$ corresponding to a hyperplane of $\mathbb{C}P^N$.

A line bundle $L \to M$ over a complex manifold M is called *ample* if for positive integer $k \gg 0$, $L^k \to M$ is very ample, where L^k is the k-th tensor product of L.

Such an ample line bundle $L \to M$ is called a *polarization* of M. The existence of an ample line bundle over M is a necessary and sufficient condition for M to be biholomorphic to a projective variety.

It is known that every compact complex manifold of dimension 1 (i.e., every compact Riemann surface) admits ample line bundles and hence is biholomorphic to a projective algebraic curve over $\mathbb{C}$. On the other hand, there are many compact complex manifolds which do not admit ample lines bundles and hence are not biholomorphic to algebraic varieties over $\mathbb{C}$.

Remark 9.1. Moduli spaces of compact complex manifolds of dimension strictly greater than 1 usually do not have desirable properties. For example, the moduli space of all complex tori of dimension $n > 1$ is not a Hausdorff space. See [18, Chapter 7] for example. Therefore, people usually study moduli spaces of polarized complex manifolds (or algebraic varieties). See [129] for example.

We note that compact Riemann surfaces are always polarizable as mentioned above. For example, when the genus g of a compact Riemann surface Σ_g is at least 2, its canonical line bundle is ample. Consequently, we define the moduli space $\mathcal{M}_g$ of compact Riemann surfaces of genus g without mentioning polarization.

The *Chern class* of an ample line bundle $L \to M$ is an integral class in $H^2(M,\mathbb{Z})$ and can be represented by a closed, positive $(1,1)$-form in the De Rham cohomology $H^2_{dr}(M,\mathbb{R})$. Conversely, by the Kodaira embedding theorem (see [56, p. 181]), every such *positive integral class* in $H^2(M,\mathbb{R})$ is the Chern class of an ample line bundle over M and hence M is polarizable.

Therefore, we have the following result.

Proposition 9.2. *A* polarization *of a complex manifold M corresponds to a* positive integral class *in $H^2(M,\mathbb{R})$. A $(1,1)$-form on M representing such a class is called a* Hodge form *of the manifold.*

9.2 Abelian varieties

For any lattice $\Lambda \subset \mathbb{C}^n$, the quotient $\Lambda\backslash\mathbb{C}^n$ is called a *complex torus*. A complex torus $\Lambda\backslash\mathbb{C}^n$ is *an abelian variety* if it is a projective variety, i.e., if it admits a polarization.

The existence of a Hodge form on the complex tori $\Lambda\backslash\mathbb{C}^n$ depends on the compatibility of the complex structure of $\mathbb{C}^n$ and the integral structure of $\mathbb{C}^n$ defined by the lattice Λ. This is reasonable since a Hodge needs both the positivity defined in terms of complex structure and the integral structure in terms of the integral homology group.

Given a basis $\lambda_1,\dots,\lambda_{2n}$ of the lattice Λ as a free abelian group of rank $2n$ and a $\mathbb{C}$-basis $e_1,\dots,e_n$ of the complex vector space $\mathbb{C}^n$, define the *period matrix* $\Omega = (\omega_{\alpha i})$ of size $n \times 2n$ by

$$\lambda_i = \sum_\alpha \omega_{\alpha i} e_\alpha.$$

Note that this period Ω measures the compatibility of these two bases.

By using the fact that every cohomology class of a complex torus (or more generally any compact Lie group) can be represented by a differential form which is invariant under translation, one can obtain the following *Riemann condition for Abelian varieties* [56, p. 306]:

Proposition 9.3. *A complex torus $M = \Lambda\backslash\mathbb{C}^n$ is an abelian variety if and only if there exists a basis $\lambda_1,\dots,\lambda_{2n}$ of Λ and a $\mathbb{C}$-basis $e_1,\dots,e_n$ of $\mathbb{C}^n$ such that the period matrix Ω with respect to them is of the form*

$$\Omega = (\Delta_\delta, Z),$$

where Δ_δ is a diagonal matrix with positive integral diagonal entries $\delta_1,\dots,\delta_n$ which satisfy the divisibility condition: $\delta_1|\delta_2,\dots,\delta_{n-1}|\delta_n$, and Z is a symmetric $n\times n$ matrix and $\mathrm{Im}(Z)$ *is positive definite.*

More precisely, let $x_1,\dots,x_{2n}$ be the coordinates of $\mathbb{C}^n$ determined by $\lambda_1,\dots,\lambda_{2n}$. Then the following form ω is an integral 2-form

$$\omega = \sum_\alpha \delta_\alpha dx_\alpha \wedge dx_{n+\alpha}, \tag{9.1}$$

since $\delta_1,\dots,\delta_n$ are integers.

Under the above condition on the period matrix Ω in Proposition 9.3, the form ω is a positive form, and hence it is a Hodge form. Therefore, ω gives a polarization of the complex torus $\Lambda\backslash\mathbb{C}^n$.

If $\delta_1 = \dots = \delta_n = 1$, then the polarization ω is called a *principal polarization*, and the abelian variety $\Lambda\backslash\mathbb{C}^n$ is called *principally polarized.*

Conversely, suppose ω is a Hodge form of $\Lambda\backslash\mathbb{C}^n$ which is invariant under translation. Then there exists a basis $\lambda_1,\dots,\lambda_{2n}$ of Λ such that ω is of the form in Equation 9.1. Now computing the form ω in the complex coordinates determined by $e_1,\dots,e_n$ shows that the condition that ω is a positive (1,1)-form is exactly equivalent to the condition on the period matrix Z stated in Proposition 9.3.

9.3 Jacobian varieties of Riemann surfaces

For a compact Riemann surface Σ_g, with $g \geq 1$, we can define its Jacobian variety as follows. Let

$$H^{0,1}(\Sigma_g) = H^0(\Sigma_g, \Omega)$$

be the space of holomorphic 1-forms on Σ_g. It is a complex vector space of dimension g. The first homology group $H_1(\Sigma_g,\mathbb{Z})$ is isomorphic to $\mathbb{Z}^{2g}$ and it can be embedded into the dual space $H^0(\Sigma_g,\Omega)^*$ of $H^0(\Sigma_g,\Omega)$,

$$\pi : H_1(\Sigma_g,\mathbb{Z}) \to H^0(\Sigma_g,\Omega)^*,$$

by integration: For every 1-cycle σ and a 1-form $\omega \in H^0(\Sigma_g,\Omega)$,

$$\pi(\sigma)(\omega) = \int_\sigma \omega.$$

Proposition 9.4. *The complex torus* $H_1(\Sigma_g,\mathbb{Z})\backslash H^0(\Sigma_g,\Omega)^*$ *admits a* canonical polarization *coming from the pairing* $H_1(\Sigma_g,\mathbb{Z}) \times H_1(\Sigma_g,\mathbb{Z}) \to \mathbb{Z}$ *and is a* principally polarized abelian variety. *It is called the* Jacobian variety *of* Σ_g.

We explain this proposition by using the above discussion of the notion of a polarization of an abelian variety.

For the complex torus $M = H_1(\Sigma_g,\mathbb{Z})\backslash H^0(\Sigma_g,\Omega)^*$, we first note that there exists a symplectic basis $a_1,b_1,\dots,a_g,b_g$ of the lattice $H_1(\Sigma_g,\mathbb{Z})$, i.e., a basis satisfying the following condition: for every pair i,j,

$$a_i \cdot b_j = \delta_{ij}, \quad a_i \cdot a_j = 0, \quad b_i \cdot b_j = 0.$$

For every such symplectic basis, there exists a normalized basis $\omega_1,\dots,\omega_g$ of $H^0(\Sigma_g,\Omega)$ such that

$$\int_{a_i} \omega_j = \delta_{ij}.$$

Note that it is true that for every j, we can find a real 1-form f_j such that $\int_{a_i} f_j = \delta_{ij}$, $i = 1,\dots,g$. On the other hand, it is not automatic that for every j there exists a holomorphic differential 1-form ω_j such that the period conditions $\int_{a_i} \omega_j = \delta_{ij}$, $i = 1,\dots,g$, are satisfied. The reason is that it is not obvious that for each cycle a_i, there is a holomorphic differential 1-form whose integral on a_i is not zero. For a detailed proof, see [56, p. 231]. Once this is known and the differential forms $\omega_1,\dots,\omega_n$ with the above specified periods on the symplectic basis are given, it is easy to see that $\omega_1,\dots,\omega_n$ are linearly independent and form a basis of $H^0(\Sigma_g,\Omega)$.

This implies that with respect to the dual basis of the complex vector space $H^0(\Sigma_g,\Omega)^*$, the period matrix (Δ_δ, Z) of the symplectic basis $a_1,b_1,\dots,a_g,b_g$ is principal, i.e.,

$$\Delta_\delta = I_g.$$

Note that the matrix $Z = (Z_{ij})$ is given by the periods on the b-cycles:

$$Z_{ij} = \int_{b_i} \omega_j.$$

The Riemann bilinear relation [56, p. 232] implies that the period Z is symmetric and that $\operatorname{Im} Z$ is positive definite. Therefore, the Riemann condition in Proposition 9.3 is satisfied by the Jacobian variety $H_1(\Sigma_g,\mathbb{Z})\backslash H^0(\Sigma_g,\Omega)^*$, and the Jacobian variety is a *principally polarized abelian variety*. This polarization on the Jacobian variety is called the *canonical principal polarization*, and the period Z is called the *canonical period* of the Riemann surface Σ_g.

The Hodge form ω for the Jacobian variety $H_1(\Sigma_g,\mathbb{Z})\backslash H^0(\Sigma_g,\Omega)^*$ is a $(1,1)$-form in the cohomology group $H^2(H_1(\Sigma_g,\mathbb{Z})\backslash H^0(\Sigma_g,\Omega)^*)$, and hence corresponds to *an integral skew-symmetric bilinear form*, i.e. a *symplectic form*, on $H_1(\Sigma_g,\mathbb{Z})$.

On the other hand, by the Poincaré duality, the intersection form on $H_1(\Sigma_g,\mathbb{Z})$,

$$H_1(\Sigma_g,\mathbb{Z}) \times H_1(\Sigma_g,\mathbb{Z}) \to \mathbb{Z},$$

is a nondegenerate skew-symmetric bilinear form. By unwrapping the definitions of the period and the symplectic form, we can prove the following.

Proposition 9.5. *For any compact Riemann surface of genus g, the integral skew-symmetric bilinear form on $H_1(\Sigma_g,\mathbb{Z})$ induced from the canonical principal polarization on the Jacobian variety of Σ_g is equal to the intersection form on $H_1(\Sigma_g,\mathbb{Z})$.*

Remark 9.6. We will see later that the Chern class of a line bundle over a smooth compact tropical curve lies in the first cohomology group of the tropical curve and that a polarization corresponds to a positive definite symmetric bilinear form instead of a skew-symmetric bilinear form. One reason is that a Riemann surface is of real dimension 2 and a tropical curve is of real dimension 1.

9.4 Moduli space of principally polarized abelian varieties $\mathscr{A}_g$

As we mentioned before, marking makes it easier to study the moduli space $\mathscr{M}_g$ of Riemann surfaces through the Teichmüller space $\mathscr{T}_g$. The same is true for moduli spaces of abelian varieties.

One analogy of Teichmüller space for abelian varieties is the Siegel upper space. For each $g \geq 1$, the *Siegel upper half space* of degree g is defined by

$$\mathfrak{h}_g = \{X + iY \mid X, Y \text{ are real symmetric } g \times g\text{-matrices}, Y > 0\}.$$

When $g = 1$, $\mathfrak{h}_g$ is equal to the upper-half plane $\mathbb{H}^2$.

It is known that the symplectic group $\mathrm{Sp}(2g,\mathbb{R})$ acts holomorphically and transitively on $\mathfrak{h}_g$, and the stabilizer of $i\mathrm{Id}_g$ can be identified with $\mathrm{U}(g)$. Therefore, we can write $\mathfrak{h}_g$ as

$$\mathfrak{h}_g \cong \mathrm{Sp}(2g,\mathbb{R})/\mathrm{U}(g).$$

Consequently, $\mathfrak{h}_g$ is a Hermitian symmetric space of noncompact type.

The above discussion shows that for every compact Riemann surface Σ_g, its canonical period Z lies in $\mathfrak{h}_g$. Different choices of symplectic bases of $H_1(\Sigma_g,\mathbb{Z})$ lead to different points in a $\mathrm{Sp}(2g,\mathbb{Z})$-orbit in $\mathfrak{h}_g$. This gives a well-defined map

$$\Pi : \mathscr{M}_g \to \mathrm{Sp}(2g,\mathbb{Z})\backslash\mathfrak{h}_g, \quad \Sigma_g \mapsto \mathrm{Sp}(2g,\mathbb{Z})Z. \tag{9.2}$$

This is called the *period map* for Riemann surfaces Σ_g.

Let $\mathscr{A}_g$ denote the moduli space of *principally polarized abelian varieties* of dimension g. Then the discussion in the previous subsection, in particular Proposition 9.3, shows the following identification.

Proposition 9.7. *The moduli space $\mathscr{A}_g$ of principally polarized abelian varieties of dimension g can be canonically identified with the Siegel modular variety,*

$$\mathscr{A}_g \cong \mathrm{Sp}(2g,\mathbb{Z})\backslash\mathfrak{h}_g,$$

where a point $\tau \in \mathfrak{h}_g$ corresponds to a principally polarized abelian variety $\mathbb{Z}^n + \tau\mathbb{Z}^n\backslash\mathbb{C}^n$.

Under the identification, the period map for compact Riemann surfaces

$$\Pi : \mathscr{M}_g \to \mathrm{Sp}(2g,\mathbb{Z})\backslash\mathfrak{h}_g$$

is equivalent to the *Jacobian map*:

$$\Pi : \mathscr{M}_g \to \mathscr{A}_g, \quad \Sigma_g \mapsto H_1(\Sigma_g,\mathbb{Z})\backslash H^0(\Sigma_g,\Omega)^*. \tag{9.3}$$

Note that each Jacobian variety $H_1(\Sigma_g,\mathbb{Z})\backslash H^0(\Sigma_g,\Omega)^*$ is given the canonical principal polarization.

The Jacobian map Π is a complex analytic map with respect to the natural complex structures on $\mathscr{M}_g$ and $\mathrm{Sp}(2g,\mathbb{Z})\backslash\mathfrak{h}_g$. It is also a morphism when $\mathscr{M}_g$ and $\mathscr{A}_g$ are given the structure of algebraic varieties.

The *Torelli theorem* [56, p. 359] says that $\Pi : \mathscr{M}_g \to \mathscr{A}_g$ is injective. As a topological map between two topological spaces, Π is an embedding. But its properties as a morphism between varieties is more subtle and complicated.

The period map Π can be lifted to holomorphic map

$$\tilde{\Pi} : \mathscr{T}_g \to \mathfrak{h}_g.$$

It is known that this lifted map $\tilde{\Pi}$ is not an immersion, i.e., its differential is singular at the locus of hyperelliptic Riemann surfaces. On the other hand, it was shown in [111] that the period map $\Pi : \mathscr{M}_g \to \mathscr{A}_g$ is an embedding as a morphism between algebraic varieties.

The Jacobian map Π and the Torelli theorem have played a fundamental role in understanding the moduli space $\mathscr{M}_g$. See for example [110].

Remark 9.8. The complex torus $H_1(\Sigma_g,\mathbb{Z})\backslash H_1(\Sigma_g,\mathbb{R})$ is usually called the *Albanese variety* of Σ_g. It will also be defined for compact smooth tropical curves. Using the duality between $H_1(\Sigma,\mathbb{R})$ and $H^1(\Sigma_g,\mathbb{R})$, which is isomorphic to $H^{0,1}(\Sigma_g)$, it can be seen that $H_1(\Sigma_g,\mathbb{Z})\backslash H_1(\Sigma_g,\mathbb{R})$ is isomorphic to the Jacobian variety $H_1(\Sigma_g,\mathbb{Z})\backslash H^{0,1}(\Sigma_g)^*$.

In general, the Albanese variety of a higher dimensional complex manifold or algebraic variety is a generalization of the Jacobian variety of a projective curve (or a compact Riemann surface).

10 Tropical abelian varieties, their moduli spaces, and Jacobians of tropical curves

10.1 Jacobian variety of tropical curves and metric graphs

As mentioned before, metric graphs considered in this book have no vertices of valence 1 or 2. By the results in Propositions 4.27 and 4.28, metric graphs can be identified with smooth tropical curves. From now on, we will *denote metric graphs by* (Γ,ℓ) *instead of* (G,ℓ) *to be consistent with the notation for tropical curves.*

Recall from Equation 2.5 and Proposition 2.23 that for any compact metric graph (Γ,ℓ), there is a positive definite quadratic form Q on the real vector space $C_1(\Gamma,\mathbb{R})$ of 1-chains on Γ with $\mathbb{R}$-coefficients, which defines an inner product on $H_1(\Gamma,\mathbb{R})$ and descends to a flat metric on the torus $H_1(\Gamma,\mathbb{Z})\backslash H_1(\Gamma,\mathbb{R})$.

Following the definitions in [82] and [34] for (combinatorial) graphs, we have the following definition.

Definition 10.1. The torus $H_1(\Gamma,\mathbb{Z})\backslash H_1(\Gamma,\mathbb{R})$ with the flat metric induced from the quadratic form Q is called the Albanese torus of the metric graph (Γ,ℓ), and denoted by $(H_1(\Gamma,\mathbb{Z})\backslash H_1(\Gamma,\mathbb{R}),Q)$.

Remark 10.2. The definition of the Albanese and Jacobian tori for graphs were introduced in [82, p. 94]. For a graph G, the torus $H_1(G,\mathbb{Z})\backslash H_1(G,\mathbb{R})$ with the flat metric where each edge is assigned length 1 is the Albanese torus, and the dual flat torus $H^1(G,\mathbb{Z})\backslash H^1(X,\mathbb{R})$ is called the Jacobian torus. These are motivated by, but slightly different from, various notions of Jacobian varieties of algebraic varieties, Kähler manifolds and Riemannian manifolds, which are defined as quotients of cohomology groups. As recalled in the previous section, for a Riemann surface Σ_g, the Jacobian variety is a quotient of the dual of the first cohomology, $H_1(\Sigma_g,\mathbb{Z})\backslash H^0(\Sigma_g,\Omega)^*$, where $H^0(\Sigma_g,\Omega)=H^{0,1}(\Sigma_g)\subset H^1(\Sigma,\mathbb{C})$.

In this metric on the space $C_1(\Gamma,\mathbb{R})$ of 1-chains (or the metric of the torus $H_1(\Gamma,\mathbb{Z})\backslash H_1(\Gamma,\mathbb{R})$) induced from the quadratic form Q, the length of an edge e of the graph Γ viewed as a 1-chain is equal to $\sqrt{\ell(E)}$. As mentioned before, this makes the definition of the metric Q seem unnatural since its dependence on the metric ℓ of the graph (Γ,ℓ) is not smooth. It might be more tempting to define Q so that the length of the 1-chain e to be $\ell(E)$ in order for it to depend smoothly on the lengths of edges of the graph (Γ,ℓ). (Of course, for graphs whose edges have normalized length equal to 1, there is no difference in the above two choices.)

In order to explain that the definition of the quadratic form in Equation 2.5 is the right one and that Q gives a *principal polarization* when $H_1(\Gamma,\mathbb{Z})\backslash H_1(\Gamma,\mathbb{R})$ is given a *natural structure of a tropical torus*, we will identify $(H_1(\Gamma,\mathbb{Z})\backslash H_1(\Gamma,\mathbb{R}), Q)$ with the tropical Jacobian variety of the tropical curve associated with (Γ,ℓ). (We note that for all metric graphs (Γ,ℓ) of genus g, $H_1(\Gamma,\mathbb{R})$ are isomorphic, and $H_1(\Gamma,\mathbb{Z})$ are isomorphic lattices. As recalled below, a tropical structure on $H_1(\Gamma,\mathbb{R})$ is determined by another lattice.)

First, we need to define differential forms on a tropical curve, and then a polarization of a tropical torus. After that, we show that the tropical Jacobian variety of a compact smooth tropical curve has a canonical principal polarization. In the rest of this subsection, we will follow [107].

Definition 10.3. Given a compact smooth tropical curve Γ, a 1-*form* ω (i.e., a tropical differential form of degree 1) on Γ is a collection of *constant real differential* 1-*forms* on the edges of Γ satisfying the balancing condition at every vertex of Γ. Specifically, let x be a vertex of valence $n+1$, and let U be a chart of x with a $\mathbb{Z}$-affine structure $\phi_U : U \to \mathbb{R}^n$. Let $e_0, \dots, e_n$ be the edges out of the vertex $\phi_U(x)$ in $\mathbb{R}^n$, and $V(e_i)$ a primitive integral vector along e_i pointing away from the vertex $\phi_U(x)$. The 1-form ω defines a constant differential 1-form on each e_i as a smooth differential manifold, and the *balancing condition* is

$$\sum_{i=0}^{n} \omega(V(e_i)) = 0. \tag{10.1}$$

It is clear that the balancing condition in the above proposition is preserved by $\mathbb{Z}$-linear maps and hence the balancing condition is well-defined.

Though we assume that all vertices have valence at least 3, the above balancing condition is satisfied at every interior point of an edge when it is viewed as a vertex of valence 2.

Let $\Omega(\Gamma)$ be the space of 1-forms on a smooth compact tropical curve Γ. This is a finite dimensional vector space.

Lemma 10.4 ([107, §6.1])**.** *If the genus of a compact tropical smooth curve Γ is equal to g, then $\Omega(\Gamma)$ is a real vector space of dimension g.*

Proof. Since Γ is homotopy equivalent to a wedge of n circles, there are g points, $p_1, \dots, p_g$, contained in the interior of edges of C such that $C - \{p_1, \dots, p_g\}$ is contractible. They are called breakpoints. From the balancing conditions at the vertices, it is clear that a global 1-form on Γ is determined by its values on primitive integral tangent vectors of the edges which contain these breakpoints $p_1, \dots, p_g$. It can also be seen that all values at the breakpoints can be achieved by some 1-forms. For example, this is true for the wedge of n-circles. By noting that the balancing conditions are preserved by contracting edges of the complement of the breakpoints, we can prove this by induction on the number of edges by starting from the wedge of circles. This proves that $\Omega(\Gamma)$ is of dimension g. □

Definition 10.5. A 1-form $\omega \in \Omega(\Gamma)$ is called integral if it takes an integral value on every integral tangent vector of every edge of Γ.

We note that the notion of integral tangent vectors of each edge of Γ can be defined using $\mathbb{Z}$-affine local charts and it is well-defined. By the same proof of Lemma 10.4, we obtain the following.

Lemma 10.6. *Let $\Omega_{\mathbb{Z}}(\Gamma)$ denote the subspace of integral 1-forms on Γ. Then $\Omega_{\mathbb{Z}}(\Gamma)$ is a lattice of the real vector space $\Omega(\Gamma)$.*

Let $\Omega(\Gamma)^*$ be the dual space of linear functionals on the real vector space $\Omega(\Gamma)$. We now define a pairing between $H_1(\Gamma,\mathbb{Z})$ and $\Omega(\Gamma)$, and hence a map

$$p : H_1(\Gamma,\mathbb{Z}) \to \Omega(\Gamma)^*.$$

(Note that p stands for period.) For any path $\gamma\colon [a,b] \to \Gamma$, the pull back $\gamma^*\omega$ is a piecewise-constant differential 1-form on $[a,b]$ (with possible discontinuity at the preimage of the vertices), and the integral $\int_\gamma \omega$ is defined to be $\int_a^b \gamma^*\omega$. This implies that for any cycle σ and any 1-form ω, we can define

$$p(\sigma)(\omega) = \int_\sigma \omega. \tag{10.2}$$

Clearly, this map $p : H_1(\Gamma,\mathbb{Z}) \to \Omega(\Gamma)^*$ is linear and can be extended to a linear transformation

$$p : H_1(\Gamma,\mathbb{R}) \to \Omega(\Gamma)^*.$$

Proposition 10.7. *The period map $p\colon H_1(\Gamma,\mathbb{Z}) \to \Omega(\Gamma)^*$ defined above is an embedding, and the linear transformation $p\colon H_1(\Gamma,\mathbb{R}) \to \Omega(\Gamma)^*$ is an isomorphism.*

Proof. If we identify a tropical curve Γ with its associated metric graph (Γ,ℓ), and on each edge e we use the length function t measured from a vertex as the parameter, then the restriction of a 1-form ω to e is given by $a\, dt$, where a is a constant. At each vertex x of the graph Γ, let $e_1,\dots,e_k$ be the edges out of x, and $t_1,\dots,t_k$ be the length functions measured from x. Suppose that ω is a 1-form, and on each edge e_i, $\omega = a_i\, dt_i$. Then *the balancing condition* at x for the 1-form ω in Equation (10.1) is equivalent to

$$\sum_{i=1}^{k} a_i = 0. \tag{10.3}$$

If we choose an orientation of an edge e, then there is a choice of a length parameter t_e such that $\int_e dt_e > 0$. Fix an orientation of all edges e of Γ and the compatible choices of the length parameters t_e. For any nontrivial cycle $\sigma = \sum_e a_e a$, define a 1-form $\omega = \sum_e a_e\, dt_e$ on Γ, which means that it restricts to $a_e dt_e$ on each edge e. The compatibility of the edges contained in σ at the vertices to make it a cycle implies that ω is a 1-form on Γ. Since

$$p(\sigma)(\omega) = \sum_e a_e^2 \int_e dt_e = \sum_e a_e^2 \ell(e) > 0,$$

this implies that $p(\sigma)$ is nonzero, and hence p is an embedding.

Since the dimensions of $H_1(\Gamma,\mathbb{R})$ and $\Omega(\Gamma)^*$ are both equal to n, the extended linear transformation is an isomorphism, and the proposition is proved. □

Remark 10.8. The above interpretation of 1-form ω on a smooth tropical curve Γ in terms of length parameters of a metric graph (Γ,ℓ) gives a different proof of Lemma 10.4. Fix an orientation of edges e of Γ and the corresponding choices of the length parameters t_e as above. We can define a linear isomorphism between $H_1(\Gamma,\mathbb{R})$ and $\Omega(\Gamma)$. For every 1-cycle $\sigma = \sum_e a_e e$, define a 1-form $\omega = \sum_e a_e\, dt_e$, which means that the restriction of ω to each edge e is equal to a_e. Conversely, for every 1-form $\omega = \sum_e a_e\, dt_e$, we have a 1-cycle $\sigma = \sum_e a_e e$. Therefore, the dimension of $\Omega(\Gamma)$ is equal to the dimension of $H_1(\Gamma,\mathbb{R})$, which is equal to g.

By Proposition 10.7, the quadratic form Q on $H_1(\Gamma,\mathbb{R})$ in Equation 2.5 defines a quadratic form on $\Omega(\Gamma)^*$. The positive definite quadratic form Q on $\Omega(\Gamma)^*$ gives an isomorphism $\Omega(\Gamma)^* \to \Omega(\Gamma)$, and together with the map $p : H_1(\Gamma,\mathbb{Z}) \to \Omega(\Gamma)^*$ in Proposition 10.7, it gives an embedding

$$i_Q : H_1(\Gamma,\mathbb{Z}) \to \Omega(\Gamma). \tag{10.4}$$

We emphasize that this embedding i_Q depends on the positive definite quadratic form Q. The following proposition explains the *reason for the choice of value* $Q(e,e)$ for an edge e in Definition (2.5) of the quadratic form Q of a compact metric graph.

Proposition 10.9. *In the above notation, the image of the embedding* $i_Q : H_1(\Gamma,\mathbb{Z}) \to \Omega(\Gamma)$ *is equal to the space of* $\Omega_{\mathbb{Z}}(\Gamma)$ *of integral 1-forms.*

Proof. We need to show that for any 1-cycle $\sigma = \sum_e a_e e$ in $H_1(\Gamma,\mathbb{R})$, the image $i_Q(\sigma) \in \Omega_{\mathbb{Z}}(\Gamma)$ if and only if $\sigma \in H_1(\Gamma,\mathbb{Z})$. For every edge e of Γ, let $V(e)$ be an integral primitive vector tangent to e. By the definition of the metric on the tropical curve, the norm of $V(e)$ is equal to 1. By definition of the integral 1-form ω, $i_Q(\sigma)$ is integral if and only if $i_Q(\sigma)(V(e))$ is integral for every edge E, which is equivalent to the fact that $\int_e i_Q(\sigma)$ is an integral multiple of $\ell(e)$. On the other hand, by definition,

$$\int_e i_Q(\sigma) = Q(e,\sigma) = a_e \ell(e).$$

Therefore, $i_Q(\sigma)$ is an integral 1-form if and only if σ is an integral 1-cycle, i.e., all coefficients a_e are in $\mathbb{Z}$. □

Remark 10.10. If we orient all edges e of Γ and pick compatible length parameters t_e for them as in Proposition 10.7, then a 1-form ω on Γ is integral if and only if for every edge e, the restriction of ω to e is an integral multiple of dt_e, and the map in Proposition 10.9 can be written down explicitly as follows: For every cycle $\sigma = \sum_e a_e e$, the restriction of the 1-form $i_Q(\sigma)$ to the edge e is equal to $a_e dt_e$, i.e., $i_Q(\sigma) = \sum_e a_e\, dt_e$. As explained in the proof of Proposition 10.9, the definition of the quadratic form Q on $H_1(\Gamma,\mathbb{Z})$ was motivated to obtain this natural equivalence.

Specifically, Q is the *unique positive definite quadratic form which gives an isomorphism* $\Omega(\Gamma) \cong \Omega(\Gamma)^*$ such that once it is composed with the isomorphism

$H_1(\Gamma,\mathbb{R}) \to \Omega(\Gamma)$ in Remark 10.8, the map $H_1(\Gamma,\mathbb{Z}) \to \Omega(\Gamma)^*$ is equal to the period map $p : H_1(\Gamma,\mathbb{Z}) \to \Omega(\Gamma)^*$ in Equation (10.2). Since both the period map and the isomorphism in Remark 10.8 are natural, the definition of Q and the induced isomorphism $\Omega(\Gamma) \cong \Omega(\Gamma)^*$ are also natural.

10.2 Tropical abelian varieties

Before continuing, we recall the definition of tropical torus according to [107, §5.1]. Compare also with §5.5.

Let $\Lambda \subset \mathbb{R}^n$ be a lattice. *A tropical torus is a real torus $\Lambda\backslash\mathbb{R}^n$ together with a $\mathbb{Z}$-affine structure on it.* A *$\mathbb{Z}$-affine structure* on $\mathbb{R}^n$ is determined by an *integral structure*, i.e., a lattice $L \subset \mathbb{R}^n$, and this integral structure on $\mathbb{R}^n$ descends to the quotient $\Lambda\backslash\mathbb{R}^n$.

For example, the *standard $\mathbb{Z}$-affine structure* on $\mathbb{R}^n$ is determined by the standard lattice $\mathbb{Z}^n$ in $\mathbb{R}^n$. Unless specified otherwise, $\mathbb{R}^n$ is given this $\mathbb{Z}$-affine structure, then for any lattice $\Lambda \subset \mathbb{R}^n$, the torus $X = \Lambda\backslash\mathbb{R}^n$ has the induced $\mathbb{Z}$-affine structure. (Note that Λ is not necessarily commensurable with $\mathbb{Z}^n$, and does not satisfy other relations with $\mathbb{Z}^n$ in general.)

More generally, suppose V is a real vector space and $L \subset V$ is a lattice. Then V has an integral and $\mathbb{Z}$-affine structure determined by L. For any lattice $\Lambda \subset V$, the torus $\Lambda\backslash V$ is a tropical torus, whose tropical structure is determined by the lattice L. Clearly, the tropical structure on the torus $\Lambda\backslash V$ depends on the lattice L.

The subspace $\Omega_{\mathbb{Z}}(\Gamma)$ of integral 1-forms defines a dual lattice $\Omega_{\mathbb{Z}}(\Gamma)^*$ in $\Omega(\Gamma)^*$, and hence defines a $\mathbb{Z}$-affine structure on $\Omega(\Gamma)^*$ and also on the torus $H_1(\Gamma,\mathbb{Z})\backslash\Omega(\Gamma)^*$. Therefore, $H_1(\Gamma,\mathbb{Z})\backslash\Omega(\Gamma)^*$ has a *canonical tropical structure*.

Definition 10.11. For a compact smooth tropical curve Γ, the tropical torus $H_1(\Gamma,\mathbb{Z})\backslash\Omega(\Gamma)^*$ with the $\mathbb{Z}$-affine structure defined above by the space $\Omega_{\mathbb{Z}}(\Gamma)$ of integral 1-forms is called the tropical Jacobian variety of the tropical curve Γ.

Remark 10.12. Clearly, $H_1(\Gamma,\mathbb{R})$ can be mapped isomorphically to $\Omega(\Gamma)^*$ by integration as in Proposition 10.7, and $H_1(\Gamma,\mathbb{Z})\backslash\Omega(\Gamma)^*$ is isomorphic to the Albanese torus $H_1(\Gamma,\mathbb{Z})\backslash H_1(\Gamma,\mathbb{R})$. Under this isomorphism, the canonical $\mathbb{Z}$-affine structure on $H_1(\Gamma,\mathbb{Z})\backslash H_1(\Gamma,\mathbb{R})$, or rather on $H_1(\Gamma,\mathbb{R})$, depends on the positive definite quadratic form Q or the metric ℓ of the graph (see Proposition 10.19). In general, the canonical $\mathbb{Z}$-affine structure on $H_1(\Gamma,\mathbb{R})$ is not the $\mathbb{Z}$-affine structure determined by the integral lattice $H_1(\Gamma,\mathbb{Z})$ in $H_1(\Gamma,\mathbb{R})$.

Besides defining a flat metric on $H_1(\Gamma,\mathbb{Z})\backslash\Omega(\Gamma)^*$, the positive definite quadratic form Q also defines a polarization, which is defined as follows.

Let X be a tropical torus. Since transition functions between trivializations of a tropical line bundle over different charts of X are affine linear functions with integral slope, a tropical line bundle over X corresponds to an element $H^1(X,\mathscr{O}^*)$, where $\mathscr{O}^* = \mathrm{Aff}$ is the sheaf of affine linear functions with integral slope.

Recall that the sheaf $\mathbb{T}^*_{\mathbb{Z}}$ of locally constant integral 1-forms is defined by the exact sequence

$$0 \to \mathbb{R} \to \mathscr{O}^* \to \mathbb{T}^*_{\mathbb{Z}} \to 0.$$

We note that the sheaf $\mathbb{T}^*_{\mathbb{Z}}$ is isomorphic to the locally constant sheaf associated with the abelian group $(\mathbb{Z}^n)^*$. This induces a map

$$c : H^1(X, \mathscr{O}^*) \to H^1(X, \mathbb{T}^*_{\mathbb{Z}}).$$

This is called the *Chern class* map.

As we recalled in the previous section, for a complex torus M, an ample line bundle L over M is called a *polarization*, or rather its first Chern class is called a polarization.

For a tropical torus X, we call *the Chern class* $c(L)$ of a tropical line bundle L over X also *a polarization of the tropical torus*, according to [107, §5.1]. It seems more natural to require also that $c(L)$ is positive in the sense defined below.

Now we unwrap the definitions and identify the Chern class of a tropical line bundle over a tropical torus X with a quadratic form. Since $X = \Lambda \backslash \mathbb{R}^n$,

$$H^1(X, \mathbb{T}^*_{\mathbb{Z}}) \cong \Lambda^* \otimes (\mathbb{Z}^n)^* \cong \mathrm{Hom}(\Lambda, (\mathbb{Z}^n)^*).$$

Then a *polarization* $c(L)$ *of a tropical torus* X is a linear map

$$c(L) : \Lambda \to (\mathbb{Z}^n)^*,$$

which is equivalent to a bilinear form

$$Q(L) : \Lambda \times \mathbb{Z}^n \to \mathbb{Z},$$

which in turn induces a bilinear form $Q(L)$ on $\mathbb{R}^n$,

$$Q(L) : (\Lambda \otimes \mathbb{R}) \times (\mathbb{Z}^n \otimes \mathbb{R}) \to \mathbb{Z} \otimes \mathbb{R}. \tag{10.5}$$

Remark 10.13. We note that for a complex torus $\Lambda \backslash \mathbb{C}^n$ to admit a polarization, its complex structure of $\mathbb{C}^n$ should be compatible with the integral structure imposed on it by Λ. Similarly, for a real torus $\Lambda \backslash \mathbb{R}^n$ to be a tropical torus with resect to the standard $\mathbb{Z}$-affine structure on $\mathbb{R}^n$ and to admit a polarization $c(L)$, the $\mathbb{Z}$-affine structure of $\mathbb{R}^n$ imposed by the standard lattice $\mathbb{Z}^n$ should be compatible with the lattice Λ in the sense that $c(L)$ induces an *integral linear map*

$$c(L) : \Lambda \to (\mathbb{Z}^n)^*.$$

Proposition 10.14 ([107, §5.1]). *For every tropical line bundle L over a tropical torus $X = \Lambda \backslash \mathbb{R}^n$, the induced bilinear form $Q(L)$ on $\mathbb{R}^n$ in Equation (10.5) is a symmetric bilinear form, and every symmetric bilinear symmetric form Q on $\mathbb{R}^n$ that is integral in the sense that it comes from a linear map $\Lambda \to (\mathbb{Z}^n)^*$ is the Chern class of a tropical line bundle over $\Lambda \backslash \mathbb{R}^n$.*

Proof. In the long exact sequence,

$$\to H^1(X,\mathscr{O}^*) \to H^1(X,\mathbb{T}^*_{\mathbb{Z}}) \to H^2(X,\mathbb{R}),$$

if we identify $H^1(X,\mathbb{T}^*_{\mathbb{Z}}) \cong \Lambda^* \otimes (\mathbb{Z}^n)^*$ and $H^2(X,\mathbb{R}) \cong \wedge^2(\mathbb{R}^n)^*$, then one can prove that the map $H^1(X,\mathbb{T}^*_{\mathbb{Z}}) \to H^2(X,\mathbb{R})$ is the restriction of the skew-symmetrization map

$$(\mathbb{R}^n)^* \otimes (\mathbb{R}^n)^* \to \wedge^2(\mathbb{R}^n)^*.$$

Then the proposition is clear. □

Definition 10.15. Given a tropical line bundle L over a tropical torus $X = \Lambda\backslash\mathbb{R}^n$, if the symmetric bilinear form $Q(L)$ associated with L in Equation (10.5) is *positive*, the tropical torus X with the line bundle L is called *a polarized tropical abelian variety*. The index of the image $c(L)(\Lambda)$ in $(\mathbb{Z}^n)^*$ is finite and called *the index of the polarization*. If the index is equal to 1, then the polarization L is called *a principal polarization* of X.

Definition 10.16. A *tropical abelian variety* is defined to be a tropical torus together with a tropical line bundle L over it whose quadratic form $Q(L)$ is positive definite.

Proposition 10.17. *The tropical torus $H_1(\Gamma,\mathbb{Z})\backslash\Omega(\Gamma)^*$ with the positive definite quadratic form Q is a principally polarized tropical abelian variety.*

Proof. The quadratic form Q on $\Omega(\Gamma)^*$ is positive definite and maps $H_1(\Gamma,\mathbb{Z})$ isomorphically to $(\Omega_{\mathbb{Z}}(\Gamma)^*)^* = \Omega_{\mathbb{Z}}(\Gamma)$ by Proposition 10.9. Therefore, by Proposition 10.14, it comes from a tropical line bundle over the tropical torus $H_1(\Gamma,\mathbb{Z})\backslash\Omega(\Gamma)^*$, and the induced polarization on $H_1(\Gamma,\mathbb{Z})\backslash\Omega(\Gamma)^*$ is principal. □

Remark 10.18. For a polarized abelian variety M, the Hodge class ω defines a non-degenerate skew-symmetric bilinear form on $H_1(M,\mathbb{Z})$ and hence a linear map

$$i_\omega : H_1(M,\mathbb{Z}) \to H_1(M,\mathbb{Z})^*.$$

It is principal if and only if the image $i_\omega(H_1(M,\mathbb{Z}))$ has index 1 in $H_1(M,\mathbb{Z})^*$. See Remark 9.6 for the case of Jacobian varieties. Therefore, the definition of a principal polarization of a tropical abelian variety is consistent with the definition of a principal polarization for complex abelian varieties.

10.3 Principally polarized tropical abelian varieties and their moduli space $\mathscr{A}_n^{tr}$

To describe moduli spaces of tropical abelian varieties, we need to identify principally polarized tropical tori with compact flat Riemmannian tori. It will turn out that this identification is also important for applications of tropical Jacobian varieties to metric graphs.

For every lattice $\Lambda \subset \mathbb{R}^n$, the standard Euclidean metric on $\mathbb{R}^n$ induces a flat metric on the torus $\Lambda\backslash\mathbb{R}^n$. More generally, for any finite dimensional real vector space V and a lattice $\Lambda \subset V$, any positive definite quadratic form Q on V defines a flat Riemannian metric on V and hence on the torus $\Lambda\backslash V$. By a change of coordinates, the more general $\Lambda\backslash V$ can be transformed to the special $\Lambda\backslash\mathbb{R}^n$. Such a flat Riemannian manifold $\Lambda\backslash V$ is called a *flat Riemannian torus.*

Proposition 10.19. *There exists a natural 1-1 correspondence between the space $\mathscr{A}_n^{tr}$ of principally polarized tropical abelian varieties of dimension n and the space $\mathscr{F}\mathscr{T}_n$ of compact flat Riemannian tori of dimension n.*

Proof. For every principally polarized tropical abelian variety $(\Lambda\backslash\mathbb{R}^n, Q)$, the positive definite quadratic form Q defines a flat Riemannian metric on $\mathbb{R}^n$ and hence also on the compact torus $\Lambda\backslash\mathbb{R}^n$. Though the integral structure $\mathbb{Z}^n$ (or $\mathbb{Z}$-affine structure) on $\mathbb{R}^n$ and $\Lambda\backslash\mathbb{R}^n$ does not show up explicitly in the compact flat torus $\Lambda\backslash\mathbb{R}^n$, it is uniquely determined by the condition that the positive definite quadratic form Q has to map Λ isomorphically to $(\mathbb{Z}^n)^*$, $\pi_Q : \Lambda \to (\mathbb{Z}^n)^*$. In other words, the lattice $\mathbb{Z}^n$ in $\mathbb{R}^n$ is uniquely determined by the pair (Q, Λ).

Conversely, for any flat compact Riemannian torus $\Lambda\backslash\mathbb{R}^n$, we claim that there is a *canonical $\mathbb{Z}$-affine structure* on $\mathbb{R}^n$ such that $\Lambda\backslash\mathbb{R}^n$ becomes a principally polarized tropical abelian variety. Let Q be the quadratic form on $\mathbb{R}^n$ which defines the flat metric of $\Lambda\backslash\mathbb{R}^n$. This defines a linear isomorphism $\pi_Q : \mathbb{R}^n \to (\mathbb{R}^n)^*$ which maps Λ to a lattice L^* in $(\mathbb{R}^n)^*$, which induces a dual lattice in $L = (L^*)^*$ in $\mathbb{R}^n$. The lattice L defines a $\mathbb{Z}$-affine structure on $\mathbb{R}^n$ and hence a $\mathbb{Z}$-affine structure on $\Lambda\backslash\mathbb{R}^n$, turning $\Lambda\backslash\mathbb{R}^n$ into a tropical torus. It can be checked directly that the symmetric form Q is integral in the sense of Proposition 10.14 and hence defines a polarization on the tropical torus $\Lambda\backslash\mathbb{R}^n$, and it is also a principal polarization by the choice of the $\mathbb{Z}$-affine structure. □

To determine the moduli space $\mathscr{A}_n^{tr}$, we need to identify the moduli space $\mathscr{F}\mathscr{T}_n$.

Proposition 10.20. *The moduli space $\mathscr{F}\mathscr{T}_n$ of compact flat Riemannian tori of dimension n can be identified with the locally symmetric space* $\mathrm{GL}(n,\mathbb{Z})\backslash\mathrm{GL}(n,\mathbb{R})/\mathrm{O}(n)$.

Proof. Let Q_0 be the standard positive quadratic form on $\mathbb{R}^n$ which induces the standard Euclidean metric on $\mathbb{R}^n$. By a suitable linear transformation, it can be shown that every compact flat torus is isometric to $(\Lambda\backslash\mathbb{R}^n, Q_0)$ for some lattices. A basic point of the geometry of numbers is that every flat torus $(\Lambda\backslash\mathbb{R}^n, Q_0)$ is isometric to $(\mathbb{Z}^n\backslash\mathbb{R}^n, Q)$ for some positive definite quadratic form. In fact, if A is the matrix formed by the column vectors which represent a basis of Λ, then $Q = A^T A$. It is clear that two flat tori $(\mathbb{Z}^n\backslash\mathbb{R}^n, Q_1)$ and $(\mathbb{Z}^n\backslash\mathbb{R}^n, Q_2)$ are isometric if and only if their quadratic forms Q_1 and Q_2 are equivalent under $\mathrm{GL}(n,\mathbb{Z})$. Since the space of equivalent classes of positive quadratic forms on $\mathbb{R}^n$ is equal to $\mathrm{GL}(n,\mathbb{Z})\backslash\mathrm{GL}(n,\mathbb{R})/\mathrm{O}(n)$, the proposition is proved. □

The combination of the above two propositions gives the following corollary, which was stated as [34, Assumption 2, p. 165] (see also [107, §6.4]).

Corollary 10.21. *The moduli space $\mathscr{A}_n^{tr}$ of principally polarized tropical abelian varieties of dimension n is equal to the locally symmetric space* $\mathrm{GL}(n,\mathbb{Z})\backslash\mathrm{GL}(n,\mathbb{R})/\mathrm{O}(n)$*:*

$$\mathscr{A}_n^{tr} \cong \mathrm{GL}(n,\mathbb{Z})\backslash\mathrm{GL}(n,\mathbb{R})/\mathrm{O}(n),$$

and $\dim \mathscr{A}_n^{tr} = \frac{n(n+1)}{2}$.

Remark 10.22. Another assumption made in [34, Assumption 1, p. 164] concerns the existence as a tropical variety of the moduli space $\mathscr{M}_n^{tr}$ of compact smooth tropical curves of genus n. This problem was discussed in some detail in §8.5. Briefly, $\mathscr{M}_n^{tr}$ is not a tropical variety, but it is a tropical orbifold since the tropical Teichmüller space $\mathscr{T}_n^{tr}$ has a natural local tropical structure.

11 The Torelli Theorem for tropical curves

We have introduced two important moduli spaces: (1) $\mathscr{M}_n^{tr}$ the moduli space of compact smooth tropical curves of genus n, and (2) the moduli space $\mathscr{A}_g^{tr}$ of principally polarized tropical abelian varieties of dimension n. In this section, we want to understand the natural map Jacobian or period between them. Unlike the Torelli theorem for compact Riemann surfaces, the tropical Torelli theorem fails. We will describe how it fails.

11.1 Period of tropical curves and contractions of graphs

For every tropical curve $\Gamma \in \mathscr{M}_n^{tr}$, by Proposition 9.3, its tropical Jacobian variety

$$(H_1(\Gamma,\mathbb{Z})\backslash\Omega(\Gamma)^*,Q)$$

is a principally polarized tropical abelian variety. By Corollary 10.21, we obtain the *tropical Jacobian map*:

$$\Pi^{tr} : \mathscr{M}_n^{tr} \to \mathscr{A}_n^{tr} = \mathrm{GL}(n,\mathbb{Z})\backslash\mathrm{GL}(n,\mathbb{R})/\mathrm{O}(n), \quad \Gamma \mapsto (H_1(\Gamma,\mathbb{Z})\backslash\Omega(\Gamma)^*,Q). \tag{11.1}$$

By Proposition 10.7,

$$(H_1(\Gamma,\mathbb{Z})\backslash\Omega(\Gamma)^*,Q) \cong (H_1(\Gamma,\mathbb{Z})\backslash H_1(\Gamma,\mathbb{R}),Q).$$

Consequently, we also denote the tropical Jacobian variety of Γ by

$$(H_1(\Gamma,\mathbb{Z})\backslash H_1(\Gamma,\mathbb{R}),Q).$$

To understand better the analogy with the Jacobian (or period) map of Riemann surfaces, $\Pi : \mathcal{M}_g \to \mathcal{A}_g$, we can interpret the quadratic form Q of a compact tropical curve Γ as the period of some normalized differential 1-forms on the tropical curve Γ.

For any smooth tropical curve of genus n, choose a basis $\sigma_1,\dots,\sigma_g$ of $H_1(\Gamma, \mathbb{Z}) \cong \mathbb{Z}^g$. For each 1-cycle σ_i, let ω_i be the corresponding 1-form as in Remark 10.8.

Proposition 11.1. *With respect to the above notations, the period of ω_j on the cycle σ_i is*

$$\int_{\sigma_i} \omega_j = Q(\sigma_i,\sigma_j),$$

i.e., the matrix of the quadratic form Q with respect to the basis $\sigma_1,\dots,\sigma_g$ is the period matrix of the forms $\omega_1,\dots,\omega_g$.

As mentioned in §4, the Jacobian map $\Pi : \mathcal{M}_g \to \mathcal{A}_g$ for compact Riemann surfaces is an embedding, but it turns out that the tropical Jacobian map $\Pi^{tr} : \mathcal{M}_n^{tr} \to \mathcal{A}_n^{tr}$ is not injective.

The failure of the injectivity of Π^{tr} is reasonable. The reason is that the Jacobian variety of a tropical curve J only depends on the flat metric Q of the torus $H_1(\Gamma,\mathbb{Z})\backslash H_1(\Gamma,\mathbb{R})$, which depends only on the lengths of 1-cycles in Γ, while as a metric graph (Γ,ℓ), Γ depends on the lengths $\ell(e)$ of all edges in Γ. Therefore, some information is lost in passing to the Jacobian variety $(H_1(\Gamma,\mathbb{Z})\backslash H_1(\Gamma,\mathbb{R}),Q)$. More precise statements are contained in Proposition 11.9 below.

On the other hand, the period of compact Riemann surface contains much more information about the Riemann surface. In fact, the Torelli theorem says that it uniquely determines the Riemann surface.

To summarize several results from [34] on the exact extent of failure of the tropical Jacobian map Π^{tr}, we need several notions on contractions of graphs.

Definition 11.2. Two metric graphs (Γ,ℓ), (Γ',ℓ') are called cyclically equivalent if there exists a bijection between their edges $\varepsilon : E(\Gamma) \to E(\Gamma')$ such that $\ell(e) = \ell'(\varepsilon(e))$, and ε induces a bijection between the cycles of Γ and the cycles of Γ'. (A cycle is a subgraph that is connected and homotopy equivalent to a circle, but does not contain separating edges.)

Definition 11.3. For $k \geq 2$, a graph Γ having at least 2 vertices is said to be k-edge connected if the graph obtained from Γ by removing any $k-1$ edges is connected.

For example, a graph is 2-edge connected if and only if it is connected and does not contain any separating edge.

To obtain highly edge-connected graphs, we introduce two types of edge contractions:

(A) contract a separating edge,
(B) contract one of two edges which form a separating pair of edges $\{e_1, e_2\}$, i.e., $\Gamma - \{e_1\}, \Gamma - \{e_2\}$ are connected, but $\Gamma - \{e_1, e_2\}$ is disconnected.

Definition 11.4. The 2-edge connectivization of a connected graph Γ is the 2-edge connected graph Γ^2 which is obtained from Γ by iterating operation (A), i.e., contracting separating edges. A 3-edge connectivization of a connected graph Γ is a 3-edge connected graph Γ^3 which is obtained from the 2-edge connectivization Γ^2 of Γ by iterating operation (B).

Note that there is no unique 3-edge connectivization of a connected graph Γ in general, but all 3-edge connectivizations are cyclically equivalent [34, Lemma 2.3.8], and they retain some information of the original graph.

To explain the above, we need the notion of $C1$-sets of Γ which corresponds to edges of Γ^3. For every subset $S \subset E(\Gamma)$ of edges of Γ, define a subgraph $\Gamma - S$ which is obtained by removing the edges in S but leaving the vertices unchanged. A new, complementary graph $\Gamma(S)$ is obtained from Γ by contracting every connected component of $\Gamma - S$ to a single point. (Different connected components are contracted to different points.)

Definition 11.5 ([34, Definition 2.3.1])**.** Let Γ be a connected graph that does not contain any separating edge. A subset $S \subset E(\Gamma)$ is called a C1-set of Γ if $\Gamma(S)$ is a cycle and $\Gamma - S$ does not contain any separating edge. Denote the collection of C1-sets of Γ by $\mathrm{Set}^1(\Gamma)$.

We discuss some examples of C1-sets. If $S = \{e\}$ and $\Gamma - \{e\}$ does not contain any separating edge, then S is a C1-set. For example, suppose that Γ has two vertices connected by more than 3 edges. If S contains more than one edges, then S is not a C1-set since $\Gamma(S)$ is homotopy equivalent to a wedge of more than one circles. In this case, every edge E is a C1-set. If Γ contains two vertices and two edges, then each edge E is not a C1-set since $\Gamma - S$ contains a separating edge, which is, in fact, the only edge of the graph $\Gamma - S$. More generally, if Γ is homeomorphic to a circle (i.e., all vertices are of valence 2), then the only C1-set S is the set of all edges due to the second condition that $\Gamma - S$ does not contain any separating edge.

For a general graph Γ, if $S = \{e_1, e_2\}$ is a separating pair of edges, then S is a C1-set if and only if $\Gamma - S$ does not contain any separating edge. The following general fact is true.

Lemma 11.6 ([34, Corollary 2.3.4]). *A graph Γ is a 3-edge connected graph if and only if it is connected and there is a bijection between $E(\Gamma)$ and* $\mathrm{Set}^1(\Gamma)$, *i.e., every edge is a C1-set and every C1-set consists of exactly one edge.*

Proposition 11.7 ([34, Lemma 2.3.8]). *Let Γ be a graph, and Γ^3 be a 3-edge connectivization of Γ. Then the following holds:*

1. *The genus of Γ^3 is equal to the genus of Γ.*
2. *There is a canonical bijection between the following three sets:* $\mathrm{Set}^1(\Gamma^3)$, $E(\Gamma^3)$, *and* $\mathrm{Set}^1(\Gamma)$.
3. *All 3-edge connectivizations of Γ are cyclically equivalent.*

We can define similar notions for metric graphs.

Definition 11.8 ([34, Definition 4.1.7]). A 3-edge connectivization of a metric graph (Γ,ℓ) is a metric graph (Γ^3,ℓ^3), where Γ^3 is a 3-edge connectivization of Γ and ℓ^3 is the length function defined by

$$\ell^3(e_S) = \sum_{e\in S} \ell(e),$$

where for each C1-set S of Γ, e_S is the corresponding edge of Γ^3 under the bijection between $\mathrm{Set}^1(\Gamma)$ and $E(\Gamma^3)$ in the above proposition.

It follows from the above definition of the metric and the above proposition that all 3-edge connectivizations of a metric graph are cyclically equivalent as metric graphs.

11.2 Tropical Torelli theorem and its failure

The Torelli theorem for metric graphs or smooth tropical curves can now be precisely stated.

Proposition 11.9 ([34, Theorem 4.1.10]). *Two metric graphs (or equivalently tropical curves) Γ and Γ' have isomorphic tropical Jacobian varieties if and only if their 3-edge connectivizations Γ^3 and $(\Gamma')^3$ are cyclically equivalent.*

Corollary 11.10. *When $n \geq 3$, the tropical Jacobian map*

$$\Pi^{tr} : \mathscr{M}_n^{tr} \to \mathscr{A}_n^{tr} = \mathrm{GL}(n,\mathbb{Z})\backslash \mathrm{GL}(n,\mathbb{R})/\mathrm{O}(n)$$

is not injective.

Proof. There are several reasons for the failure of the injectivity of Π^{tr}. If Γ contains separating edges, then all separating edges are contracted in Γ^3. We can move one bridge to any other place without changing the image. On the other hand, we note that as metric graphs, the lengths of these separating edges matter, and that the lengths of these separating edges do affect the quadratic form on $H^1(\Gamma,\mathbb{Z})$.

For graphs which do not contain any separating edges, the tropical Jacobian map can still fail to be injective. Take a graph Γ such that some C1-set S of Γ contains at least 2 edges, $e_1,\dots,e_k$, $k\geq 2$. Now in the 3-edge connectivization Γ^3, the edge e_S corresponding to the set S has length $\ell^3(e_S)=\ell(e_1)+\ldots+\ell(e_k)$. By Proposition 11.9, the Jacobian variety of (Γ,ℓ) is the same as the Jacobian variety of its 3-edge connectivization (Γ^3,ℓ^3), and it depends on the sums $\sum_{e\in S}\ell(E)$ for C1-sets S. On the other hand, as metric graphs, they depend on the lengths of edges $\ell(e)$, $e\in S$. There are certainly infinitely many nonisomorphic metric graphs with the same value of $\ell^3(e_S)=\ell(e_1)+\ldots+\ell(e_k)$.

Some examples of C1-sets that contain more than one element are explained after Definition 11.5 (see also Remark 11.11 below). In particular, we see that for every $g\geq 3$, there exist metric graphs of genus g which do not contain any separating edge but contain C1-sets S with cardinality of at least 2, for example, separating pairs of edges. This proves Corollary 11.10. □

Remark 11.11. To construct a graph Γ which contains a large C1-set S, we start with a graph Γ_1 which is homeomorphic to the circle and has $k\geq 3$ edges. But the valence of every vertex of Γ_1 is equal to 2. To overcome this, we attach a connected graph whose vertices have valence of at least 3 at every vertex of Γ_1. Denote the new graph by Γ. Take S to consist of the edges of Γ_1. Then it can be seen that S is a C1-set of Γ with k edges. This explains that there are sets of metric graphs of arbitrarily large dimension which are mapped to a point under the tropical Jacobian map Π^{tr}.

12 Complete invariant metrics on outer space X_n

In this section, we solve the problems on constructing intrinsic complete geodesic metrics on the outer space X_n which are invariant under $\mathrm{Out}(F_n)$, i.e., Problems 1.1 and 1.3. There are several different ways to construct them. But the identification of metric graphs with tropical curves and the tropical Jacobian map are crucial. These different constructions are treated in different sections below.

12.1 Tropical Jacobian map and invariant complete geodesic metrics on X_n

As discussed in the previous section, the tropical Jacobian variety of tropical curves defines the tropical Jacobian map in Equation (11.1):

$$\Pi^{tr}:\mathscr{M}_g^{tr}\to\mathscr{A}_n^{tr}\cong \mathrm{GL}(n,\mathbb{Z})\backslash\mathrm{GL}(n,\mathbb{R})/\mathrm{O}(n).$$

By the identification between smooth tropical curves and compact metric graphs in Propositions 4.27 and 4.28, there is a natural inclusion

$$X_n \to \mathscr{T}_g^{tr},$$

which factors through the action of $\mathrm{Out}(F_n)$ and gives an embedding

$$\mathrm{Out}(F_n)\backslash X_n \to \mathscr{M}_g^{tr} = \mathrm{Out}(F_n)\backslash \mathscr{T}_n^{tr}$$

and hence a *tropical period map* for metric graphs

$$\Pi^{tr} : \mathrm{Out}(F_n)\backslash X_n \to \mathscr{A}_n^{tr} = \mathrm{GL}(n,\mathbb{Z})\backslash \mathrm{GL}(n,\mathbb{R})/\mathrm{O}(n).$$

This tropical period map can be lifted to the equivariant period map in Equation (2.7)

$$\Pi = \Pi^{tr} : X_n \to \mathrm{GL}(n,\mathbb{R})/\mathrm{O}(n), \tag{12.1}$$

with respect to the natural homomorphism $\mathrm{Out}(F_n) \to \mathrm{GL}(n,\mathbb{Z})$ and the actions of these groups on X_n and $\mathrm{GL}(n,\mathbb{R})/\mathrm{O}(n)$. For simplicity of notation, we denote the *tropical period map* from now on by Π instead of Π^{tr}.

This map Π can be explained more directly as follows. We can fix an isomorphism

$$H_1(R_n,\mathbb{Z}) \cong \mathbb{Z}^n,$$

where R_n is the rose with n petals. Then for each marked metric graph (Γ,ℓ,h) in X_n, the marking h induces a canonical isomorphism

$$h_* : H_1(\Gamma,\mathbb{Z}) \cong \mathbb{Z}^n,$$

and hence an isomorphism

$$h_* : H_1(\Gamma,\mathbb{R}) \cong \mathbb{R}^n$$

by linear extension. By Proposition 10.7, we have

$$\Omega(\Gamma)^* \cong H_1(\Gamma,\mathbb{R})$$

under the period map in Equation 10.2. This implies that the positive quadratic form Q on $\Omega(\Gamma)^*$ associated with the metric graph Γ induces a positive definite *matrix* on $\mathbb{R}^n$ with respect to the standard basis of $\mathbb{R}^n$ and consequently gives a point in the symmetric space

$$\mathrm{GL}(n,\mathbb{R})/\mathrm{O}(n)$$

of positive definite $n \times n$-symmetric matrices, which is also the *Teichmüller space of flat Riemannian tori of dimension n.*

We also note that by Proposition 10.19, the tropical structure and the principal polarization Q on the tropical torus

$$H_1(\Gamma,\mathbb{Z})\backslash\Omega(\Gamma)^* \cong H_1(\Gamma,\mathbb{Z})\backslash H_1(\Gamma,\mathbb{R})$$

are all uniquely determined by the positive definite quadratic form Q.

Note that $\mathrm{GL}(n,\mathbb{R})$ is a reductive Lie group, $\mathrm{O}(n)$ is a maximal compact subgroup of $\mathrm{GL}(n,\mathbb{R})$, and consequently $\mathrm{GL}(n,\mathbb{R})/\mathrm{O}(n)$ is a symmetric space of nonpositive curvature, though not of noncompact type. Fix an invariant Riemannian metric on $\mathrm{GL}(n,\mathbb{R})/\mathrm{O}(n)$, and let d_{inv} be the induced Riemannian distance function on $\mathrm{GL}(n,\mathbb{R})/\mathrm{O}(n)$.

If the map Π in Equation 12.1 were injective, then the pull-back metric $\Pi^* d_{inv}$ would give an $\mathrm{Out}(F_n)$-invariant metric on the outer space X_n. Unfortunately, it is not injective due to the infinite kernel of $\mathrm{Out}(F_n) \to \mathrm{GL}(n,\mathbb{Z})$ as pointed out in §2, or by Corollary 11.10 when $n \geq 3$.

On the other hand, $\Pi^* d_{inv}$ is a pseudo-metric, i.e., it satisfies all the conditions of a metric except for the positivity condition: $\Pi^* d_{inv}(x,y) > 0$ when $x \neq y$.

Recall that X_n admits the canonical $\mathrm{Out}(F_n)$-invariant incomplete simplicial metric d_0 (Proposition 2.10). The *simple key idea* to solve the above problem is to combine $\Pi^* d_{inv}$ with the metric d_0. For every two points $x, y \in X_n$, define a *distance* $d_1(x,y)$ between them by

$$d_1(x,y) = d_0(x,y) + d_{inv}(\Pi(x), \Pi(y)). \tag{12.2}$$

Proposition 12.1. *The function* $d_1 : X_n \times X_n \to \mathbb{R}_{\geq 0}$ *is an* $\mathrm{Out}(F_n)$*-invariant metric on* X_n.

Proof. Since the map Π in Equation (12.1) is equivariant with respect to the homomorphism $\mathrm{Out}(F_n) \to \mathrm{GL}(n,\mathbb{Z})$, d_{inv} is invariant under $\mathrm{GL}(n,\mathbb{Z})$, and d_0 is invariant under $\mathrm{Out}(F_n)$, it is clear that d_1 is invariant under $\mathrm{Out}(F_n)$. Since d_0 is a metric, it is also clear that d_1 is a metric on X_n. □

Proposition 12.2. *The metric space* (X_n, d_1) *is a complete locally compact metric space.*

Proof. Fix a basepoint $y_0 \in X_n$. Since X_n is a locally finite simplicial complex with some simplicial faces missing, it suffices to prove that for any sequence of marked metric graphs $y_j = (\Gamma_j, \ell_i, h_i)$ in X_n, if y_j converges to a missing simplicial face of a simplex of X_n, then $d_1(x_0, y_j) \to +\infty$ as $j \to +\infty$. We note that when y_j approaches a missing simplicial face, the length of the shortest loop (or nontrivial cycle) of the metric graph (Γ_j, ℓ_i, h_i) goes to 0 as $j \to +\infty$. This means that the shortest closed geodesic of the flat tori

$$H_1(\Gamma_j,\mathbb{Z})\backslash H_1(\Gamma_j,\mathbb{R}) \cong H_1(R_n,\mathbb{Z})\backslash H_1(R_n,\mathbb{R}) \cong \mathbb{Z}^n\backslash\mathbb{R}^n$$

goes to zero, i.e., the minimum value of the positive definite matrix $\Pi(y_j) \in \mathrm{GL}(n,\mathbb{R})/\mathrm{O}(n)$ on nonzero integral points in $H_1(\Gamma_j,\mathbb{Z}) \cong \mathbb{Z}^n$ goes to 0. By the Mahler compactness theorem for quadratic forms (or lattices), this implies that $\Pi(y_j)$ is a divergent sequence in the symmetric space $\mathrm{GL}(n,\mathbb{R})/\mathrm{O}(n)$. Since the invariant metric d_{inv} on the symmetric space $\mathrm{GL}(n,\mathbb{R})/\mathrm{O}(n)$ is complete, $d(\Pi(y_0),\Pi(y_j)) \to +\infty$ as $j \to +\infty$. Since $d_1(y_0,y_j) \geq d_{inv}(\Pi(y_0),\Pi(y_j))$, the proposition is proved. □

Remark 12.3. The Mahler compactness theorem for lattices used in the above proof is a basic result in the geometry of number. See [36] for example. It can be explained easily in terms of the moduli space of flat Riemannian tori. The moduli space n-dimensional flat Riemannian tori of volume 1 can be identified with the locally symmetric space $\mathrm{SL}(n,\mathbb{Z})\backslash\mathrm{SL}(n,\mathbb{R})/\mathrm{SO(n)}$. Then a sequence of flat Riemannian tori $\Lambda_j\backslash\mathbb{R}^n$ in $\mathrm{SL}(n,\mathbb{Z})\backslash\mathrm{SL}(n,\mathbb{R})/\mathrm{SO(n)}$ diverges if and only only if the length of the shortest geodesic of $\Lambda_j\backslash\mathbb{R}^n$ goes to 0 as $j \to \infty$. Note that the length of the shortest geodesic of $\Lambda_j\backslash\mathbb{R}^n$ is equal to the minimal norm of nonzero vectors of the lattice Λ_j.

Remark 12.4. The above result shows an important difference between the tropical period map Π in Equation (12.1) and the period map for marked compact Riemann surfaces of genus g,

$$\Pi : \mathscr{T}_g \to \mathfrak{h}_g,$$

where $\mathscr{T}_g$ is the Teichmüller space of compact surfaces of genus g and $\mathfrak{h}_g = \mathrm{Sp}(2g,\mathbb{R})/\mathrm{U}(g)$ is the Siegel upper half space. When we pinch a family of hyperbolic Riemann surfaces Σ_g along a separating geodesic, i.e., a geodesic which cuts the surface into two connected components, its period matrix does not go to the infinity of $\mathfrak{h}_g$. Instead it converges to a point in

$$(\Omega_1,\Omega_2) \in \mathfrak{h}_{g_1} \times \mathfrak{h}_{g_2} \subset \mathfrak{h}_g,$$

where the limit hyperbolic surface is $\Sigma_{g_1,1} \cup \Sigma_{g_2,1}$, $g_1+g_2=g$, and Ω_1 is the period of Σ_{g_1}, and Ω_2 is the period of Σ_{g_2}. In particular, the period does not detect the punctures in $\Sigma_{g_1,1}$ and $\Sigma_{g_2,1}$. This implies that the pulled back metric on $\mathscr{T}_g$ by Π from the invariant metric of the Hermitian symmetric space $\mathfrak{h}_g$ is not complete. This metric on $\mathscr{T}_g$ is usually called the Satake metric, probably due to the fact that a compactification of $\mathscr{M}_g = \mathrm{Mod}_g\backslash\mathscr{T}_g$ was first constructed via Π from the Satake compactification of the Siegel modular variety $\mathrm{Sp}(2g,\mathbb{Z})\backslash\mathfrak{h}_g$. We note that metric graphs in the outer space X_n, whenever a closed path (or loop) is pinched, the metric graphs will go to the infinity of X_n and their images in the period space $\mathrm{GL}(n,\mathbb{R})/\mathrm{O}(n)$ also go to infinity.

Recall that a *geodesic segment* in a metric space (X,d) is an isometric embedding

$$i : (a,b) \to (X,d).$$

Sometimes its image is also called a geodesic segment. A metric space (X,d) is called a *geodesic metric space* if every pair of points are connected by a geodesic segment. For many applications, it is important to obtain geodesic metrics.

For every path connected metric space (X,d), there is a natural length structure (X,d_ℓ) [57, p. 3, 1.4.(a)] [31, §2.3]. Briefly, for every path $\gamma\colon [a,b] \to X$, define its length

$$\ell(\gamma) = \sup_{t_0<\ldots<t_N} \{\sum_{i=1}^{N} d(\gamma(t_{i-1}),\gamma(t_i))\},$$

where the sup is over all partitions $a = t_0 < t_1 < \ldots < t_N = b$ of the interval $[a,b]$. For every two points $x,y \in X$, the length metric $d_\ell(x,y)$ is given by

$$d_\ell(x,y) = \inf_\gamma \ell(\gamma),$$

where the inf is over all paths γ connecting x,y.

In general, a length metric space is not necessarily a geodesic space. Let $d_{1,\ell}$ be the length metric on X_n induced from the metric d_1 in Proposition 12.2.

Theorem 12.5. *The length metric $d_{1,\ell}$ on X_n is an* $\mathrm{Out}(F_n)$*-invariant complete geodesic metric.*

Proof. Since d_1 is invariant under $\mathrm{Out}(F_n)$, it is clear that $d_{1,\ell}$ is also $\mathrm{Out}(F_n)$-invariant. Since X_n is a locally finite simplicial complex with some simplicial faces missing, and d_1 is a complete metric on X_n, it is clear that (X_n,d_1) is a complete locally compact metric space. It is also clear from the definition of d_1 and the local finiteness of the simplicial complex X_n that for every two points $x,y \in X_n$, $d_{1,\ell}(x,y) < +\infty$. By definition, $d_{1,\ell} \geq d_1$, and hence $(X_n,d_{1,\ell})$ is also a complete locally compact length space. By [31, Proposition 2.5.22], $(X_n,d_{1,\ell})$ is a geodesic length space. □

From the above discussion, it is clear that we can also construct other $\mathrm{Out}(F_n)$-invariant complete geodesic metrics on X_n. For example, we can define a metric d_2 on X_n by

$$d_2(x,y) = \sqrt{d_0(x,y)^2 + d_{inv}(\Pi(x),\Pi(y))^2},$$

and obtain a corresponding *complete geodesic metric $d_{2,\ell}$*. Similarly, we can define

$$d_\infty(x,y) = \max\{d_0(x,y), d_{inv}(\Pi(x),\Pi(y))\},$$

and obtain a *complete geodesic metric $d_{\infty,\ell}$*.

As mentioned in the introduction, we would like to construct some complete metrics on X_n which resemble Riemannian metrics, for example, its restriction to each open simplex of X_n is a Riemannian metric. It is not clear whether the above geodesic metrics $d_{1,\ell}, d_{2,\ell}, d_{\infty,\ell}$ on X_n enjoy this property.

In the next two subsections, we construct two such metrics, which might offer better analogies of the Riemannian metrics on symmetric spaces and Teichmüller spaces.

12.2 Tropical Jacobian map and a complete pesudo-Riemannian metric on X_n

In the previous subsection, we use the period map of metric graphs in Equation (12.1) to pull back the Riemannian *distance* d_{inv} on the symmetric space $\mathrm{GL}(n,\mathbb{R})/\mathrm{O}(n)$ in order to define geodesic metrics on X_n. In this subsection, we will construct geodesic metrics on X_n which are Riemannian metrics on each simplex of X_n.

Recall that $\mathrm{GL}(n,\mathbb{R})/\mathrm{O}(n)$ is a Riemannian symmetric space of nonpositive curvature. Let ds^2_{inv} be a Riemannian metric tensor on it which is invariant under $\mathrm{GL}(n,\mathbb{R})$. Equivalently, let $\langle\cdot,\cdot\rangle_{inv}$ be the invariant Riemannian metric on $\mathrm{GL}(n,\mathbb{R})/\mathrm{O}(n)$.

When restricted to each open simplex Σ of X_n, the period map

$$\Pi : \Sigma \to \mathrm{GL}(n,\mathbb{R})/\mathrm{O}(n)$$

is a smooth map. It is natural to pull back the Riemannian metric ds^2_{inv} by Π. But the problem is that Π is not an embedding, and not even a local immersion when $n \geq 3$, by Corollary 11.10. (As seen in the proof of Corollary 11.10, it can map subsets of positive dimension of X_n to one point.) Therefore, $\Pi^*(ds^2_0)$ is not positive definite, but only semi-positive definite.

To overcome this difficulty, we note that each open simplex Σ of X_n has a canonical flat Riemannian metric $\langle\cdot,\cdot\rangle_0$ such that each edge has length $\sqrt{2}$. Let ds^2_0 be the metric tensor of this flat Riemannian metric.

Define a bilinear form on the tangent bundle of Σ by

$$\langle u,v\rangle_\Sigma = \langle d\Pi(u), d\Pi(v)\rangle_{inv} + \langle u,v\rangle_0, \tag{12.3}$$

where u,v are tangent vectors of Σ at any point. Equivalently, its associated symmetric metric tensor is

$$ds^2 = \Pi^*(ds^2_{inv}) + ds^2_0. \tag{12.4}$$

Lemma 12.6. *The bilinear form $\langle\cdot,\cdot\rangle_\Sigma$ in Equation* 12.3 *defines a Riemannian metric on every open simplex Σ of X_n.*

Proof. Since ds^2_0 is positive definite, and $\Pi^*(ds^2_{inv})$ is semi-positive definite, their sum $ds^2 = \Pi^*(ds^2_{inv}) + ds^2_0$ is positive definite. Since Π depends smoothly on the simplicial coordinates of Σ, this implies that ds^2 defines a smooth Riemannian metric on Σ. □

Since X_n is a simplicial complex and is a disjoint union of open simplices Σ, the metrics ds^2 on the open simplices Σ give a *stratified Riemannian metric* on X_n. We denote it by ds^2 or $\langle\cdot,\cdot\rangle$.

A simplicial complex is a stratified space with a *smooth structure* in the sense of [120, Chapter 1], and we want to show that the stratified Riemannian metric ds^2

gives a smooth Riemannian metric on X_n in the sense of [120, §2.4], or rather more directly it is a piecewise smooth Riemannian metric on X_n.

For each simplex Σ of X_n, let $\mathbb{R}^i$ be the linear space of the same dimension that contains Σ, or is spanned by Σ. Let $\overline{\Sigma}$ be the closure of Σ in X_n, which is contained in $\mathbb{R}^i$ and but still not a closed simplex. A function f or a tensor T on $\overline{\Sigma}$ is called *smooth* if there exists an open subset U of $\mathbb{R}^i$ which contains $\overline{\Sigma}$ such that f or T can be extended to a smooth function or a tensor on U.

Proposition 12.7. *For every simplex Σ of X_n, the Riemannian metric ds^2 can be extended to a smooth Riemannian metric on $\overline{\Sigma}$.*

Proof. It suffices to prove that for every point $p \in \Sigma$, there is a neighborhood U_p of p in $\mathbb{R}^i$ such that the period map $\Pi : \Sigma \to \mathrm{GL}(n,\mathbb{R})/\mathrm{O}(n)$ extends to a smooth function on

$$U_p \cup \Sigma \to \mathrm{GL}(n,\mathbb{R})/\mathrm{O}(n).$$

Let (Γ,h) be a marked graph whose corresponding simplex $\Sigma_{(\Gamma,h)}$ in X_n is Σ. Then the point p corresponds to a marked metric graph (Γ,ℓ_p,h). When p lies on the boundary faces of σ, some edge lengths $\ell(e)$ are equal to 0, but for every 1-cycle, its total length is positive. By changing all edge lengths slightly (allowing them to be positive, zero or negative) under the condition that the sum of all edges is equal to 1, we obtain a neighborhood of U_p in $\mathbb{R}^i$ such that for every 1-cycle of the graph Γ, its length is positive. Since the period Q of a metric graph (Γ,ℓ,h) only depends on the lengths of 1-cycles, Q can be extended to U_p and the extended Q is positive definite. This gives the required extension of the period map π and the metric $\Pi^*(ds^2_{inv}) + ds_0^2$ to $U_p \cup \Sigma$. □

Proposition 12.8. *For every two simplices Σ_1, Σ_2 with $\overline{\Sigma_1} \cap \overline{\Sigma_2} \neq \emptyset$, the extended Riemannian metrics $\langle\cdot,\cdot\rangle_{\Sigma_1}$ on $\overline{\Sigma_1}$ and $\langle\cdot,\cdot\rangle_{\Sigma_2}$ on $\overline{\Sigma_2}$ agree on the intersection $\overline{\Sigma_1} \cap \overline{\Sigma_2}$. Therefore, the stratified metric ds^2 on X_n is a piecewise smooth Riemannian metric on X_n as a simplicial complex.*

Proof. Let Σ_3 be a common simplicial face of $\overline{\Sigma_1}$, $\overline{\Sigma_2}$. For every point $p \in \Sigma_3$, let $\langle\cdot,\cdot\rangle_p$ be the inner product associated with the metric tensor $\Pi^*(ds^2_{inv}) + ds_0^2$ on Σ_3. Then for every tangent vector v to Σ_3 at p, it is clear from the previous proposition that

$$\langle v,v\rangle_{\Sigma_1} = \langle v,v\rangle_p, \quad \langle v,v\rangle_{\Sigma_2} = \langle v,v\rangle_p.$$

Therefore, $\langle\cdot,\cdot\rangle_{\Sigma_1}$ on $\overline{\Sigma_1}$ and $\langle\cdot,\cdot\rangle_{\Sigma_2}$ agree on Σ_3. □

With this piecewise smooth Riemannian metric on the stratified space X_n, we will introduce a length metric on X_n, following the general setup of [120].

Specifically, we consider piecewise smooth curves in X_n, i.e., continuous maps $c : [a,b] \to X_n$ such that there is a partition $a = t_0 < t_1 < \ldots < t_n = b$, where each piece $c : [t_{i-1}, t_i]$ is a smooth curve in a simplex $\overline{\Sigma_i}$. Define the length of c by

$$\ell(c) = \sum_{i=1}^{n} \int_{t_{i-1}}^{t_i} \langle c'(t), c'(t)\rangle_{\Sigma_i} dt.$$

For every pair of points $p, q \in X_n$, define a length function

$$d_{\ell,R}(p,q) = \inf_c \ell(c),$$

where c ranges over all piecewise smooth curves in X_n connecting p and q. We note that the subscript R stands for a Riemannian metric.

Proposition 12.9. *The function $d_{\ell,R}$ defines a complete, locally compact length metric on X_n which is invariant under* $\mathrm{Out}(F_n)$ *and restricts to a Riemannian distance on each simplex Σ of X_n.*

Proof. Since X_n is a locally finite simplicial complex with a piecewise smooth Riemann metric, and since every two points p, q of X_n are connected by a piecewise smooth curve, it is clear that $d_{\ell,R}(p,q) < +\infty$, and if $p \neq q$, then $d_{\ell,R}(p,q) > 0$, and that $d_{\ell,R}$ defines the original topology of X_n.

Since the Riemannian metric ds^2 is invariant under $\mathrm{Out}(F_n)$, the length function $d_{\ell,R}$ is also invariant under $\mathrm{Out}(F_n)$. We need to show that it is complete. This follows from Lemma 12.10 below. □

Lemma 12.10. *Let $c : [0,1) \to X_n$ be a continuous path such that when $t \to 1$ the point $c(t)$ approaches a missing simplicial face of X_n, i.e., the total length of some cycle (or loop) of the marked metric graph represented by $c(t)$ goes to 0. Then the image $\Pi(c(t))$ goes to the infinity of the symmetric space* $\mathrm{GL}(n,\mathbb{R})/\mathrm{O}(n)$ *when $t \to 1$. In particular, the length of the curve $\Pi(c(t))$, and hence of the curve $c(t)$, $t \in [0,1)$, is equal to infinity.*

Proof. When $c(t)$ goes to a missing simplicial face of X_n as $t \to 1$, the period matrix $Q(t)$ of the metric graph corresponding to $c(t)$ converges to the subspace of semipositive definite singular symmetric matrices. This implies that as $t \to 1$, $\Pi(c(t))$ goes to the infinity of the symmetric space $\mathrm{GL}(n,\mathbb{R})/\mathrm{O}(n)$. Since the symmetric space $\mathrm{GL}(n,\mathbb{R})/\mathrm{O}(n)$ is a complete Riemannian manifold, the length of the curve $\Pi(c(t))$, $t \in [0,1)$, is equal to infinity.

When we compute the length of c using $\Pi^* ds^2_{inv}$, it is equal to the length of $\Pi(c(t))$. Since the length element of ds^2 is greater than or equal to $\Pi^* ds^2_{inv}$, the length of the curve $c(t)$, $t \in [0,1)$, is also equal to infinity. □

Theorem 12.11. *The outer space X_n with the length metric $d_{\ell,R}$ of the piecewise Riemannian metric ds^2 is a complete geodesic space, and every two points are connected by a geodesic which is a piecewise smooth curve in X_n.*

Proof. Since X_n is locally compact and $d_{\ell,R}$ is a length distance, it follows from [31, Proposition 2.5.22] and [120, Theorem 2.4.15] that $(X_n, d_{\ell,R})$ is a geodesic space, i.e., every two points are connected by a geodesic segment. Since $(X_n, d_{\ell,R})$ is the length structure induced from a piecewise smooth Riemann metric of a locally finite simplicial complex, it follows from the usual argument in Riemannian geometry (see [87] for example) that every geodesic segment in X_n is a piecewise smooth curve. □

Remark 12.12. Once we have endowed X_n with an invariant piecewise Riemannian metric, it is natural to study some geometric invariants. For example, we can study the growth of balls in X_n, and the spectral theory of X_n and $\mathrm{Out}(F_n)\backslash X_n$ with respect to suitable defined Laplacian operators. As discussed in [120], there is little known for analysis on general stratified spaces.

12.3 Complete pseudo-Riemannian metric on X_n via lengths of pinching loops

In this subsection, we construct another Riemannian metric on the outer space X_n whose asymptotic behaviors are explicitly given. This is motivated by the McMullen metric for Teichmüller space in [97].

Fix a sufficiently small $\varepsilon > 0$, for example, $\varepsilon < \frac{1}{6n}$. Define a modified edge length of 1-cycles σ of metric graphs (Γ, ℓ): $\ell_\varepsilon(\sigma)$ is a positive smooth function of $\ell(\sigma)$ satisfying the conditions

$$\ell_\varepsilon(\sigma) = \begin{cases} \varepsilon, \text{ if } \ell(\sigma) \geq 2\varepsilon, \\ \ell(\sigma), \text{ if } \ell(\sigma) \leq \varepsilon. \end{cases} \tag{12.5}$$

For each simplex Σ in X_n, define a Riemannian metric by

$$ds_\varepsilon^2 = ds_0^2 + \sum_{\sigma \text{ with } \ell(\sigma)<\varepsilon} \left(\frac{d\ell_\varepsilon(\sigma)}{\ell_\varepsilon(\sigma)}\right)^2, \tag{12.6}$$

where the sum is over all cycles whose lengths are less than ε.

It is clear that for each simplex Σ, ds_ε^2 defines a Riemannian metric. This defines a stratified Riemannian metric on X_n. We note that a sequence of marked metric graphs converges to some missing simplicial complexes of X_n if and only if there are cycles (or loops) σ whose lengths $\ell(\sigma)$ converge to 0. Such loops are called *pinching loops* of metric graphs and they correspond to pinching geodesics on hyperbolic surfaces.

By the proof of Proposition 12.7, the length function $\ell(\sigma)$ and hence the modified length function $\ell_\varepsilon(\sigma)$ is a smooth function on the closure $\overline{\Sigma}$ in X_n, and the Riemannian metric ds_ε^2 extends to a smooth Riemannian metric on $\overline{\Sigma}$. Then by the proof of Proposition 12.8, we can prove the following proposition.

Proposition 12.13. *The Riemannian metric ds_ε^2 on the stratified space X_n defines an* $\mathrm{Out}(F_n)$*-invariant piecewise smooth Riemannian metric on X_n.*

By arguments similar to those of the proofs of Proposition 12.9 and Theorem 12.11, we can prove the following.

Theorem 12.14. *The piecewise smooth Riemannian metric ds_ε^2 on the outer space X_n defines a complete geodesic metric $d_{\ell,R,\varepsilon}$ on X_n which is invariant under* $\mathrm{Out}(F_n)$ *such that every two points are connected by a piecewise smooth geodesic.*

12.4 Finite Riemannian volume of the quotient $\mathrm{Out}(F_n)\backslash X_n$

For a Riemannian manifold, a natural invariant is its volume. As mentioned before, with respect to many natural metrics on $\mathcal{T}_g$, the quotient $\mathrm{Mod}_g\backslash\mathcal{T}_g$ has finite volume. Note that the locally symmetric space $\mathrm{GL}(n,\mathbb{Z})\backslash\mathrm{GL}(n,\mathbb{R})/\mathrm{O}(n)$ has infinite volume. Since elements of $\mathrm{GL}(n,\mathbb{Z})$ have determinant equal to ± 1, $\mathrm{GL}(n,\mathbb{Z})$ also acts on $\mathrm{SL}(n,\mathbb{R})/\mathrm{SO}(n,\mathbb{Z})$, which is a symmetric space of noncompact type, and the quotient space $\mathrm{GL}(n,\mathbb{Z})\backslash\mathrm{SL}(n,\mathbb{R})/\mathrm{SO}(n,\mathbb{Z})$ has finite volume. Or equivalently, we note that $\mathrm{GL}(n,\mathbb{Z})$ contains $\mathrm{SL}(n,\mathbb{Z})$ as a subgroup of finite index, and the quotient $\mathrm{SL}(n,\mathbb{Z})\backslash\mathrm{SL}(n,\mathbb{R})/\mathrm{SO}(n,\mathbb{Z})$ has finite volume.

If we pursue the analogy between the three pairs of transformation groups $(\mathrm{Out}(F_n),X_n)$, $(\mathcal{T}_g,\mathrm{Mod}_g)$, $(\mathrm{SL}(n,\mathbb{Z}),\mathrm{SL}(n,\mathbb{R})/\mathrm{SO}(n))$, then we expect that $\mathrm{Out}(F_n)$ is a lattice and the following conjecture seems natural.

Conjecture 12.15. *The volume of* $\mathrm{Out}(F_n)\backslash X_n$ *with respect to either of the Riemannian metrics ds^2, ds_ε^2 defined in the previous section is finite.*

By definition, the volume of a simplicial complex with respect to a piecewise Riemannian metric is the sum of volumes of the top dimensional simplices. By Proposition 2.16, there are only finitely $\mathrm{Out}(F_n)$-orbits of simplices. Therefore, it suffices to decide whether for each top dimensional simplex Σ of X_n, the volume of Σ with respect to ds^2 or ds_ε^2 is finite.

Proposition 12.16. *When $n=2$, the volume of every 2-simplex Σ in X_n with respect to the Riemannian metric ds^2 is finite, and hence the volume of* $\mathrm{Out}(F_n)\backslash X_n$ *with respect to the Riemannian metric ds^2 is finite.*

Proof. To prove this proposition, we need to compute $\Pi^*(ds_{inv}^2)$. Let (Γ,h) be a marked graph corresponding a 2-dimensional simplex Σ of X_2. Then Γ has 3 edges e_1,e_2,e_3 and 2 vertices, with each edge connecting these two vertices. Denote the edge lengths by a,b,c. Then they are characterized by the following conditions: (1) $a,b,c\geq 0$, (2) $a+b+c=1$, (3) $a+b,a+c,b+c>0$. In particular, the vanishing of one edge length is allowed, but not two edges simultaneously. Therefore, Σ has 3 missing simplicial faces of dimension 0, i.e., missing vertices, with each corresponding to the vanishing of two edge lengths. It suffices to show that the area of the corner near each missing vertex is zero. Assume that $a,b=0$ at this corner. Let ε be a small positive number and define a corner Ω_ε by $a+b\leq\varepsilon$. The period matrix of the marked metric graph is

$$Q = \begin{pmatrix} a+b & b \\ b & 1-a \end{pmatrix},$$

and its inverse is

$$Q^{-1} = \frac{1}{ab+(a+b)(1-a-b)} \begin{pmatrix} 1-a & -b \\ -b & a+b \end{pmatrix}.$$

By [126, §4.1.2], the invariant Riemannian metric of the symmetric space $X = \mathrm{GL}(n,\mathbb{R})/\mathrm{O}(n)$ is given by the metric tensor

$$ds_{inv}^2 = \mathrm{Tr}((Y^{-1}dY)^2),$$

where $Y \in X$ is a symmetric matrix in X. Therefore, the induced Riemannian metric on Σ is $ds^2 = ds_0^2 + \mathrm{Tr}((Q^{-1}dQ)^2)$. We estimate the growth of ds^2 when $a,b \to 0$. The term ds_0 is bounded and we ignore it. Since $\frac{1}{ab+(a+b)(1-a-b)}$ is comparable with $\frac{1}{a+b}$, and

$$dQ = \begin{pmatrix} ad+db & db \\ ab & -da \end{pmatrix},$$

by a direct computation, the area form of the Riemannian metric ds^2 is bounded from above by a multiple of $\frac{1}{(a+b)^{3/2}} da \wedge db$. By dividing the region Ω_ε into two subregions according to $(1): a \leq b$, $(2): b \leq a$, we can show easily that the area of Ω_ε with respect to ds^2 is finite. □

Remark 12.17. In the above estimate, the assumption that $n = 2$ was used for the explicit computation. It is conceivable that a more detailed computation will allow one to prove the general case.

Using the explicit form of the metric ds_ε^2, we can also show that in the corner Ω_ε of a simplex Σ as above, ds_ε^2 is comparable with $\frac{(da+db)^2}{(a+b)^2} + da^2 + db^2$, and that the area form is bounded by a multiple of $\frac{1}{a+b} ad \wedge db$. Therefore, we can prove that the area of the corner is finite with respect to the Riemannian metric ds_ε^2 and prove the following result.

Proposition 12.18. *When $n = 2$, the volume of every 2-simplex in X_n with respect to the Riemannian metric ds_ε^2 is finite, and hence the volume of* $\mathrm{Out}(F_n)\backslash X_n$ *with respect to the Riemannian metric ds_ε^2 is finite.*

Remark 12.19. When $n = 2$, there is a natural identification of X_2 with the Poincare disk [133]. It is probably true that the metric ds^2 is quasi-isometric to the Poincare metric.

In this book, we have introduced several geodesic metrics $d_{1,\ell}$, $d_{2,\ell}$, $d_{\infty,\ell}$, $d_{\ell,R}$, $d_{\ell,R,\varepsilon}$ on X_n. We note that $d_{1,\ell}$, $d_{2,\ell}$, $d_{\infty,\ell}$ are clearly quasi-isometric to each other. The following problem seems to be natural and tempting.

Problem 12.20. *Decide whether all the above* $\text{Out}(F_n)$*-invariant, complete geodesic metrics* $d_{1,\ell}$, $d_{\ell,R}, d_{\ell,R,\varepsilon}$ *on* X_n *are quasi-isometric.*

References

[1] D. Abramovich, L. Caporaso, S. Payne, *The tropicalization of the moduli space of curves*, Ann. Sci. Éc. Norm. Supr. (4) 48 (2015), no. 4, 765–809.

[2] N. A'Campo, L. Ji, A. Papadopoulos, *On the early history of Moduli and Teichmüller spaces*, in *Lipman Bers: A Life in Mathematics*, edited by L. Keen, Editor), I. Kra, R. Rodriguez, AMS, 2015, pp. 175–262.

[3] A. Adem, J. Leida, Y. Ruan, *Orbifolds and stringy topology*, Cambridge Tracts in Mathematics, 171. Cambridge University Press, Cambridge, 2007. xii+149 pp.

[4] D. Alessandrini, *A compactification for the spaces of convex projective structures on manifolds*, arXiv:0801.0165.

[5] Y. Algom-Kfir, *Strongly contracting geodesics in outer space*, Geom. Topol. 15 (2011), no. 4, 2181–2233.

[6] Y. Algom-Kfir, M. Bestvina, *Asymmetry of outer space*, Geom. Dedicata 156 (2012), 81–92.

[7] C. Athorne, D. Maclagan, I. Strachan, *Tropical geometry and integrable systems*, Proceedings of the conference held at the University of Glasgow, Glasgow, July 38, 2011. Contemporary Mathematics, 580. American Mathematical Society, Providence, RI, 2012. x+155 pp.

[8] O. Baker, *The Jacobian map on outer space*, Thesis, Cornell University, 2011.

[9] D. Burago, Y. Burago, S. Ivanov, *A course in metric geometry*, Graduate Studies in Mathematics, 33. American Mathematical Society, Providence, RI, 2001. xiv+415 pp.

[10] G. Bergman, *The logarithmic limit-set of an algebraic variety*, Trans. Amer. Math. Soc. 157 (1971) 459–469.

[11] M. Bestvina, *A Bers-like proof of the existence of train tracks for free group automorphisms*, Fund. Math. 214 (2011) 1–12.

[12] M. Bestvina, *The topology of* $\text{Out}(F_n)$, Proceedings of the International Congress of Mathematicians, Vol. II (Beijing, 2002), 373–384, Higher Ed. Press, Beijing, 2002.

[13] M. Bestvina, *Geometry of outer space*, Geometric group theory, 173–206, IAS/Park City Math. Ser., 21, Amer. Math. Soc., Providence, RI, 2014.

[14] M. Bestvina, M. Feighn, *The topology at infinity of* $\text{Out}(F_n)$, Invent. Math. 140 (2000) 651–692.

[15] R. Bieri, J. Groves, *The geometry of the set of characters induced by valuations*, J. Reine Angew. Math. 347 (1984), 168–195.

[16] R. Bieri, W. Neumann, R. Strebel, *A geometric invariant of discrete groups*, Invent. Math. 90 (1987), no. 3, 451–477.

[17] R. Bieri, R. Strebel, *Valuations and finitely presented metabelian groups*, Proc. London Math. Soc. (3) 41 (1980), no. 3, 439–464.

[18] C. Birkenhake, H. Lange, *Complex tori*, Progress in Mathematics, 177. Birkhäuser Boston, Inc., Boston, MA, 1999. xvi+251 pp.

[19] F. Block, *Counting algebraic curves with tropical geometry*, *Tropical geometry and integrable systems*, 41–54, Contemp. Math., 580, Amer. Math. Soc., Providence, RI, 2012.

[20] A. Borel, L. Ji, *Compactifications of symmetric and locally symmetric spaces*, Mathematics: Theory & Applications. Birkhäuser Boston, Inc., Boston, MA, 2006. xvi+479 pp.

[21] S. Brannetti, M. Melo, F. Viviani, *On the tropical Torelli map*, Adv. Math. 226 (2011) 2546–2586.

[22] M. Bridson, *Geodesics and curvature in metric simplicial complexes*, in *Group theory from a geometrical viewpoint*, 373–463, World Sci. Publ., River Edge, NJ, 1991.

[23] M. Bridson, K. Vogtmann, *Automorphism groups of free groups, surface groups and free abelian groups, Problems on mapping class groups and related topics,* 301–316, Proc. Sympos. Pure Math., 74, Amer. Math. Soc., Providence, RI, 2006.

[24] M. Bridson, K. Vogtmann, *On the geometry of the automorphism group of a free group*, Bull. London Math. Soc. 27 (1995), no. 6, 544–552.

[25] E. Brieskorn, H. Knörrer, *Plane algebraic curves*, Translated from the German original by John Stillwell. [2012] reprint of the 1986 edition. Modern Birkhäuser Classics. Birkhäuser/Springer Basel AG, Basel, 1986. x+721 pp.

[26] J. Brock, B. Farb, *Curvature and rank of Teichmüller space*, Amer. J. Math. 128 (2006), no. 1, 1–22.

[27] E. Brugallé, *Some aspects of tropical geometry*, Eur. Math. Soc. Newsl. No. 83 (2012), 23–28.

[28] E. Brugallé, I. Itenberg, G. Mikhalkin, K. Shaw, *Brief introduction to tropical geometry, Proceedings of the Gökova Geometry-Topology Conference 2014,* 1–75, Gkova Geometry/Topology Conference (GGT), Gökova, 2015.

[29] E. Brugallé, G. Mikhalkin, *Enumeration of curves via floor diagrams*, C. R. Math. Acad. Sci. Paris 345 (2007), no. 6, 329–334.

[30] E. Brugallé, K. Shaw, *A bit of tropical geometry*, Amer. Math. Monthly 121 (2014) 563–589.

[31] D. Burago, Y. Burago, S. Ivanov, *A course in metric geometry*, Graduate Studies in Mathematics, 33. American Mathematical Society, Providence, RI, 2001. xiv+415 pp.

[32] L. Caporaso, *Algebraic and tropical curves: comparing their moduli spaces, Handbook of moduli.* Vol. I, 119–160, Adv. Lect. Math. (ALM), 24, Int. Press, Somerville, MA, 2013.

[33] L. Caporaso, J. Harris, *Counting plane curves of any genus*, Invent. Math. 131 (1998), no. 2, 345–392.

[34] L. Caporaso, F. Viviani, *Torelli theorem for graphs and tropical curves*, Duke Math. J. 153 (2010), no. 1, 129–171.

[35] E. Casas-Alvero, *Singularities of plane curves*, London Mathematical Society Lecture Note Series, 276. Cambridge University Press, Cambridge, 2000. xvi+345 pp.

[36] J.W.S. Cassels, *An introduction to the geometry of numbers*, Corrected reprint of the 1971 edition. Classics in Mathematics. Springer-Verlag, Berlin, 1997. viii+344 pp.

[37] M. Chan, *Combinatorics of the tropical Torelli map*, Algebra Number Theory 6 (2012), no. 6, 1133–1169.

[38] M. Chan, M. Melo, F. Viviani, *Tropical Teichmüller and Siegel spaces*, *Algebraic and combinatorial aspects of tropical geometry*, 45–85, Contemp. Math., 589, Amer. Math. Soc., Providence, RI, 2013.

[39] A. Chambert-Loir, *Mesures et équidistribution sur les espaces de Berkovich*, J. Reine Angew. Math. 595 (2006), 215–235.

[40] D. Cooper, M. Culler, H. Gillet, D. Long, P. Shalen, *Plane curves associated to character varieties of 3-manifolds*, Invent. Math. 118 (1994), no. 1, 47–84.

[41] M. Culler, P. Shalen, *Varieties of group representations and splittings of 3-manifolds*, Ann. of Math. (2) 117 (1983), no. 1, 109–146.

[42] M. Culler, K. Vogtmann, *Moduli of graphs and automorphisms of free groups*, Invent. Math. 84 (1986), no. 1, 91–119.

[43] M. Culler, K. Vogtmann, *The boundary of outer space in rank two*, in *Arboreal group theory*, 189–230, Math. Sci. Res. Inst. Publ., 19, Springer, New York, 1991.

[44] G. Daskalopoulos, R. Wentworth, *Classification of Weil-Petersson isometries*, Amer. J. Math. 125 (2003) 941–975.

[45] M. Einsiedler, M. Kapranov, D. Lind, *Non-Archimedean amoebas and tropical varieties*, J. Reine Angew. Math. 601 (2006), 139–157.

[46] B. Farb, D. Margalit, *A primer on mapping class groups*, Princeton Mathematical Series, 49. Princeton University Press, Princeton, NJ, 2012. xiv+472 pp.

[47] S. Fomin, G. Mikhalkin, *Labeled floor diagrams for plane curves*, J. Eur. Math. Soc. 12 (2010), no. 6, 1453–1496.

[48] S. Francaviglia, A. Martino, *Metric properties of outer space*, Publ. Mat. 55 (2011) 433–473.

[49] G. Francois, S. Hampe, *Universal families of rational tropical curves*, Canad. J. Math. 65 (2013), no. 1, 120–148.

[50] S. Garoufalidis, *The degree of a q-holonomic sequence is a quadratic quasipolynomial*, Electron. J. Combin. 18 (2011), no. 2, Paper 4, 23 pp.

[51] A. Gathmann, *Tropical algebraic geometry*, Jahresber. Deutsch. Math.-Verein. 108 (2006), no. 1, 3–32.

[52] A. Gathmann, M. Kerber, H. Markwig, *Tropical fans and the moduli spaces of tropical curves*, Compos. Math. 145 (2009), no. 1, 173–195.

[53] A. Gathmann, H. Markwig, *Kontsevich's formula and the WDVV equations in tropical geometry*, Adv. Math. 217 (2008), no. 2, 537–560.

[54] A. Gathmann, H. Markwig, *The Caporaso-Harris formula and plane relative Gromov-Witten invariants in tropical geometry*, Math. Ann. 338 (2007), no. 4, 845–868.

[55] S. Gersten, *The automorphism group of a free group is not a CAT(0) group*, Proc. Amer. Math. Soc. 121 (1994), no. 4, 999–1002.

[56] P. Griffiths, J. Harris, *Principles of algebraic geometry*, Pure and Applied Mathematics. Wiley-Interscience [John Wiley & Sons], New York, 1978. xii+813 pp.

[57] M. Gromov, *Metric structures for Riemannian and non-Riemannian spaces*, With appendices by M. Katz, P. Pansu and S. Semmes. Translated from the French by Sean Michael Bates. Progress in Mathematics, 152. Birkhäuser Boston, Inc., Boston, MA, 1999. xx+585 pp.

[58] M. Gross, *Tropical geometry and mirror symmetry*, CBMS Regional Conference Series in Mathematics, 114. Published for the Conference Board of the Mathematical Sciences, Washington, DC; by the American Mathematical Society, Providence, RI, 2011. xvi+317 pp.

[59] W. Gubler, *Tropical varieties for non-Archimedean analytic spaces*, Invent. Math. 169 (2007), no. 2, 321–376.

[60] W. Gubler, *The Bogomolov conjecture for totally degenerate abelian varieties*, Invent. Math. 169 (2007), no. 2, 377–400.

[61] W. Gubler, *A guide to tropicalizations*, *Algebraic and combinatorial aspects of tropical geometry*, 125–189, Contemp. Math., 589, Amer. Math. Soc., Providence, RI, 2013.

[62] P. Hacking, S. Keel, J. Tevelev, *Stable pair, tropical, and log canonical compactifications of moduli spaces of del Pezzo surfaces*, Invent. Math. 178 (2009), no. 1, 173–227.

[63] M. Handel, L. Mosher, *The expansion factors of an outer automorphism and its inverse*, Trans. Amer. Math. Soc. 359 (2007), no. 7, 3185–3208

[64] M. Handel, L. Mosher, *Axes in outer space*, Mem. Amer. Math. Soc. 213 (2011), no. 1004, vi+104 pp.

[65] J. Harer, *The cohomology of the moduli space of curves*, Theory of moduli (Montecatini Terme, 1985), 138–221, Lecture Notes in Math., 1337, Springer, Berlin, 1988.

[66] I. Itenberg, G. Mikhalkin, *Geometry in the tropical limit*, Math. Semesterber. 59 (2012), no. 1, 57–73.

[67] I. Itenberg, G. Mikhalkin, E. Shustin, *Tropical algebraic geometry*, Second edition, Oberwolfach Seminars, 35. Birkhäuser Verlag, Basel, 2009. x+104 pp.

[68] N. Ivanov, *Mapping class groups*, Handbook of geometric topology, 523–633, North-Holland, Amsterdam, 2002.

[69] L. Ji, *A tale of two groups: arithmetic groups and mapping class groups*, Handbook of Teichmüller theory. Vol. III. Edited by Athanase Papadopoulos. IRMA Lectures in Mathematics and Theoretical Physics, 17. European Mathematical Society (EMS), Zürich, 2012. pp. 157–295.

[70] L. Ji, *Buildings and their applications in geometry and topology, Differential geometry*, 89–210, Adv. Lect. Math. (ALM), 22, Int. Press, Somerville, MA, 2012.

[71] L. Ji, *The story of Riemann's moduli space*, in *The legacy of Bernhard Riemann after one hundred and fifty years*, Higher Education Press and International Press, 2016, pp. 307–359.

[72] L. Ji, E. Leuzinger, *The asymptotic Schottky problem*, Geom. Funct. Anal. 19 (2010), no. 6, 1693–1712.

[73] L. Ji, A. Papadopoulos, *Sophus Lie and Felix Klein: the Erlangen program and its impact in mathematics and physics*, IRMA Lectures in Mathematics and Theoretical Physics, 23. European Mathematical Society, Zürich, 2015. xviii+330 pp.

[74] M. Jonsson, *Degenerations of amoebae and Berkovich spaces*, Math. Ann. 2016 (364) 293–311.

[75] M. Joswig, *Essentials of Tropical Combinatorics*, Springer, 2014.

[76] T. Kajiwara, *Tropical toric geometry*, (English summary) Toric topology, 197-207, Contemp. Math., 460, Amer. Math. Soc., Providence, RI, 2008.

[77] E. Katz, *A tropical toolkit*, Expo. Math. 27 (2009), no. 1, 1–36.

[78] S. Kerckhoff, *The Nielsen realization problem*, Annals of Mathematics. 117 (1983) 235–265.

[79] N. Koblitz, *p-adic numbers, p-adic analysis, and zeta-functions*, Second edition, Graduate Texts in Mathematics, 58. Springer-Verlag, New York, 1984. xii+150 pp.

[80] M. Kontsevich, Y. Manin, *Gromov-Witten classes, quantum cohomology, and enumerative geometry*, Comm. Math. Phys. 164 (1994), no. 3, 525–562.

[81] M. Kontsevich, Y. Soibelman, *Affine structures and non-Archimedean analytic spaces*, *The unity of mathematics*, 321–385, Progr. Math., 244, Birkhäuser Boston, Boston, MA, 2006.

[82] M. Kotani, T. Sunada, *Jacobian tori associated with a finite graph and its abelian covering graphs*, Adv. in Appl. Math. 24 (2000) 89–110.

[83] D. Kozlov, *Moduli spaces of tropical curves of higher genus with marked points and homotopy colimits*, Israel J. Math. 182 (2011), 253–291.

[84] D. Kozlov, *The topology of moduli spaces of tropical curves with marked points*, Asian J. Math. 13 (2009) 385–403.

[85] S. Krstic, K. Vogtmann, *Equivariant outer space and automorphisms of free-by-finite groups*, Comment. Math. Helv. 68 (1993), no. 2, 216–262.

[86] A. Lagerberg, *Super currents and tropical geometry*, Math. Z. 270 (2012) 1011–1050.

[87] J. Lee, *Riemannian manifolds, An introduction to curvature*, Graduate Texts in Mathematics, 176. Springer-Verlag, New York, 1997. xvi+224 pp.

[88] G. Litvinov, *Idempotent and tropical mathematics; complexity of algorithms and interval analysis*, Comput. Math. Appl. 65 (2013) 1483–1496.

[89] G. Litvinov, *Tropical mathematics, idempotent analysis, classical mechanics and geometry*, in *Spectral theory and geometric analysis*, 159–186, Contemp. Math., 535, Amer. Math. Soc., Providence, RI, 2011.

[90] G. Litvinov, *The Maslov dequantization, and idempotent and tropical mathematics: a brief introduction*, J. Math. Sci. (N. Y.) 140 (2007) 426–444.

[91] G. Litvinov, S. Sergeev, *Tropical and idempotent mathematics*, Papers from the International Workshop TROPICAL-07 held at the Independent University of Moscow, Moscow, August 25-30, 2007. Contemporary Mathematics, 495. American Mathematical Society, Providence, RI, 2009. x+382 pp.

[92] K. Liu, X. Sun, S.T. Yau, *Canonical metrics on the moduli space of Riemann surfaces*, I. J. Differential Geom. 68 (2004) 571–637.

[93] M. Luxton, Z. Qu, *Some results on tropical compactifications*, Trans. Amer. Math. Soc. 363 (2011) 4853–4876.

[94] D. Maclagan, *Introduction to tropical algebraic geometry*, *Tropical geometry and integrable systems*, 1–19, Contemp. Math., 580, Amer. Math. Soc., Providence, RI, 2012.

[95] D. Maclagan, B. Sturmfels, *Introduction to Tropical Geometry*, preprint of a book, 2014.

[96] W. Magnus, A. Karrass, D. Solitar, *Combinatorial group theory. Presentations of groups in terms of generators and relations*, Second revised edition, Dover Publications, Inc., New York, 1976. xii+444 pp.

[97] C. McMullen, *The moduli space of Riemann surfaces is Kähler hyperbolic*, Ann. of Math. (2) 151 (2000) 327–357.

[98] C. McMullen, *Thermodynamics, dimension and the Weil-Petersson metric*, Invent. Math. 173 (2008) 365–425.

[99] G. Mikhalkin, *What is ... a tropical curve?* Notices Amer. Math. Soc. 54 (2007) 511–513.

[100] G. Mikhalkin, *Introduction to Tropical Geometry (notes from the IMPA lectures in Summer 2007)*, arXiv:0709.1049.

[101] G. Mikhalkin, *Tropical curves and their Jacobians*, a preprint posted at http://www.mathcenter.spb.ru/nikaan/tropic/2.ps

[102] G. Mikhalkin, *Moduli spaces of rational tropical curves*, *Proceedings of Gökova Geometry-Topology Conference 2006*, 39–51, Gökova Geometry/Topology Conference (GGT), Gökova, 2007.

[103] G. Mikhalkin, *Enumerative tropical algebraic geometry in* $\mathbb{R}^2$, J. Amer. Math. Soc. 18 (2005) 313–377.

[104] G. Mikhalkin, *Amoebas of algebraic varieties and tropical geometry*, *Different faces of geometry*, 257–300, Int. Math. Ser. (N. Y.), 3, Kluwer/Plenum, New York, 2004.

[105] G. Mikhalkin, *Tropical geometry and its applications*, International Congress of Mathematicians. Vol. II, 827–852, Eur. Math. Soc., Zürich, 2006.

[106] G. Mikhalkin, J. Rau, *Tropical Geometry*, a preprint of a book, 2014.

[107] G. Mikhalkin, I. Zharkov, *Tropical curves, their Jacobians and theta functions.* in *Curves and abelian varieties*, pp. 203–230, Contemp. Math., 465, Amer. Math. Soc., Providence, RI, 2008

[108] J. Morgan, P. Shalen, *Valuations, trees, and degenerations of hyperbolic structures*, I. Ann. of Math. (2) 120 (1984), no. 3, 401–476.

[109] J. Mueller, W. Schmidt, *On the Newton polygon*, Monatsh. Math. 113 (1992) 33–50.

[110] D. Mumford, *Geometric invariant theory*, Ergebnisse der Mathematik und ihrer Grenzgebiete, Neue Folge, Band 34 Springer-Verlag, Berlin-New York 1965 vi+145 pp.

[111] F. Oort, J. Steenbrink, *The local Torelli problem for algebraic curves, Journées de Géometrie Algébrique d'Angers*, Juillet 1979/Algebraic Geometry, Angers, 1979, pp. 157–204, Sijthoff & Noordhoff, Alphen aan den Rijn–Germantown, Md., 1980.

[112] M. Passare, D. Pochekutov, A. Tsikh, *Amoebas of complex hypersurfaces in statistical thermodynamics*, Math. Phys. Anal. Geom. 16 (2013) 89–108.

[113] M. Passare, H. Rullgard, *Amoebas, Monge-Ampére measures, and triangulations of the Newton polytope*, Duke Math. J. 121 (2004) 481–507.

[114] M. Passare, A. Tsikh, *Amoebas: their spines and their contours*, *Idempotent mathematics and mathematical physics*, 275–288, Contemp. Math., 377, Amer. Math. Soc., Providence, RI, 2005.

[115] L. Pachter, B.Sturmfels, *Tropical geometry of statistical models*, Proc. Natl. Acad. Sci. USA 101 (2004), no. 46, 16132–16137.

[116] S. Payne, *Analytification is the limit of all tropicalizations*, Math. Res. Lett. 16 (2009) 543–556.

[117] R.C. Penner, *Tropical lambda lengths, measured laminations and convexity*, Journal of Differential Geometry (2013) 343–365.

[118] R. Penner, J. Harer, *Combinatorics of train tracks*, Annals of Mathematics Studies, 125. Princeton University Press, Princeton, NJ, 1992. xii+216 pp.

[119] J. Pin, Tropical semirings, *Idempotency* (Bristol, 1994), 50–69, Publ. Newton Inst., 11, Cambridge Univ. Press, Cambridge, 1998.

[120] M. Pflaum, *Analytic and geometric study of stratified spaces*, Lecture Notes in Mathematics, 1768. Springer-Verlag, Berlin, 2001. viii+230 pp.

[121] M. Pollicott, R. Sharp, *A Weil–Petersson type metric on spaces of metric graphs*, Geometriae Dedicata, 172 (2014) 229–244.

[122] J. Richter-Gebert, B. Sturmfels, T. Theobald, *First steps in tropical geometry*, in *Idempotent mathematics and mathematical physics*, 289–317, Contemp. Math., 377, Amer. Math. Soc., Providence, RI, 2005.

[123] D. Speyer, *Parameterizing tropical curves I: Curves of genus zero and one*, Algebra Number Theory 8 (2014) 963–998.

[124] D. Speyer, B. Sturmfels, *Tropical mathematics*, Math. Mag. 82 (2009) 163–173.

[125] J. Tevelev, *Compactifications of subvarieties of tori*, Amer. J. Math. 129 (2007) 1087–1104.

[126] A. Terras, *Harmonic analysis on symmetric spaces and applications*, II. Springer-Verlag, Berlin, 1988. xii+385 pp.

[127] S. Tillmann, *Boundary slopes and the logarithmic limit set*, Topology 44 (2005) 203–216.

[128] S. Tillmann, *Degenerations of ideal hyperbolic triangulations. Math*, Z. 272 (2012) 793–823.

[129] E. Viehweg, *Quasi-projective moduli for polarized manifolds*, Ergebnisse der Mathematik und ihrer Grenzgebiete (3), 30. Springer-Verlag, Berlin, 1995. viii+320 pp.

[130] O. Viro, *From the sixteenth Hilbert problem to tropical geometry*, Jpn. J. Math. 3 (2008) 185–214.

[131] O. Viro, *Dequantization of real algebraic geometry on logarithmic paper*, *European Congress of Mathematics*, Vol. I (Barcelona, 2000), 135–146, Progr. Math., 201, Birkhäuser, Basel, 2001.

[132] F. Viviani, *Tropicalizing vs. Compactifying the Torelli morphism*, *Tropical and non-Archimedean geometry*, 181–210, Contemp. Math., 605, Amer. Math. Soc., Providence, RI, 2013.

[133] K. Vogtmann, *Automorphisms of free groups and outer space*, Geom. Dedicata 94 (2002) 1–31.

[134] K. Vogtmann, *The cohomology of automorphism groups of free groups*, International Congress of Mathematicians. Vol. II, 1101–1117, Eur. Math. Soc., Zürich, 2006.

[135] T. White, *Fixed points of finite groups of free group automorphisms*, Proc. Amer. Math. Soc. 118 (1993) 681–688.

[136] S. Wolpert, *Geodesic length functions and the Nielsen problem* J. Differential Geom. 25 (1987) 275–296.

[137] S. Zhang, *Equidistribution of small points on abelian varieties*, Ann. of Math. (2) 147 (1998)159–165.

Index

Surveys of Modern Mathematics

www.intlpress.com/smm

SMM 14: *Introduction to Moduli Spaces of Reimann Surfaces and Tropical Curves*
by Lizhen Ji and Eduard Looijenga

SMM 13: *Gauss–Manin Connection in Disguise: Calabi–Yau Modular Forms*
by Hossein Movasati

SMM 12: *Differential Geometry & Integral Geometry: Selected papers & lecture notes of Shiing-Shen Chern*
edited by Shiu-Yuen Cheng and Lizhen Ji

SMM 11: *Topics in Differential Geometry: Selected papers & lecture notes of Shiing-Shen Chern*
edited by Shiu-Yuen Cheng and Lizhen Ji

SMM 10: *Finite Groups: An Introduction*
by Jean-Pierre Serre

SMM 9: *Compressible Flow and Euler's Equations*
by Demetrios Christodoulou and Shuang Miao

SMM 8: *Lie-Bäcklund-Darboux Transformations*
by Y. Charles Li and Artyom Yurov

SMM 7: *Introductory Lectures on Manifold Topology: Signposts*
by Thomas Farrell and Yang Su

SMM 6: *Open Problems and Surveys of Contemporary Mathematics*
Lizhen Ji, Yat-Sun Poon and Shing-Tung Yau, eds.

SMM 5: [no volume published]

SMM 4: *An Introduction to Rota-Baxter Algebra*
by Li Guo

SMM 3: *Application of Elementary Differential Geometry to Influence Analysis*
Yat-Sun Poon and Wai-Yin Poon, eds.

SMM 2: *Lie Theory and Representation Theory*
Naihong Hu, Bin Shu and Jianpan Wang, eds.

SMM 1: *Analytic Methods in Algebraic Geometry*
by Jean-Pierre Demailly

Made in United States
North Haven, CT
04 November 2026

21724631R00128